Virtual Geographies

Virtual Geographies explores the possibilities and the dangers brought about by the revolution in communication technologies. Outlining how these technologies are being used to produce new geographies and new types of space, this book de-mystifies the hype over 'cyberspace' and its consequences, and reveals how new communication technologies can open up whole new vistas.

Contributors drawn from a wide range of disciplines, including geography, sociology, English and philosophy, investigate how particular visions of cyberspace have been constructed and articulated, exploring the influence of literature, gender, how the experience of online interaction is expressed and of the interest groups or rhetoric that shape many developments. As scepticism of the consequences of cyberspace emerges, the contributors critically assess the status of virtual environments and geographies, how they interact with everyday spaces, and how they can reshape how we think and write about the world.

Mike Crang is a lecturer in Geography, Durham University; **Phil Crang** is a lecturer in Geography, University College London and **Jon May** is a lecturer in Human Geography, University of Sussex.

Sussex Studies in Culture and Communication

Series Editors: Roger Silverstone, London School of Economics, Craig Clunas, University of Sussex and Jane Cowan, University of Sussex

Books in this series express Sussex's unique commitment to interdisciplinary work at the cutting edge of cultural and communication studies. Transcending the interface between the social and the human sciences, the series explores some of the key themes that define the particular character of life, and the representation of life, at the end of one millennium and the beginning of the next.

Our relationships to each other, to our bodies and to our technologies are changing. New concepts are required, new evidence is needed to advance our understanding of these changes. The boundaries between disciplines need to be challenged. Through monographs and edited collections the series will explore new ways of thinking about communication, performance, identities and the continual refashioning of meanings, messages, and images in space and time.

Virtual Geographies

Bodies, space and relations

Edited by
Mike Crang, Phil Crang and Jon May

London and New York

First published 1999
by Routledge
11 New Fetter Lane, London EC4P 4EE

Simultaneously published in the USA and Canada
by Routledge
29 West 35th Street, New York, NY 10001

Typeset in Garamond 3 by Keystroke, Jacaranda Lodge, Wolverhampton
Printed and bound in Great Britain by T.J. International Ltd, Padstow, Cornwall

British Library Cataloguing in Publication Data
A catalogue record for this book is available from the British Library

Library of Congress Cataloging in Publication Data
Virtual geographies : bodies, space & relations / [edited by] Mike
Crang, Phil Crang & Jon May.
p. cm.
1. Computer networks—Psychological aspects. 2. Computer
networks—Social aspects. 3. Virtual reality—Psychological
aspects. 4. Virtual reality—Social aspects. I. Crang, Mike.
II. Crang, Phil. III. May, Jon. IV. Series.
ZA4375.V57 1999
303.48'33—dc21 98-30450

ISBN 0–415–16827–9 (hbk)
ISBN 0–415–16828–7 (pbk)

Contents

Illustrations

Figures

Plate

Contributors

Nick Bingham is currently a Research Associate at the University of Sheffield working on an ESRC funded project under Dr Gill Valentine on the use of Cyberspace by children.

Laura Chernaik is Lecturer in American Studies at the University of Nottingham. She has published articles in *Gender, Place and Culture*, *Feminist Review*, *Renaissance and Modern Studies and Letterature d'America*. She is completing a book, *Social and Virtual Space: transnationalism and the New Social Movements*.

David B. Clarke is Lecturer in Human Geography and Affiliate Member of the Institute of Communications Studies at the University of Leeds; he is also an ESRC Research Fellow. He has published widely on consumption, the media, modernity, postmodernity and space. Having recently edited *The Cinematic City* (Routledge 1997), he is currently working on a book on the spatiality of the consumer society, *Commodity, Sign and Space* (Blackwell forthcoming).

Marcus A. Doel is a Lecturer in Human Geography at the Department of Geography, Loughborough University. He has written widely on poststructural and postmodern approaches to spatial science, and on the geographies and cultural politics of film and literature, subjectivity, and the Holocaust. He is presently writing a book entitled *Poststructuralist Geography*.

Oliver Froehling is a Doctoral Candidate at the Department of Geography, University of of Kentucky, and is currently finishing a dissertation on Indigenous Autonomy in Oaxaca, Mexico. He is also affiliated with the Centro Intercultural de Encuentros y Diálogos (CIED) in Oaxaca.

Stephen Graham works in the Centre for Urban Technology at Newcastle University's Department of Town and Country Planning. His research addresses the interrelationships between telecommunications and information technologies and urban theory, development and policy. He is co-author of the books *Telecommunications and the City: Electronic Spaces, Urban Places* (1996) and *Splintering Networks, Fragmenting Cities: Infra-*

structure and Urban Development in a Global-Local Age (1999) (both with Simon Marvin and published by Routledge).

Ken Hillis is an Assistant Professor in the Department of Communication Studies, The University of North Carolina at Chapel Hill. He received his PhD in Human Geography from the University of Wisconsin, Madison, and has published in *Urban History Review*, *Cartographica*, *Ecumene*, *Progress in Human Geography*, and in the anthologies *Cultures of Internet* and *Mapping the Body*. His book, *Digital Sensations: Space, Identity and Embodiment in Virtual Reality* (University of Minnesota Press) is forthcoming.

Otto Imken completed a PhD in the Department of Philosophy at the University of Warwick. He has organised conferences around the possibilities of cyberspace and is interested in the transformations of thought and experience rendered possible through this.

Michael Joyce teaches in the English Department at Vassar College. His most recent hyperfictions include *Twelve Blue* for the Web and *Twilight, a symphony* on CD ROM.

James Kneale is a Lecturer in Human Geography at the University of Exeter. His research interests include geographies of the media and popular culture and particularly popular fiction.

Wendy Larner is a Lecturer in Sociology at the University of Auckland, Aotearoa/New Zealand. She has published various articles and book chapters on globalization, restructuring and identity in Aotearoa/New Zealand. Her current projects include a book manuscript on market governance in the telecommunications industry.

Jennifer S. Light is finishing her doctoral work at Harvard University in the History of Science. She holds prior degrees from Harvard College and Cambridge University. Her dissertation examines the recent history of domestic technologies including home security systems, 'smart house' automation devices, and environmental control units for persons with disabilities. The essay that appears here grew out of a course she taught at the University of Edinburgh in autumn 1996.

Christopher Ray is a Research Fellow in the Centre for Rural Economy, University of Newcastle upon Tyne. Using sociological and anthropological approaches, he undertakes research into participative forms of local development and into the relationship between cultural identity and rural development.

Jeremy Stein is a Research Fellow in the Innovation Management Research Unit (IMRU) in the Business School at the University of Birmingham, England. His research mainly focuses on connections between modernity and the social embedding of technology in nineteenth- and twentieth-century British and Canadian cities. He is currently working on a

Leverhulme funded project, 'Dilemmas of a Maturing Technology', examining technological and organisational innovation, knowledge management and business failure in Britain's nineteenth-century steam power engineering industry.

Hilary Talbot is the Research Manager at the Centre for Rural Economy at the University of Newcastle upon Tyne. Her work with organisations in the North of England to devise a strategy for the introduction of telematics in the rural region forms the case study for the 'rural' chapter of this book. She has also recently completed a study of rural community information needs linked to the opportunities to improve information provision electronically.

Nina Wakeford is the Foundation Fund Lecturer in Sociology and a member of the Digital World Research Centre [http://www.surrey.ac.uk/dwrc/] at the University of Surrey. She currently holds an ESRC Fellowship and an ESRC grant on women's experiences of computer-mediated communication. She is co-editing with Peter Lyman of the University of California, Berkeley a volume on virtual methodologies, *Analysing Virtual Society: New Directions in Methodology* (Sage, forthcoming), and is currently completing a book entitled *Networks of Desire: Gender, sexuality and computing culture*, to be published by Routledge.

Acknowledgements

We would like to register our thanks to all at Routledge, especially Sarah Lloyd and Sarah Carty. Phil apologises for the protracted labour at the end. Thanks also to Roger Silverstone for recruiting us into the Sussex series; Steve Pile for being on board in the very early stages; and the Social and Cultural Geography Research Group of the RGS–IBG for sponsoring the conference session from which some of the contributions are drawn and which provided the impetus for this book. We acknowledge the agreement of Pion Ltd to allow Stephen Graham to reproduce some passages from his paper in Environment and Planning D: Society and Space.

Mike, Phil and Jon
(Durham, London and Brighton)

1 Introduction

Mike Crang, Phil Crang and Jon May

The collection

Whether framed through the more generalised notion of cyberspace, or the more specific phenomena of the Internet, the World Wide Web, Virtual Reality, hypertext and genres of science fiction such as cyberpunk, it is hard to miss the proliferating debates over the social and geographical significance of new technologies of computer mediated communication. For some, these technologies are seen as facilitating, if not producing, a qualitatively different human experience of dwelling in the world; new articulations of near and far, present and absent, body and technology, self and environment (for a collection of essays mostly in this spirit see Featherstone and Burrows 1995). For others, emphasis is laid on the capacity of digitalisation to integrate previously separate operations such as computation, communication and surveillance, with the consequent emergence of new informational networks and 'spaces of flows', with associated morphologies of connection and disconnection (see Castells 1996). In either case, what is at stake is, at its starkest, the suggestion that computer mediated communication technologies are 'generating an entirely new dimension to geography . . . Virtual Geography' (Batty 1997b: 339).

This collection of fourteen essays is provoked by such claims. Provoked, in that we endorse calls to take the development and use of these technologies seriously, to subject them to careful conceptual and empirical scrutiny, and to be open to the possibility that they embody different kinds of spatialities to those hegemonic within theorisations of 'non-virtual' worlds; but also provoked by worries about the danger of falling into what Otto Imken, in his contribution to this volume, calls 'cyperbole', an overdrawn opposition of the real and the virtual, whether this be through the reproduction of the (self-) promotional rhetorics of committed cyber enthusiasts and marketers or the dystopian visions of cyberpessimists. Instead, this collection, whilst not without some claims for a radical transformation of social life as constituted through virtual technologies – Imken's own chapter is a wonderfully engaging example – seeks to approach the virtual in ways that allow a serious analysis of particular socio-technical developments but also avoid their fetishisation as unambiguous locations of social good or evil. Collectively, four main elements to this approach can be identified.

The first is to *eschew any simple technological determinism*. In part this is based on a recognition that virtual technologies and virtual geographies are not synonymous. This collection's understanding of 'virtual geographies' is therefore that they include virtual technologies but are also constituted by the social relations, discourses and sites in which these technologies are embedded. Technologies are not self-contained entities that impact on the social. To use Nina Wakeford's terminology in her contribution on an Internet café, technologies cannot be considered in isolation from the 'landscapes of translation' in which they are encountered, used and for which they may be designed. One impact of this is that technologies have to be seen as socialised. Crucially, this is not a question of already existing technologies being reworked in a social realm that we can locate as exterior to them. No technology can come into being without its socialisation; and this socialisation is an ongoing process throughout the circuits linking technological production, distribution and usage. An example of work in this vein is that on the constructed and contested genderings of technologies, an issue pursued in this collection through considerations of the gendered forms of communicative interaction associated with technologies such as the Internet, through the gendered forms of communality they facilitate, through the gendered forms of expertise associated with computational and communicative technologies (and note the potential for multiplicity here), and through the gendered qualities of the spatial textures users are involved in (see the contributions of Wakeford and Joyce in particular). Moreover, as the case of gender exemplifies, technologies are not just socialised as technologies. They are also socialised as a variety of other entities – for example as commodities, as property and infrastructure, as the objects of attention for workers and consumers, as tools for economic and regional development, as items of interior decoration, as genres of literature – and in consequence are constituted through a range of social dimensions that, at first sight, might seem to have little to do with technology or indeed virtuality (see also Silverstone *et al.* 1992). Of course, at the same time it is important to recognise that technologies are also in part constitutive of these social dimensions. Just as technology does not come into being outside of the social, so the social does not come into being outside of the technological. To that extent, and although the language is difficult here, one can speak of the power of technology to affect social relations, without succumbing to technological determinism. One way to express this is to speak of a dialectic composed of the social shaping of technology and the technological building of the social (Bijker *et al.* 1987; Bijker and Law 1992) (see also Stein's contribution to this volume). Or pushing this a fraction further, one arrives at a conceptual emphasis on the 'technosocial' (see Bingham 1996 and this volume). Associated in particular with the writings of Michel Serres and Bruno Latour, this involves a recasting of the social so that it no longer excludes non-human actants, and a consequent conceptualisation of technological formations as 'co-productions between people and things' (Bingham this volume: 253). The question of technological determinism therefore becomes redundant, as the very entities of the technological and the social are themselves deconstructed.

We will have more to say on this, and in particular on the alternative questions and forms of analysis suggested by this deconstruction, when we come to introduce the individual contributions. For now, though, let us turn to a second feature that characterises the collection as a whole: its emphasis on *historical geographies of virtuality*. Given the aura of futurology that surrounds discussions of the virtual this may seem paradoxical. However, it is important, we would argue, to counteract or at least question the associations with novelty and epochal transition that cling to the subject of virtual geographies. For instance, one can point to how contemporary celebrations and worries about computer mediated communications are paralleled in debates over 'past' virtual technologies, such as the television, the telephone (see Stein this volume) or even the camera obscura (see Hillis this volume). (We place the now obligatory quote marks around 'past' to signal that the associations of technologies with modernity are contingent not only historically but also geographically; for much, indeed most of the world, the telephone is still thoroughly new and modern.) For example, as Jennifer Light argues in her essay on cityspace and cyberspace, 'while statements such as "something unusual is happening today in the relation between the real and the imaginary" (Soja 1996: 242) . . . may be applicable today, they would also have characterised observations of fifty and one hundred years ago' (this volume: 124). It is also important to emphasise the dangers of falling into either an uncritical celebration of the new as new (the ideology of modernisation beloved of the technophile) or an unproductive lamentation for all that is perceived to be threatened by it (an ideology of nostalgia, mobilised by many a technophobe, but often as detrimental to a critical understanding of the past as it is to any account of the present and future). And at a more substantive level, one needs to recognise how new technologies are in part socialised via their shaping according to old, existent technologies. None of this should be taken as denying the possibility of contemporary virtual geographies being in some important respects new and different. However, it does highlight how our investigations of these contemporary virtual geographies can benefit from a historical perspective. First, through the sharper focus this gives on just what is different and new about them. Second, through an insistence on moving beyond simplistic periodisations of a virtual future, a partly virtual present, and a non-virtual past, and towards more textured understandings of the varying forms of virtuality worked through different technologies in different times and places. And third by conceptualising and responding to contemporary changes through the lens of longer running processes of virtuality, for example in terms of histories of vision (Hillis), space-time distanciation (Bingham), civic life (Light), mediatised communication (Froehling), or writing, reading and representation (Joyce).

The presence in the previous sentence of a number of different longer histories through which the contemporary virtual geographies of computer mediated communication can be framed makes even more obvious something we have so far been skirting around, but no doubt readers have already been pondering. The question of just what we or anyone else means by the virtual in virtual geographies. The collection addresses this at two levels, both of which also contribute to the ethos of toning down the epochal hyperbole attached to virtuality. Substantively, both within and between essays there is an emphasis on *the heterogeneity of material–semiotic practices constituting virtual geographies.* There is a diversity to both the constitution and character of virtual geographies. Hence, for example, we can see how 'the production and consumption of ideas of cyberspace take place in many very different contexts' (Kneale this volume: 205) including: generically conventionalised technologies and techniques of writing and reading (the term comes from William Gibson's short stories and novels) (see Kneale and Joyce this volume), as well as practices of and on the net (see Wakeford and Froehling this volume), and the more strategic visions of planners and place developers (see Light this volume). Indeed one potential conclusion is that cyberspace is in fact produced through the interrelationships between these practices, in so far as planners read (or at least quote) William Gibson, or William Gibson's readers use the Internet and use it to make sense of what they read, or the Internet is marketed through auras drawn from cyberpunk science fiction and/or virtual urban planning. There is then no single version of cyberspace, but a plurality of networked conceptions, each associated with particular generic and geographical sites, and translated for others with varying degrees of success. It is these sites, networks and their translations that we need to understand if we are to investigate the production of cyberspace or other virtual geographies (Ray and Talbot use this approach in analysing rural telematics).

More generally, the work collected here emphasises how unhelpful it is to seek or proclaim a singular character to virtual geographies. This is worth stressing as, perhaps because of the limited volume and depth of work on virtuality, there is still a tendency to conduct debate in these terms: to ask whether virtual worlds are democratising or controlling; whether they are a liberation from the structures and drags of the non-virtual world, or marked by an intensification of existing, 'real world' relations of inequality and domination; and so on. This totalising logic produces a bluntness of analytical judgement, fetishises and over-simplifies both the virtual and the non-virtual (of which more below), and allows little room for expressing variations across virtual time and space. It is hardly surprising then that, in contrast, arguments can be found in this volume for virtual geographies as being both de-centred and centralising, masculinist and feminist, rationally ordered and inconceivable to the rational mind. In part, this is a reflection of fundamental ambiguities which are constitutive of particular virtual geographies. For example, Ken Hillis' account of virtual reality points to its paradoxical combination of self-centred rationalist realism – through an emphasis on viewers making sense of what they are immersed in – and of de-centred magical transcendence – through an abandonment to this received virtual environment. As he puts it:

> VR thereby achieves a cultural point of purchase with subjects who seek to maintain control over their individual production of meaning even as they might play with the spectre of abandoning the formal maintenance of modern identity to external sources such as VR and the 'performativity' it encourages.
>
> (Hillis this volume: 31)

But more prosaically the problems with any 'once and for all verdicts' on the character of virtual geographies are a result of the many kinds of virtual geographies in existence and yet to come into being. So, for example, the Internet is, unsurprisingly, far from homogeneous. Thus Nina Wakeford's chapter on the gendering of the net mentions in passing the varying moral norms with regard to styles of communication found in different MUDS and MOOS, and between discussion lists and newsgroups, to the extent that 'on-line landscapes cannot be characterised by a single set of conventions relating to gender' (this volume: 187). And of course the net cannot be taken as a generalisable model for all computer mediated communications, let alone contemporary virtual geographies. The matrices of integrated telematics have a number of different arms. Stephen Graham's chapter, for example, examines integrations of surveillance, computational and simulatory technologies (themselves as diverse as digital CCTV, electronic tracking and geo-positional technologies, and home teleservices) and, in contrast to the emphasis on the difficulties of controlling the operations of the Internet (see for example Shade 1996), draws out their role in intensified state and commercial practices of surveillance. The character of virtual geographies is dependent on what sort of virtual technosocialities one chooses to focus on.

Substantively, then, we have to unpack the category of virtual technologies, both through identifying the different forms in which computer mediated communicational networks are being developed and, as we suggested earlier, through recognising the much longer histories of virtuality that have preceded and coexist with these technologies. Even more importantly, this needs to be paralleled by *a conceptual unpacking of virtuality*, an exploration of the dimensions through which this empirical diversity is structured. Four such conceptual dimensions, we think, can be seen to be highlighted and developed in the essays that follow: we term these *simulation*, *complexity*, *mediazation* and *spatiality*. These are far from mutually exclusive, but each signifies a rather different approach to virtuality, and stimulates slightly different questions about it. The first three emphasise the virtual constitution of our human geographies, the last highlights the geographical constitution of the virtual. We will now say a little about each in turn.

Dimensions of virtuality

Simulation is perhaps the dominant dimension through which virtual technologies are popularly conceived. Notwithstanding their influence in understandings of the virtual, here we use the term not as a sign of adherence to Jean Baudrillard's particular formulations (Baudrillard 1994a), but to signal a looser set of understandings in which the virtual is positioned in relationship to the real, indeed as its Other, in an oppositional imaginative geography which at the same emphasises the mutual constitution of reality and virtuality (Said 1978). A number of potential approaches to virtual geographies results. First, as Marcus Doel and David Clarke argue in their insightful concluding essay, this oppositional framing frequently positions the virtual as a copy, always striving towards but never quite achieving a mimetic replication of the real. Hence the presence of a two-sided cult of authenticity around the virtual, which generates both the promotional celebrations of each step made closer to this replication (whether within popular culture of the new virtual realities and digital special effects that are even more real, or in academic culture of the computational geographies that deliver an ever more exact portrait of reality 'on the ground'), but also the criticisms of the virtual as being a retreat from, and poor substitute for, real life (for example the stigmatic stereotype of Internet chat users as rather sad individuals, either unable to cope with or increasingly divorced from face-to-face social interaction; or the broader social critique of virtuality for a retreat into idealised fantasy worlds rather than facing up to real world issues).

Second, despite the dominant rhetorics of mimesis, there is the potential within the dimension of simulation to argue for the continuing representational nature of virtual geographies; to emphasise that despite some claims made on their behalf, computer mediated communications do not institute a post-symbolic order. They do not pull off what Donna Haraway calls the 'God trick'. Paralleling the forms of critical analysis applied to literary and artistic representation, studies in this vein might involve an exposition of the poetics and politics of the virtual worlds and subjects created in virtual space: the bodily picturings of avatar personae, the social and geographical landscapes of games and other simulations, and so on. Perhaps more provocatively, it can also stimulate an investigation of the different representational economies of varying virtual technosocialities. In this volume, for example, Ken Hillis' chapter provides just such an analysis of VR, at least in terms of its emphasis on visuality, teasing out the complex and often ambivalent conceptions of optics and truth that it mobilises. More implicitly, Michael Joyce's evocation of hypertextuality is a highly suggestive thinking through of the reconfigurations of representational authority that hypertext can produce, as it reconfigures the compositions and locations of authors, texts and readers and the relations between them.

A third approach to the virtual–real opposition can also be detected here: the recognition not just of virtual realities but of 'real virtualities' (Castells 1996), operating in parallel to the 'real-real'. The virtual not as copy or representation but as alternative. Sometimes these virtual alternatives are conceived as problematic; sometimes as socially and politically progressive. The anthropologist Danny Miller has recently developed a strong argument of the former sort in an analysis of the regressive socio-political implications of the virtual, abstract worlds of economic modellers, public service auditors and postmodern theorists (Miller 1998). The juxtapositions here are themselves provocative, but let us just stay with the first of these. Miller's contention, part of a longer running opposition to the dominant intellectual project of contemporary economics, is twofold: that economists' understandings of economics are woefully detached from the real, lived worlds of the producers and consumers involved in the practice of lived capitalism; but, and here is the rub, that they also have the authority to impose their own detached visions on to those real producers and consumers. Structural adjustment is the epitome of this brutal virtualism. Miller's verdict is worth quoting at length:

> Structural adjustment . . . comprises a series of procedures and models devised by groups of economists working within some of the key institutions that were set up following the epochal meeting of Bretton Woods. These models, fostered by the IMF, World Bank and their ilk are purely academic models, in the sense that they seem to pay no attention whatsoever to local context . . . they are simply idealised and abstract models that represent the university departments of economics engaged in academic modelling. . . . So while capitalism as a process by which firms seek to reproduce and increase capital through the manufacture and trade in commodities has become increasingly contextualised, complex and often contradictory (Miller 1997), another force has arisen which has become increasingly abstract. These are academics, paid for by states and international organizations and given the freedom to rise above context to engage in highly speculative processes of modelling. . . . While capitalism was forced to engage with the world and was thus subject to the transformations of context, economics remains disengaged. . . . Social scientists may not think of academics as particularly powerful – but then they are not economists. . . . economics has a form of power that again surpasses early capitalism, that is the legitimate authority to transform the world into its own image. . . . In short, in every case where the existing world does not conform to the academic model the onus is not on changing the model . . . but on changing the world . . . , and the very power of this new form of abstraction is that it can indeed act to eliminate

> the particularity of the world as a series of distortions which prevent the world from working as the model predicts it should.
>
> So it is not that the principles of the market represent capitalism, but that capitalism is being instructed to transform itself into a better representation of the model of the market.
>
> (Miller 1998: 8–9)

Miller's argument, then, is that structural adjustment is not a product of the inherent and unyielding logic of capitalism, but of the abstractions and decontextualisations of a particular form of virtualism: academic economic modelling. Whilst Miller is tentative about its wider applicability – he writes 'I do not pretend to be clear myself as to how far the argument works in relation to other phenomena such as . . . the virtualism of virtual reality within computer technologies' (1998: 2) – we think his analysis raises important issues that can be usefully extended into other forms of virtuality than those he directly addresses. One is the possibility of the economy of real and virtual being reversed, so that the real becomes the poor imitation of the virtual rather than vice versa, and it is the real that has to 'adjust'. Another is the potential of the virtual, through abstraction, to mobilise and produce forms of highly located power which operate precisely through their determined denial of the contextuality of practice and knowledge (including their own).

Others, though, point to a different politics, and a different political geography, that real-virtualities can perform: the liberation of those who enter into them from the very contexts of the 'real-real' that Miller wishes to defend. At one level such claims for the virtual are framed in terms of a de-localisation of the visibility of events, and hence also of experience (see Wark 1994b for an example). Thus the World Wide Web is promoted as a learning tool, as a space that allows access to extra-local knowledges and encounters. At another level, this is linked to a progressive, if always somewhat anxious, strand of the modern condition; de-traditionalisation and the opening up of a realm of the possible to supplement that of the customary (for a critical review see Heelas *et al.* 1996). And at another level again, the value of the real-virtual is seen in terms of the particular qualities of the spatial architectures, textures and inhabitations it facilitates; for example through its potential to reconfigure dominant regimes of human embodiment, to allow a greater fluidity and play than the more viscous and dichotomously regulated real-real (Plant 1995). Of course, all these claims for the positive possibilities of the real-virtual can be, and need to be, scrutinised. The local moral regulatory norms of its interactions and performances, and their relationship to those operating in the real-real, need to be recognised and investigated rather than assumed away (see for instance Wakeford's contribution on the gendering of interaction on the net). The ways in which users actually engage with and make sense of the 'expanded terrain' they access through 'the telecommunications networks criss-crossing the globe' (Wark 1994b: vii) is crucial, as studies of television viewing have shown (Silverstone 1994) and as Deborah Lupton's writings on popular conceptions of personal computers also emphasise (Lupton 1994, 1995). And, of course, the importance of just who has access to these terrains, and in what capacities, cannot be overstated (see for example Dyrkton 1996 on e-mail in Jamaica). Nick Bingham's contribution includes the following jolting, if on a moment's reflection unsurprising, statistic: up to half the world's population are over two hours' travel from the nearest telephone. The virtual world has very definite geographies of inclusion and exclusion. Nonetheless, one cannot simply dismiss the auras of possibility that surround virtual geographies. Care needs to be taken in finding out the extent to which they are fulfilled and for whom but in so far as virtual geographies do become an arena in which the possible is located, they can operate as a field very different to that critiqued by Miller; not as a homogenising imposition of singular histories, but as a space-time characterised by contingency and a tangible sense of openness to change.

Indeed, these qualities are symptomatic of a second thread in the conceptualisation of virtuality. Here, the complexity of virtual geographies is a recurrent refrain, whether in more generalised form through an emphasis on their inconceivability and 'unthinkable complexity' (Gibson 1984: 67) or through more detailed, but still often rather mantric, recitations of their non-linearity, self-organisation, emergent orders, chaos, fractal multidimensional spatiality and autopoetic character. Again, though, this complexity is not a fixed quality. It marks out a domain of conceptual and material contestation. For instance, in some conceptions it is suggestive of a cybernetic spatiality which exceeds the rationalist mappings of masculinist worlds. As Otto Imken puts it in his exploration of the global matrix, a spatiality that is characterised by 'multiple, non-interchangeable dimensions . . . [that] cannot be easily mapped or traversed and require alternative means of navigation and construction, ones rooted in an event-oriented situationist approach' (this volume: 95). A spatiality in which new configurations of self and world are materialised, in which one can lose one's (real-real) self through the processes of getting lost and living in the situation. A spatiality which means that, despite the origins of so much of the virtual in masculinist cultures of technology and militarism, '[c]yberspace is . . . perhaps even the place of woman's affirmation . . . , not of her own patriarchal past, but what she is in a future that has yet to arrive but can nevertheless already be felt' (Plant 1995: 60). However, one might also point to how, not least through a dominant emphasis on user-friendly interfaces and search engines, such possibilities are being marginalised; how the dominant spatialities of cyberspace are characterised less by wanderings than by namings which reproduce real world and common-place structures of intelligibility. One might, like Imken, attribute this to the operation of power seeking and powerful interests, for example those who wish the virtual to model itself on the mall and its consumerist ideologies. Or one might emphasise the desire of many users to make sense of the virtual, to envision it in intelligible and rationalisable forms (as James Kneale argues in his contribution on readings of William Gibson). In all cases, though, what is clear is that the complexity of virtual geographies is not innate; it is subject to multiple constructions and reconstructions.

If virtuality as simulation and virtuality as complexity are two common and interrelated ways of conceiving virtual geographies, a third is virtuality as *mediazation*. We borrow this term from John Thompson's social theoretical analysis of modern media (see Thompson 1990, 1995) and use it to emphasise the potential for understanding virtual technosocialities as recent developments in much longer histories of mediated and distanciated communication. Thompson's argument has four main elements. At the most general level he approaches the media as involved in the 'cultural transmission of symbolic forms', but this semiotic function is also thoroughly materialised through attention to the 'institutional apparatuses' and 'material substrata' or 'technical media' of communication, and through an emphasis on the interweaving of the cultural with other forms of power – economic, political and military (Thompson 1990: 13). Second, through sketching out a history spanning from the European printing presses of the fifteenth century to the communication conglomerates of the present day, Thompson positions communication media as key constituent components of the modern world: 'if we wish to understand the nature of modernity – that is, of the institutional characteristics of modern societies and the life conditions created by them – then we must give a central role to the development of communication media and their impact' (Thompson 1995: 3).

In particular, and this is the third strand of his approach, for Thompson the most important aspect of the modernity of mass communications is their facilitation of 'space-time distanciation':

> The transmission of a symbolic form necessarily involves the detachment, to some extent, of this form from the original context of its production: it is distanced from this context, both spatially and temporally, and inserted into new contexts which are located at different times and places. . . . [P]art of what constitutes modern societies as 'modern' is the fact that the exchange of symbolic forms is no longer restricted primarily to the contexts of face-to-face interaction, but is extensively and increasingly mediated by the institutions and mechanisms of mass communication.
>
> (Thompson 1990: 13,15)

The mediazation of modern culture therefore 'transforms the spatial and -temporal organization of social life' (Thompson 1995: 4), in particular reconfiguring its contextuality. For instance, Thompson draws on the work of the social psychologist Joshua Meyrowitz (1984) to point to mass communication's role in permeating some of the boundaries between different contexts, increasing the visibility of arenas of social life previously distant and concealed. Others emphasise the recontextualisations made both through the montage effects of media vectors (see Wark 1994b) and the situated consumptions of particular readers, listeners or viewers (see for example Liebes and Katz 1990). The overall effect is to problematise both bounded senses of the local and homogenised portraits of the global. The geographies of mediazation produce globalised locals and localised globals. Fourth, and finally, Thompson examines the implications of this reconfiguration of the contextuality of social life. His particular concerns are with its impacts on the character of social action and interaction (where he contrasts face-to-face interaction with less dialogic mediatised 'quasi-interaction'); the effects of a profusion of mediated materials for processes of self-formation; and the forms of public life that modern media do and might facilitate. Underlying all these is a belief that communicational media do more than simply transmit information between people and places. Rather, 'the use of communication media involves the creation of new forms of action and interaction in the social world, new kinds of social relationship and new ways of relating to others and to oneself' (Thompson 1995: 4).

The kinds of questions posed by treating virtuality as mediazation are made fairly clear by Thompson's synoptic account: what are the institutional apparatuses and technical substrata that characterise computer mediated communications; in what ways, and through what sorts of visibility, do they reconfigure the contextuality of social life; what forms of space-time distanciation do they embody; what forms of interaction take place through them, and in particular what degrees and forms of dialogue are apparent; how do they position the subjects involved in them; and what forms of publicness, and indeed privacy, can they constitute? Contributions in this book provide both a conceptual foundation for such concerns (see for example Bingham's essay which argues for conceiving of cyberspace as a 'message-bearing system') and some more detailed analyses of particular cases (perhaps most fully in Froehling's account of the mediazation of the Mexican Zapatista movement through the Internet, which argues for the net's potential to present events through a dialogic economy of geographical knowledge that promotes not just 'distance learning' but distanciated coalitions).

Thompson's account of mediazation also highlights a fourth take on virtuality that recurs throughout this volume: *the virtual as spatial*. This spatial constitution of the virtual is, we hope, already implicitly apparent from the preceding passages. Nonetheless, it is worth emphasising that this collection's title signals not only the need for studies of human geography to take virtuality seriously but, equally importantly, the need for studies of virtuality to place questions of geography centre stage. This argument has three main supports. First, if overgeneralised portraits of the virtual are to be avoided, the distribution of virtuality must be addressed: through documenting the differentials in access to virtual forms (for example, by tackling the obvious but still less than fully answered question of who uses and is able to use the Internet); and by analysing the contexts within which virtual technologies are put into practice (for example asking why and how, in particular times and places, the Internet is used). Second, and building on this last point, the (re)territorialisations produced through the incorporation of the virtual into other geographies need to be examined. In this collection, for example, Jeremy Stein shows how the early development of the telephone, at least in London, was shaped particularly strongly by the existing political-economic geographies of urban land holding and property rights. Christopher Ray and Hilary Talbot analyse how ideas and practices of telematics get worked through multiple territorialisations when incorporated within 'development' discourses. Third, the other dimensions of virtuality we have identified – mediazation, complexity and simulation – are themselves all fundamentally geographical. Thus one gets Thompson's understanding of mediazation as the reconfiguration of contextuality; as well as others' arguments for how 'message bearing systems' do more than move information, but produce spatial entities, networks of connection, and sites of centrality and peripherality (see Bingham's review of the work of Latour and Serres). The form of this mediatised spatiality may be debated, and indeed may be far from uniform – contrast for example Stephen Graham's account of how virtual technologies are used to intensify centripetal panoptical geographies of surveillance with Michael Joyce's emphasis on the centrifugal, decentralised 'pantopical' (everywhere at the same time) geographies of hypertextuality – but the geographical constitution of the virtual-as-mediazation remains a constant theme. The same is true of portraits of the virtual in terms of complexity. After all, the auras of complexity that surround specific virtual assemblages are conceived in fundamentally spatial ways (see for examples Imken on the 'global matrix' and Joyce on the 'boundfulness' of hypertextuality). Finally, understandings of the virtual as Other to the real are equally, if sometimes less explicitly, concerned with space. Marcus Doel and David Clarke's concluding chapter makes this argument at a general conceptual level. In theorising virtual worlds they suggest that:

> what is at stake . . . is not just how we think about reality, virtuality, and virtual reality: it is also how we figure space-time itself. . . . It is the need to rethink space-time, rather than any new-fangled technologies, which poses the most pressing challenge.
>
> (Doel and Clarke this volume: 297)

In a neat symmetry, Ken Hillis' opening contribution on Virtual Reality develops along a parallel path, for example by exploring its positioning of subjects (for example as immersed in and/or distanced from the world) and its locatings of reality and enlightenment. Virtuality, then, is not just something which operates through and across space. It is at its heart a spatial phenomenon.

The essays

As the preceding sections illustrate, this collection is deliberately wide-ranging both in substance and style. Partly because of this, whilst we have already woven into our introductory remarks some comments on the essays that follow, we want now to say a little about each in turn. The book begins with three contributions that emphasise the historical geographies of the virtual. Ken Hillis' essay, 'Toward the light "within" . . .', sets out to engage critically with the influential hype surrounding new communicational and computational technologies, and to that end contextualises them 'within longer histories of optics, light, envisioning and mediation'. Focusing primarily on Virtual Reality (VR) technologies, Hillis argues that they, like other contemporary instances of telematics, not only combine digital computation with telecommunications but also put great emphasis on vision and the experiences it offers, especially as mediated through the screen. This is more than a technical feature. Rather, much of the power attributed to these technologies to recast the world and the self and to provide a new source of enlightenment is bound up with their visualities. Weaving a commentary on VR through an analysis of 'prefigurative technologies' of seeing such as the camera obscura, the magic lantern, the panorama and the stereoscope, Hillis draws out four dimensions through which these visualities can be understood: their positioning of the subject in relationship to the world; their locating of knowledge and enlightenment; their spatialisation of the real and the image; and their metaphorical understanding of light and its relationships to truth. The overall effect is to make some of the hyperbolic claims for VR both understandable – for instance through an explanation of how light and truth become located and localised in technologies – and open to cautionary criticism.

Jeremy Stein's chapter on 'The telephone, its social shaping and public negotiation . . .' shares Hillis' concern with historical contextualisation, albeit pursuing it with less broad brush strokes through a focus on late nineteenth-century and early twentieth-century London. Stein points out that contemporary debates over the transformations of urban space brought about by the emergence of informational and networked cities were paralleled nearly 100 years ago in debates over urban telephony. Reviewing debates in the national press over what became known as 'the telephone question' he shows how much the same issues dominated: the effect of this new communicational technology on social life in general, and face-to-face interaction in particular; its relations to existing moral codes of privacy and social status; the social patterning of access to the telephone; its economic and geographical impacts; its symbolic role as a sign of personal, organisational and municipal status.

The first contribution of Stein's discussion is therefore to stimulate some scepticism over the 'newness' of our contemporary new technologies. Moreover, Stein's approach to the telephone is to emphasise its (ongoing) social production rather than its intrinsic powers. He argues that 'the development of the telephone was literally "grounded" in London's unique political and institutional geography, in its system of land structure and in a range of political discourses' (this volume: 45). He pays particular attention to the crucial role played by existing territorial property relations – above all the legal difficulties telephone providers had in achieving 'wayleave powers' to facilitate the building of the telephonic infrastructure – and institutional ideologies – where professional and entrepreneurial ethoses clashed. The overall emphasis is on how 'the way the telephone system developed in Britain was not inevitable'. This virtual technology did not emerge fully formed, but developed, and continues to develop, in particular contexts and in relationship to other technologies, institutional and economic ideologies, and existing configurations of space.

Wendy Larner's chapter on 'Consumers or workers?: restructuring telecommunications in Aotearoa/New Zealand' marks a shift of period, to the 1980s, and place but in many ways it extends the analytical approach developed by Stein. Examining the dramatic shift made in New Zealand's telecommunications from 'state governance' to neo-liberal 'market governance' – which in a much milder form is prefigured in Stein's discussion of the arguments over the National Telephone Company in London 80 years earlier – Larner focuses on the relations between communication technologies, institutional forms and social identities. In particular, she examines how the figure of 'the consumer' becomes a hegemonic identity, supplanting that of 'the worker' which is reduced to a purely economic rather than a more fully social lexicon. In passing it is worth pointing out, as Larner does, that both the consumer and the worker are, of course, themselves somewhat virtual constructions; after all, real consumers are often workers too, and much that is done in their name is therefore highly detrimental to them (Miller 1995). But Larner's central argument is that telecommunications technologies need to be contextualised in specific times and places, and in terms of the institutional forms and social identities they are constituted by. In turn, a fascinating question for further consideration becomes the extent to which those (socially shaped) technologies themselves facilitate, both discursively and practically, the development of these institutional forms and social identities. So, for example, in what ways does market governance draw on telecommunicational symbolism and materiality to construct its emphases on a surpassing of the state by globalisation?

The chapter from Laura Chernaik, 'Transnationalism, technoscience and difference', introduces a rather different style of analysis. Drawing in particular on the writings of Deleuze and Guattari, Haraway, and Latour, Chernaik highlights a number of productive concepts for thinking through the virtual. Perhaps, particularly important is her emphasis, taken from

Haraway, on surpassing oppositions of the material and the semiotic through an emphasis on 'material–semiotic practices'. Given the dual and paradoxical tendency to conceive of the virtual as the ephemeral, and technology as somehow outside the social and its webs of meaning, this is an important corrective on two fronts. A related contribution is suggested for Deleuze's and Guattari's concept of 'machines', which far from viewing them as bounded things treats them as collectivities, bringing together entities from a host of different categories (both human and non-human, organic and inorganic), and only through this assemblage, rather than through an intrinsic, narrowly technical capacity, having the effective power to 'construct a new way of being and thinking'. Otto Imken's essay on 'the convergence of virtual and actual in the Global Matrix' certainly takes such an approach. It opens: 'I think it best to drop all trepidation and begin to treat the global telecommunications Matrix like a new artificial life-form: not a mere organism . . . but a nonlinear, asymmetrical, chaotically-assembled functionality with much more potential freedom than that of an entity encased in skin or limited to being an agglomeration of discrete organs'. Imken, then, offers an alternative to cautious appraisal, emphasising how, through their interconnection into a broader set of networks and assemblages, virtual technologies are implicated in a qualitative shift of human geographies into a 'matrix' form. Indeed, for Imken, attempts to mask this qualitative difference, for example by constructing user-friendly interfaces that present the Matrix in familiar, everyday guises are inherently problematic. He argues, 'if we want to understand anything about the Matrix, we must see it on its own terms . . .'. However, Imken emphasises how the politics of the Matrix are far from fixed. A key reason why we must recognise its nature and significance is so we can be active agents rather than passive participants within it. In particular, Imken stresses the need to work at developing the Matrix's potential for a decentralised and possible geography, one in which something original can be made to happen, not least through the making of unexpected links and connections. Whilst recognising the corporate and institutional pressures attempting to limit this, Imken at least sees the hope that a creative situationist ethos can be kept alive in, and indeed fostered by, the Matrix world of virtual geographies.

Jennifer Light's critical review of debates over the impacts of cyberspace on urban life, 'from cityspace to cyberspace', is likewise critical of overly pessimistic portraits. She reviews how two parallel critiques of virtual urban geographies have grown up and mirrored each other: one lamenting the loss of authentic urban life under the (non) weight of simulated and managed spaces (malls, walkways and so on); the other characterised by a 'cyber-pessimism' on the impacts of information technology. Together, these two critiques construct and use a dichotomy of the virtual and the real in which the 'real city' is authentic, organic, public and communal and the 'virtual city' is inauthentic, commodified, privatised and individualised. All the elements of this dichotomy are, Light suggests, questionable, something she demonstrates through a brief analysis of the ways in which community organisations in Chicago use the Internet. Real and virtual cities are, then, neither oppositionally different nor unconnected. In consequence new communication technologies, far from signalling the death of the city, may actually provide 'exciting opportunities to revitalise civic engagement in new ways'.

The complexity and multiplicity of virtual geographies is suggested by Stephen Graham's subsequent contribution on the 'geographies of surveillant-simulation'. As we have already noted earlier in this introduction, Graham rightly points out that contemporary developments in integrated telematics have often been conceived in overly narrow ways, with the Internet in particular coming to stand as a model for all. Drawing in part on the ideas of William Bogard (1996), he counteracts this by focusing on the connections being established between technologies and practices of surveillance (marked by the monitoring and attempted disciplining of behaviour), computation (with its construction of data bases) and simulation (with its real-time representations of behaviour and data). Three main examples are highlighted: digital CCTV and electronic tracking, as promoted for crime control; home teleservices and cyber-shopping, and in particular the 'transactionally generated information' and 'digital consumer personas' thereby produced; and 'road transport informatics', or the development of 'smart, digitally controlled highways'. Without subscribing to any form of technological determinism – Graham points out that surveillance technology can be used and socially organised in more progressive ways – the overall emphasis is on an intensification of surveillance, and a domination by corporate and institutional concerns with profit, flexibility and the effective targeting of subjects.

As the chapters by Stein, Light and Graham demonstrate, the virtual's encounters with 'real' geographies are often analytically located in the city. Christopher Ray's and Hilary Talbot's account of 'Rural telematics?: the information society and rural development' therefore provides an important empirical complement. However, perhaps the chapter's chief contribution is to supplement the 'grounded' analysis of virtuality undertaken by Stein and Larner, in so far as Ray and Talbot's concern is with how telematics, and its auras of a de-territorialising conquest of space, are re-territorialised through their subsumption within a number of rural development agendas. In part this combination of de-territorialisation and re-territorialisation is constituted within the discourses of telematics itself, at least as they are formulated by the likes of the European Union, through an emphasis on their ability both to transcend space and distance and, at the same time, to revalorise place, for example through the consequent ability of rural people to remain in and work from their local areas, or to promote their community to the wider world. However, the dialectic of de- and re-territorialisation is not quite this neat, for there is no single re-territorialisation being put into operation. Using a case study of the North of England, Ray and Talbot show

how telematics discourses are bound up with the network-building efforts of a range of differently territorialised institutions and organisations: European, national, regional and rural. In the end, what is at stake in their negotiations and conflicts is the sense and materiality of where telematics come from and what kinds of geographies they are productive of.

Oliver Froehling's chapter on 'Internauts and guerrilleros: the Zapatista rebellion in Chiapas, Mexico and its extension into cyberspace' continues this examination of the dialectic of de- and re-territorialisation. Froehling's focus is on the 're-scaling' of the Zapatista rebellion movement produced through its coverage and elaboration on the Internet. This has played a crucial role in converting what the Mexican government wished to locate as a local issue in the region of Chiapas into an internationally known and supported struggle. To some extent, then, this essay restates Jennifer Light's earlier optimism for the possibilities of the Internet to enact an engaged public consciousness and practice. However, it goes further. For Froehling's contention is not just that the Zapatistas have used the net as part of their campaign, but, especially given that the majority of sites devoted to the Zapatistas are 'located' outside of Chiapas and indeed Mexico, that this re-scaling has involved complex processes of translation, mediazation, and coalition building. For example, a key factor in the popularity of the Zapatistas is their 'net-friendly' politics and presentation: their non-hierarchical ethos can be seen as paralleling users' own conceptions of the politics of the net; and much of their networking was fronted by Subcommandante Marcos, who with his irreverence and humour became an iconic cyber personality. In turn, the face-to-face 'international encounters' that took place in 1996 and 1997 and set up the supportive 'International of Hope' maintained a focus on Chiapas but developed a broader global opposition to neo-liberalism. Froehling signals both the positive possibilities and inevitable tensions here, but his primary concern is with emphasising how one therefore needs to understand the Internet as part of a wider mediascape, fulfilling much the same role as newspapers and television, but with important differences in organisation (in this case including less Mexican state control), interactional character (having the potential to be more explicitly dialogic and participative), and 'audiences' and ethoses of affiliation (so that the arguments and rhetorics that work to build coalitions on the net are different to those that work in Chiapas or in other media).

The following chapter by Nina Wakeford stays on the topic of the Internet and its forms of communality, as it discusses 'Gender and the landscapes of computing in an Internet cafe'. It works from and through two theoretical starting points. The first is to approach gender and technology as 'mutually constituted', such that not only is technology gendered but it also plays an important role in the making of embodied gender orders and practices itself. The second is to identify three different landscapes of computing and the net: 'on-line landscapes' (the visual and textual landscapes interacted with when logged on); 'expert landscapes of the machine' (the technical expertise that establishes the net and is necessary for the inhabitation of the other net landscapes); and 'translation landscapes of computing' (the sites where 'the Internet is produced and interpreted for "ordinary people"'). These three Internet landscapes and their genderings are then examined through the case of an Internet cafe in London; a place where Internetting goes with a good cup of coffee, a slice of cyberpunk style, feminised ideologies of hospitality and service, and, despite the efforts of the cafe itself, masculinised ideologies of technical competence. Wakeford therefore provides an example of how to approach communicational technologies not as pre-given entities but as incorporated into particular places and practices, whilst at the same time presenting the Internet as constituted by multiple and contested genderings.

James Kneale's chapter on 'The virtual realities of technology and fiction: reading William Gibson's Cyberspace' continues the emphasis on what is done with the virtual, and its emplacement in the fabric of users' wider lives. It does this by examining readers' understandings of William Gibson's evocations of cyberspace, evocations that Kneale argues provide 'a complex and ambiguous fictional space for readers to explore, one which is rationally ordered [for example through the grid-like ordering of the matrix] but also open to fantastic uncertainty'. Substantively, Kneale's research with readers suggested a recognition of this uncertainty, but a trend to reduce rather than embrace it, both through playing up fragments of scientific realism and through bringing to bear other representations and experiences of cyberspace. More generally, this argument endorses and extends Wakeford's notion of landscapes of translation, through an analysis of the generic conventions and situated engagements that comprise the translational activity of reading.

Another aspect of Kneale's contribution is to recognise the constitution of cyberspace through technologies of writing and reading. This theme is taken up by Michael Joyce in his evocation of hypertextuality, 'On boundfulness: the space of hypertext bodies'. The precise nature of hypertextuality is of course not fixed, but by way of definition we can say that at the least it usually emphasises the production of a computerised, networked (hyper)text that comprises a 'navigable space' with no authorised route through it (unlike a linear narrative), and quite often combines this with interaction between readers and text and between different readers. Joyce tries to evoke something of this in his chapter. As he himself explains, 'in an attempt to approximate the space of hypertextuality in this linear form this essay is written in a series of overlays'. Indeed, through embodying the tensions of writing 'about' hypertextuality in a non-hypertextual volume – our own editorial requests rebound back on us here – Joyce's chapter brings into fascinating relief just what may be at stake in hypertext: its reconfigurations of authority, readership and textual spatiality. Taking these in reverse order, hypertextuality is understood as fundamentally spatial. This spatiality is characterised in terms of 'boundfulness' rather than boundedness, that is as 'a space that ever makes itself, slice by slice, section by section, contour by contour . . .'. More specifically, this boundfulness therefore involves: a 'metonymic imagination' which emphasises the relations between parts; contrapuntal aesthetics, productive of both disjuncture and plenitude; and quoting Marcos Novak, a potential 'condition of being in all places at one time', or at least of reworking the ontological status of there, here and now. (Joyce writes of one of his subject figures being 'no places at many times and though nowhere all at once elsewhere here'!) There is a recognition that such spatialities can be differently explored by different readers and readings: for example Joyce contrasts the 'young men on the flying trapezoids [who] move from link to link' with the evocation of a student of his, Samantha Chaitkin, who says she'd 'rather jump up in the air and let the ground re-arrange itself'. However, hypertextuality does facilitate, he argues, a particular way of experiencing textual space, a way of 'passing by', something he suggests through the notion of 'hypertextual contour', a 'sense of changing change across the surface of a text, . . . something less isobaric than erotic . . .'.

The following chapter by Nick Bingham begins to move us back towards concluding reflections. Entitled 'Unthinkable complexity? Cyberspace otherwise' it offers a strong argument against approaching virtual technologies as locations of the 'sublime', that is the experience of an-other entity which is awesome, powerful, unknowable, and in consequence desirable. This is not a question, we think, of denying the kinds of experiences of other spaces evoked by Joyce. It is, though, to emphasise the importance of not converting them into more general conceptions of the virtual, a virtual which is then treated as a technologically based totality impacting in awesome ways on non-virtual societies and selves. Rather, Bingham draws on the writings of Michel Serres and Bruno Latour to advocate an understanding of 'cyberspace as a message-bearing system'; that is, as a recent form of a much longer history of sociotechnical developments including writing, print, money, postal systems, cartography and telephony that bring together things and people so as to produce and link different spaces of human life. The effect, he argues, is to open up an approach to cyberspace which does not position it as a singular entity, characterised in terms of the usual 'cyber-cliches', but as a complex of technosocial networks, connections and disconnections.

The final essay, 'Virtual worlds: simulation, suppletion, s(ed)uction and simulacra' by Marcus Doel and David Clarke, approaches these questions of virtual possibilities through an interrogation of ideas of the virtual and the real. Both, they argue, are often poorly grasped. In particular, Doel and Clarke work through a critique of two positions: one that understands the virtual as a copy of the real, as a 'duplicitous xerography'; and another that reverses this logic to see the virtual as a resolution of the real, as its hyperreal improvement and solution (another manifestation of Bingham's technological sublime perhaps). Both, they suggest, leave the really pressing analytical task largely untouched; a task which, in a move that inverts the logic of most contributions to this collection with their emphasis on the grounded reality of the virtual, Doel and Clarke see as the rethinking of space-time in ways that address the virtuality of the real. This is a fitting end. For if any single lesson can be drawn from the diverse approaches contained here, it is that virtual geographies, far from being a specialised concern of interest only to net nerds, mad modellers or dedicated followers of intellectual fashion, are implicated in much wider questions of human life, human geography and human reality.

Part I

Embedding the virtual

2 Toward the light 'within'

Optical technologies, spatial metaphors and changing subjectivities

Ken Hillis

But of all the sciences Optics is the most fertile in marvelous expedients.
(Sir Daniel Brewster, *Letters on Natural Magic*, 1832: 5)

Introduction

In *City of Bits*, William J. Mitchell's (1996) paean to information technologies and their reconfiguration of urban social life, the author argues that immersion in the simulated environments that Virtual Reality (VR) technology makes possible constitutes a change in subjective identity. If previously the subject or viewer stared at a rectangular screen, with VR she or he becomes an inhabitant, moving from voyeur to engaged participant (p. 20).[1] Mitchell's observation, though part of the unproblematised hype surrounding VR and other 'new media', nonetheless points to a changed spatial milieu within and across which *experience* is gained, negotiated, transacted, and increasingly constituted by interactions with visual representations made possible by optical and digital technologies. Written from a more critical stance, this chapter examines some of the ways by which new technologies of 'seeing' such as VR are bound up with new spatialities and subjectivities.

VR, like the World Wide Web (WWW), the Internet, intranets, and Internet Relay Chat (IRC) technology, is an instance of telematics which not only merges telephony with digital computation but also relies on vision and the experience of it, in part mediated through the cathode ray tube or LED screen. Mitchell's assertion of engaged participation in VR relies on the seeming ability of VR, the 'wearable technology', to shrink the distance between the conceptions it proposes via the flat plane of the computer screen users wear and individual users' perceptions. Just as its position tracking device is able to synthesise something approaching a seamless continuity between body motility and screen images, VR's employment of a wrap-around stereoscopic display shrinks to almost nothing the line of sight between the technology's interface and the sight/vision of its user.

The positioning of VR as a *new* technology, the *next* thing, expresses a transcendent yearning to deny history and the contested circumstances of its making, and the necessary limits that attend material realities and their accompanying forms. The extreme spatialisation of VR, and users' relative ability to reformulate the virtual environment at will, seem to confirm the efficacy of denying history as a linear narrative, while at the same time the technology's reliance on software and codes affirms the social constructionist argument that 'all the world's a text', our bodies included. If one concurs that the world and reality are always already socially constructed, then it would seem 'only natural' to redirect a long-standing Western desire for transcendence towards technologies such as VR where limits appear to be constrained only by the imagination and the cultural contexts within which they operate. Avital Ronell (1994) notes that the so-called 'death of God' has dispersed sacred meaning, part of which has come to reside within technology. Modern individuals, charged with producing meaning and organised as the source of their own 'truth', may well approach VR with desires to don or 'perform' new identities as a kind of transcendence partaking of technology's power. In their popular account of VR, promoters Sherman and Judkins suggest that the technology 'is neither uncritically functional nor tackily quasi-scientific . . . it is poetic, mysterious, elusive . . .' (1993: 38).

The positioning of new technologies as the next thing reflects an ongoing quest for technological progress rooted partly in a belief that links human perfectibility to protheses, that enlightenment and a moral core of goodness is within us as subjects, if we work with the tools at hand to find it. Optical technologies relying on light, such as the camera obscura, the magic lantern, and VR, along with the metaphors by which these technologies are discursively and strategically positioned within cultures, are applied to making this exalted task of getting in touch with the light less onerous, and this alone becomes adequate moral justification for the current focus on virtual transcendence machines. Despite their differences of scale, both tools and technology extend our grasp. Earlier tools such as light metaphors and geometries were concepts used for accessing Godly truth and understanding. These tools were both bridging mechanisms between humans and their highest, most mysterious, cultural technology and ways to contemporise and reinvigorate God-as-highest-concept. For later Enlightenment theorists postulating a light within, the fixed source of Absolute light above was not fully extinguished. Though for many today the God behind this light is missing, lost, cancelled out, or written off, optical technologies may be seen to offer a labour-saving substitute: to allow individuals to communicate with one another as participants within an ideal sphere of continuously circulating communication. The wish to achieve such a state is what drives Kevin Kelly's (1994a) synthetic vision of telematics, metaphysics, and politics. Kelly is editor of *Wired*, the successful mass circulation magazine promoting the telematic reality of a coming wired world. He proposes we join together as 'dumb terminals' in an ecstatic unity via a rhizome-like cybernetic net to achieve a state he identifies as 'Hive Mind'. Hive Mind is a techno-humanist version of the ancient metaphysical notion of 'World Soul' first set forth by Plotinus. For Plato, World Soul is the animating principle of all things.

A delicate balancing act is at work in VR. Just as vision and sight have histories, so too do technologies of vision. Though VR and other optical devices are 'part of a sweeping reconfiguration of relations between an observing subject and modes of representation' (Crary 1994: 1), the technology, with its seemingly novel ability to suggest that subjectivity might be re-imagined in radically disembodied fashion, extends aspects of earlier optical technologies and draws from a number of philosophies of space, vision and light whose 'discursive formations' bear traces thousands of years old.

The account that follows, organised as a discussion of the continuities and breaks that inform immersive VR technology, helps explain why the technology is gaining cultural acceptance as an *idea*. I contextualise new communication and information technologies within longer histories of optics, light, envisioning and mediation. Promotions of current technological developments emphasise their novelty. I work through the issue of novelty by providing a history that focuses on VR as influenced by technologies of vision and metaphors of light, all of which have a pre-history. Thus, I examine the 'perfect vision' believed possible within previous technologies of light and vision, questions of immersion and fantasy through an examination of the camera obscura, stereoscope, panorama and magic lantern.

VR extends aspects of earlier optical technologies; but in doing so it also both renews and conflates long-standing Western notions of space related to a number of metaphors of light and vision. This chapter also therefore traces important changes (themselves partly depending upon the introduction of technologies that once were new) in metaphors of light and how they influence the understanding of light itself. These shifts confirm that though technologies have trajectories, histories and agency, these are not inevitable but contingent. For the Ancients, light is first on high, in the sky like God and the sun. Later, as nature materially recedes from cultural purviews, light is theorised as having relocated *to* the sphere of culture and then even inside ourselves *as if* we were Gods. As optical technologies become more powerful, able to correspond *progressively* to a 'perfect copy' of reality, we are able to reposition this light to within technologies that confirm the 'naturalness' of the inner light of individuated subjectivity. I relate the history of metaphors of light to how the different, fluid, and interdepending ways they connect subjectivity to light and transcendence are manifested in and help operationalise virtual technologies, which themselves reflect the resiliency of utopic thinking in the West. VR suggests that we 'see visions' when we close our eyes to the outside world. These visions, I argue, are Ideal ones confirming the 'correctness' of neo-Platonic associations between light, 'vision', and an originary source of truth. This truth source, once widely associated with a 'god on high', is now understood as part of technology's praxis; as a result, emerging optical technology itself becomes truth and, for some, even a god.

Smoke and mirrors: casting new Light on 'The Subject'

As with most new technologies, VR both extends and disrupts intertwining histories, in this case, histories of optical technologies, of light and vision, and of relationships between viewer/subject/user and machine. With this in mind, it is useful to discuss four 'prefigurative' optical technologies to suggest how their discursive positioning and repositioning over time inflects current optical practices informing VR. Because of its lengthier history, and its being made a metaphor for different, even oppositional, theories of subjectivity, the camera obscura in particular suggests ways VR extends and disrupts these earlier metaphors. I then introduce the magic lantern, stereoscope and panorama to provide a context for a larger discussion of how these devices, particularly the camera obscura and magic lantern, are reflected within the confines of VR.

Camera obscura

It is customary to credit Renaissance Neapolitan Giovanni Battista della Porta with the invention of the camera obscura,[2] some time before 1558. The thrust within American VR research is to have VR increasingly *correspond* to the natural world, to achieve through technical means a 'perfect copy' of reality that would be indistinguishable from that which it represents (see Bryson 1983; Coyne 1994). Belief in the eventual attainability of such correspondence, in part, extends and is informed by the dynamic underlying the Renaissance Doctrine of Signatures. This Doctrine infuses Porta's conceptualisation of the lived world and his theorisation of how the camera obscura might be used to represent and even double that world. The Doctrine of Signatures asserted that imprints and signs of inner meaning are everywhere to be found in the natural world, and they reflect or communicate a use or intention which can be read and acted upon (see Sack 1976). The Doctrine is also consonant with medieval belief that paintings were animated by the real world of nature, of which everything, including the paintings, formed a part. As I have argued elsewhere, for medieval people depiction in painting was literally true, and understood as coeval with the material or imaginative reality being represented (Hillis 1994: 4). Under the Doctrine of Signatures, referent and reference become the same; all things are linked regardless of time, place, scale or (im)materiality. Yet although representation has since replaced the similitudes posited by the Doctrine, it is a technology of thought to the degree that it renders *thought processes* more representational. Looking at the shape of a walnut's meat, for example, and linking this to the notion that walnuts must therefore benefit the brain, as medieval people believed, suggests a kind of dialectical thinking or juxtaposition on the part of the observer who gives meaning to patterns and shapes revealed to sight (see Manovich 1992). Walter Benjamin, in noting that the sphere of life that formerly seemed 'governed by the law of similarity was comprehensive'

(1979: 160), suggests that the ability or gift of producing similarities or what he terms 'natural correspondences' is also the gift of recognising them. Such correspondences awaken our mimetic faculty. They seem innate and yet their manifestations reflect the cultural and historical contexts in and through which they take place. Extending Benjamin, looking at a walnut is not the same as staring into a computer-generated world if only because the imaginative engagement posited by the Doctrine of Signatures has been, in a sense, built or factored into VR's technological practices. Put another way, VR may issue from the awakening of mimetic imagination suggested by earlier forms of correspondence.

Porta's description of the camera obscura anticipates something of the potential for VR touted by its promoters:

> in a dark Chamber . . . one may see as clearly and perspicuously, *as if they were before his eyes*, Huntings, Banquets, Armies of Enemies, Plays and all things that one desireth. Let there be over against that Chamber, where you desire to represent these things, some spacious Plain, where the sun can freely shine: upon that you shall set trees in Order, also Woods, Mountains, Rivers and Animals, that are really so, or made by Art, of Wood, or some other matter . . . Let there be Horns, Cornets, Trumpets sounded: those that are in the Chamber shall see Trees, Animals, Hunters Faces, and all the rest so plainly, that they cannot tell whether they be true or delusions . . . Hence it may appear to Philosophers, and those that study Opticks, *how vision is made*.
>
> (Porta 1658: 364–365, emphases added)

If the Doctrine of Signatures, as a cultural technology, assumed a correspondence of meaning or a similitude between things that looked alike, the exponential increase in the power of twentieth-century optical technologies to suggest the empirical truth of the illusion of reality they present not only supplants such earlier imaginative conceptions but also, ironically, suggests that recent technologies are truly novel and diverge from earlier concepts and devices. The technologies under review here raise issues of how the physiology of sight works. This was an ongoing concern of early modern science. Though VR relies on earlier scientific discoveries pertaining to sight, the development of VR is 'as a tool to re-present and manage data in a world where levels of information have increased exponentially' (Bleeker 1992: 11–12). VR *is* a novel form of 'training ground' on and in which users learn to overcome what would have been until recently resistance to the incoherent proposal that they might occupy the space of an image. This learning, however, is abetted by a lingering residue of belief in similitude: though today we claim to distinguish fully between images and referents, not only do users understand that images themselves are real, but they may choose to allow the image to stand in for the reality it represents. In this, the technology's sophistication is critical; however, this choice made by users speaks to the essense of Baudrillard's *sim*ulacra and reflects the underacknowledged cultural capital still invested in *sim*iltude. Aspects of magical thinking are alive in such technologised practices. The aforementioned twentieth-century divergence towards data and away from physiology, however, could only have taken place within a mode of thinking privileging the eye as a detached optical device. And in this way, though VR implicates users' bodies, it also suggests a visual home for a disincorporated or excised optical subjectivity already present in Descartes' study of the camera obscura, *La Dioptrique*, published in 1637. Porta clearly shows his own synthesis of idea and technology and his positioning of the camera obscura as both a scientific and magical device. Porta can be profitably read against Michael Benedikt's description of the polyvalency in VR. Benedikt suggests that the following hypothesis reflects a staggering increase in the future possibilities for rational communication in VR:

> You might reach for a cigarette that in my world is a pen, I might sit on a leather chair that in your world is a wooden bench. She appears to you as a wire whirlwind, to me as a ribbon of color. While I am looking at a three-dimensional cage of jittering data jacks, you can be seeing the same data in a floating average, perhaps a billowing field of 'wheat'.
>
> (Benedikt 1991: 180)

Porta's camera obscura is a mechanism used by individuals for seeing and comprehending a shared external world given by God. Benedikt sees VR as allowing access to a subjectively given world that, despite his claims to the contrary, cannot be shared precisely because each user's world can be so different. In effect, Benedikt is proposing a reality that celebrates pluralism expressed as consubstantial *difference* along with a technology of representation that trades on the already-noted lingering or resurgent belief in similitude. Though the forms and cultural contexts of the camera obscura and VR differ, all address an ongoing Western desire for transcendence from 'this earthly plane', and each suggests that this might be obtained, if only virtually, through the fusion of images and reality, and abandonment of the embodied constraints of real places.

Two possible objections to my argument should be addressed. Understood as a confluence of social practices conjoining in their form and using a host of desires, goals, ideologies, labour practices, discursive strategies and so forth, optical technologies clearly have 'utilitarian' applications; however, in no way do these preclude their being positioned within 'the social imaginary' as transcendent devices. Second, though one might argue, for example, that VR is in its infancy, like the Ford Model T, the contemporary automobile is more like the Model T than dissimilar. At base the technology remains the same. Though each device opens new possibilities, the camera obscura and VR, as theorised by Porta and Benedikt, both offer imaginary access to a parallel world in which, as if by magic, users might become the creator of their own ontological ground. I am not suggesting that transcendence has a universal signifier. For some it is escape from the body, for others the planet, for others both. For some the route follows a path toward celestial or outer space, for others that 'space' is ironically interior, whether the cyberspace on the 'other side' of the computer's interface, or the 'world' of the imagination, or both conjoined by the hybrid 'space' of an immersive virtual environment.

Technologies are not neutral. As material components of ideologies they help constitute as well as exemplify social processes and interests. The Renaissance camera obscura confirmed belief in the consubstantiality of all things, a belief I am suggesting is ironically updated, for example, in Benedikt's assertions about the benefits of extreme polyvalency of form within VR. In Enlightenment thought, the camera obscura was positioned as a model of visual truth, confirming the subjective interiority of viewers (Crary 1994). VR mixes and matches Renaissance understandings of the camera obscura as confirming the equivalence of simulation and reality with an Enlightenment understanding of the device as confirming the truth of individual subjective vision, hence Benedikt's belief that VR will augment communication by making each user's vision available to other users. Any communication in his model is between or among radically relative subjectivities who believe that total control over images-as-identities is key to a more direct communication with other images, machines, and (presumably) people. In other words, Benedikt hopes for a machine that delivers on the wish expressed in the phrase, 'If only you could see what I mean'. Such a wish forgets that visual symbols and images, like language, are always culturally inflected. It also promotes the wishful thinking that VR as communicatory 'space' would somehow obviate the need for discourse and negotiation of meaning.

Magic lanterns, panoramas and stereoscopes

The magic lantern, or *Phantasmagoria* as it was often called during the nineteenth century, likely invented in 1646 by a Dane (Godwin 1979: 83), though often credited to its populariser, Jesuit Athanasius Kircher (1602–1680), used a projection booth in which an artificial light source was refracted by and through a series of lenses, each with an image superimposed on it. Light passes through the images and projects them on to a wall or screen (sometimes formed of vapour or smoke) in front of relatively immobile viewers, who much as in the cinema, are in a darkened chamber between the projection device and the image.

The *Panorama* was a 360° cylindrical painting, which, when viewed from the centre, offered a sense of a simulated world that both surrounded the viewer and placed her or him at the centre of its finite display. Designed by Irishman Robert Barker, and demonstrated in a commercial setting in London's Leicester Square in 1792, the device also provided an experience of spatial and temporal mobility. Unlike the magic lantern, viewers depended on body movement in order to take in the 'finite but unbounded' surroundings. Painted landscapes of an earlier, bucolic countryside brought something of the country to the town, and of the past to the present (Friedberg 1993: 22). And the receding of the frame, one of the main features of immersive VR, is already present in the panorama experience.

Charles Wheatstone's invention of the stereoscopic display in 1833 revealed by instrumental means the importance of binocular vision in depth perception (Schwartz 1994: 40). The *Stereoscope* and stereoscopic photography are products of the sharp increase in study of physiology taking place between 1820 and 1840 (Bleeker 1992). The stereoscope is based on separate dual images, each depicting the same scene from slightly different vantage points, which together mimic the distance between our eyes (Rheingold 1991: 65). The Viewmaster is premised on Wheatstone's invention, and like VR, requires the viewer's intimate physical contact with the device. When presented separately to a user's left and right eyes, his or her visual sense merges the two disparate views into a single 3-D scene. The stereoscopic display created within twin video display terminals (VDT) built into contemporary head mounted displays (HMD) operate in a similar way in creating its illusion of immersion into virtual space.

Joining the dots . . .

While eighteenth-century discourse on the camera obscura positions a radical disjuncture between exterior world and subject, at base both the camera obscura and VR are immersive, and I would argue that this immersivity also inspires Porta's pre-Enlightenment, pre-modern vision of 'the light within'. His vision is one of opening up the world to exploration and representation in novel ways. The notion of explorer/exploration requires a freeing up of the encrusted medieval imagination which is, however, for Porta, not yet yoked to the Calvinist weight of responsibility that attends the requirement that individuals produce their own meaning. Jonathan Crary (1994: 38–40) argues that from the late 1500s, the camera obscura becomes *the* site of subjective individuation. The observer is isolated, enclosed, autonomous. From within the interiorising and privatising confines of the device he or she witnesses the mechanical representation of an objective world and determines appropriate distinctions between this world and the visual representations inside the machine (ibid.: 41). This adjudication or politicised aesthetics, flows, in part, from a desire to exclude disorder and privilege reason. The concept of a shared external world given by God is not so much rejected as it is supplanted by a growing awareness of an interior conscious (confirmed by using the device) increasingly focused on how it produces meaning and *orders* the world around it. The same technology that once confirmed God's plan now facilitates individuated perception of the world by the Cartesian *cogito* and users may place themselves in a sovereign position analogous to God's eye. By the eighteenth century, the camera obscura will have been repositioned to confirm the superiority of an interiorised individual producing meaning on 'his' own, fully in concordance with the Enlightenment's 'discovery' that a light within the modern individual can be cultivated through reason, taste, and hard work (Taylor 1994: 27–30).

Crary (1994: 33) argues against making links between the camera obscura and magic lantern. His important enquiry into the nineteenth-century subject opposes Enlightenment arguments about the camera obscura's relationship to an interiorised truth and a modernising (Protestant) subjectivity against the counter-Reformation context within which Kircher popularises the magic lantern. Such oppositions, however, are always partially dependent on spatio-temporal contexts. Though Crary notes the centrality of all things optical to the twentieth century, his project does not specifically address how earlier technologies of vision collectively contribute, in various partial ways, to current optical inventions and processes. I have already suggested connections between the camera obscura and VR, and it is equally possible to theorise how the magic lantern spectacle prefigures the transcendent luminosity and the ghostly or 'uncanny' illusions of today's virtual worlds. The magic lantern's nineteenth-century commercial success, moreover, depended less on associations with an originary source of divine illumination and more on a separate, sinister association with the spiritual. The deployment of magic lanterns for popular entertainment after 1802, in contradistinction to how the technology was positioned by scientists such as Sir David Brewster to dispel mysticism and the obscured mechanisms of illusion, *confirmed* the experience of spectres, ghosts and the spirit world (Crary 1994; Castle 1995).[3] If the camera obscura and magic lantern once reflected oppositional religious and ideological strategies of subjectivity and relationships of the subject to truth production, VR borrows aspects from any earlier optical technology that contains precursive mechanisms desirable to and confirming of a fracturing subjectivity seeking transcendence. VR thereby achieves a cultural point of purchase with subjects who seek to maintain control over their individual production of meaning even as they might play with the spectre of abandoning the formal maintenance of modern identity to external sources such as VR and the 'performativity' it encourages. Stated otherwise, VR is a world of images and data into which users insert themselves in search of greater productivity, enhanced subjectivity, escape, or combinations thereof. Whether positioned as a transcendence machine or a utilitarian prosthesis enhancing thought, VR reflects a desire for a return to either a pre-linguistic or a pre-lapsarian state, or both (Hillis 1998).

VR, with its brilliant interior of images that needs bear little relationship to the exterior world save for the socially inflected conceptions of reality of its software designers and users, is a world of artificial light; any 'objective' world it models is contained within a computer program. The technology, therefore, not only sets aside the spatio-temporal hierarchy between outside object and inside image, but also suggests that causal links between real world references and virtual environments are less necessary than once might have been judged to be the case. In other words, although VR dispenses with the dialectical model of clarity that Enlightenment thought believed was modelled by the camera obscura's relationship between exterior object (the real) and interior representation, by repositioning this object–subject binary entirely within its purview in a similar fashion to the panorama, VR seems to maintain the distinctions between environment and users (or space and subject). These distinctions, however, are culturally constructed. In the tradition of the camera obscura the dialectic between self and world appears to be confirmed, even as image, language, and referentiality stand in for the real. So, VR maintains distinctions between an 'anterior real' and its referents, even as it repositions these distinctions away from an observer who uses the technology to confirm these distinctions to one in which users themselves are inserted into the dialectic of the technology in order to confirm the reality of the illusions it presents, images of themselves included. VR thereby suggests that the interiority or 'blackbox' of computer software and hardware can operate in an adequate fashion to suggest an exteriority in opposition to users, who nonetheless, in Cartesian fashion, must imaginatively set their bodies aside to enter a virtual world and in a sublatory, almost re-medievalised fashion, merge with the display.

This ironic sense of merger also relies on the kind of immersive visual education provided by the stereoscope. Crary (1994: 40) notes that this device advanced the conflation between real and optical. The reduction of the idea of vision this implies is wholly embraced by many members of the American VR community and reflected in arguments that we will soon see what we mean. On a related note, like the stereoscope, VR relies on the saturation of gradients and surfaces users encounter with the kinds of visual details that Crary notes filled nineteenth-century stereoscopic images. A sense of flatness is at once confirmed yet denied through engaging the eye's attention with detail so that the implicit isotropic sense of space manifested in current immersive VR technology seems relaxed or given more of the attributes of a place experienced in extreme close-up. Unlike the panorama, at times VR can produce a parallax effect. This quasi-hallucinatory quality further deflects attention from the inherent flatness of the 2-D screen and images.

In her discussion of nineteenth-century *Phantasmagoria* technology, Terry Castle (1995: 141) notes that 'something external and public', the spectral illusions produced by the device, 'has now come to refer to something wholly internal or subjective: the phantasmic imagery of the mind'. The polyvalent world anticipated by Benedikt, wherein you are a ribbon of colour and I am a jittering data jack, reproduces the modern belief outlined by Castle that we 'see' figures and scenes in our minds, are haunted by our thoughts, which can 'materialize before us like phantoms in moments of hallucination, waking dream, or reverie' (ibid.: 143). Belief that we see such materialisations reflects the ongoing saliency of certain Stoical understandings of *phantasiai*, presentations or manifestations of what the soul seeks to see or believe (Goldhill 1996: 23). VR suggests the marriage not only of viewing and desire, but also of its own externality (and the publicness that networked applications may provide) to the interiority of the human imagination 'extended' to 'engage' with the privatised interior datascapes of sacred and profane commerce and pleasure generated within the machine. In 1931, Benjamin wrote that 'every day the need to possess the object in close-up in the form of a picture, *or rather a copy*, becomes more imperative' (1979: 250, emphasis added). As Benedikt's elegy suggests, in a virtual environment we each can be the controllers of our own phantasmagoria as we pursue individual combinations of 'truths' once available via the camera obscura, escapes provided by the magic lantern, and the uncanny sense of possessing familiar objects via manipulating their images in stereoscopes. Immersive VR further combines these kinds of pleasurable controls with the illusion that users might inhabit something like 'the space of a dream', and coexist and co-mingle therein with copies of their 'inner' thoughts, imaginations, and fancies. All of this relies on the play of light in virtual worlds. VR is interior illumination incarnate, subjective illumination conjoined to the machine in a hybrid or cyborg exorcism of interior subjectivity, even as the technology also confirms the light of a discrete inner subjectivity. VR is artificial light, a culturally-generated illumination that blends light and image to suggest it can be the source of 'natural' light, if only virtually. VR seems to suggest that it is only natural that light as a source of truth would be wholly artificial in nature. It is the genesis of this apparent dichotomy, or even incoherency, that is the subject of the following section.

Light metaphors and virtual technology

The West's understanding of the relationship between vision, sight, and light has been shaped by contradictory and complementary concepts often expressed through metaphors of light. The interplay between vision, sight, and light constitutes the essence of experience within a virtual environment. In the pages below I look at ways that three metaphors of light[4] (re)position spatial relationships between seeker/viewer/subject and light understood as a source of truth. I also suggest how these metaphors inform VR. The links between meaning and image, however, are fluid; maintained, policed, fought over and subject to change. To ignore the power of metaphors would be to take all images as literal expressions (Bal and Bryson 1994: 193), an idealism ironically fostered by much of the hype surrounding VR which suggests implicitly that we are 'hard wired' to perceive reality directly through vision, and that this new form of technologically-enhanced vision will lead to the promised land of 'post-symbolic communication' (Biocca and Lanier 1992: 160–161). Though the contexts and meaningful uses of metaphor evolve, aspects of specific connections suggested in early Hellenistic light metaphors between space, light, and the subject retain ongoing salience. Virtual technologies draw physical sight and metaphors of vision together. In so doing, they participate in a metaphysics of light as old as Plato's cave.

It is useful to think of concepts, for example, the spatial concept of being in the light, as metaphors whose origins we have forgotten. The earliest light metaphor I examine situates seers as being *in* the light that shines on high. In the second, after light is made to recede conceptually from the earth as its home, humans are no longer in the light, but look *into* it at a distance. Finally, the modern subject is both *in* and *of* the light; in addition to looking into the light, a separate inner light is posited as illuminating the individual's rational search for enlightenment.

In the light

Light and its associations with the day have been central to many cultures' metaphors of transcendence, the good, truth, and power. The early Greek philosopher Parmenides believed that darkness was overcome in the essence of light (Blumenberg 1993: 32). The concept of light originates in the primordial view of the world as darkness and light. Enmity between these forces generates awareness that nothing is self-evident, including truth. This does not mean that the dark is denied its due. Everything has a place in ancient Hellenistic expectations, the dark included (Walter 1988: 185). In the essence of light, darkness is overcome, and intellectuality surmounts material actuality. Light is the 'wherein' of nature and not a component part. Light is only visible when reflected by objects and is transcendent because it is not *of* the matter it reveals. Rather, like space, light articulates relations between this and that, here and there. Early classical thought understood humans as being *in* the light. In a similar way, we may think of ourselves as objects arrayed *in* space, a space in and by which we relate to other people and things.[5]

Plato's cave metaphor, however, transforms light into an *idea* of the good. For the Ancients, light, which gives all else visibility, does not have the character of an object. Ironically perhaps, abstracting light in this fashion turns 'a way of expressing the naturalness of truth into its opposite: truth becomes "localized" in transcendency' (Blumenberg 1993: 33). Light becomes a metaphysical truth, and partially because of this, light, along with the truth it carries, is conceptually withdrawn from the *kosmos* or world. Furthermore, despite Plato's identification of the 'eye of the soul' and the 'light of reason', and Aristotle's connections between vision, desire for knowledge and sensual delight, no Greek thinker really explained which *material* properties of sight might qualify it for such 'supreme philosophical honors' (Jonas 1982: 135). Plato, writing about *vision*, is most often using a metaphor of insight or pathway to knowledge and enlightenment. His implication of *sight per se* occurs through his use of visual metaphors that trade on mechanisms of seeing.

Parmenides's influence on Plato is considerable, and Plato's cave allegory does not deny the existence of dark places so much as suggests the natural connection between Being, light, and truth. The cave is a place metaphor for the *kosmos*. It is also a 'doctrine of the restriction of human knowledge imposed by the body', which does not allow us to grasp truth, but only shadows and echoes (Couch and Geer 1961: 496). People trapped in the cave learn to love the illusions 'projected on the walls of the dungeon of the flesh' (Heim 1993: 88), and it is also here that light is seized, exhausted and *lost*. Freed of the temptations of this limited earthly realm, those formerly trapped in the cave can ascend to the realm of active thought. However, few mortals are equal to this task despite the classical imperative that being in touch with God or the Idea of the Good 'was essential to full being'[6] (Taylor 1994: 28). This moral conundrum provides a second reason why light is detached from the earthly realm and metaphorised into salvation and immortality. Further, 'light, now otherworldly and pure . . . demands extraordinary, ecstatic attention, in which fulfilling contact and repellent dazzling become one' (Blumenberg 1993: 34). With the cave metaphor light is already withdrawn, in a kind of 'cosmic flight', from a connection with (human) nature to a more supernatural realm. Any former prisoner of the cave who might ascend toward the pure light would look back with compassion on those left in ignorance below. As if prefiguring the twentieth-century ideal dynamic of endless circulation within digital realms of information, such an illuminated, cosmopolitan individual would never return to the cave or a life among the (embodied) shadows, even though complete wisdom or virtue would forever elude her or his grasp (see Kitto 1964: 498–499).

Cicero (106 BC–43 BC) made Greek insight available to Roman culture with his translations of Greek philosophers. Amalgamating different theories of light, Cicero developed the concept of 'natural light', linking 'the metaphor of light with inner moral self-evidence' (Blumenberg 1993: 35), thereby somewhat reorienting a metaphysics of light. This *naturalis lux* would eventually filter down to inform Enlightenment assertions that humans also constitute a light source by virtue of access to this inner light as a gestating source for the self. In earlier Greek thought light articulated a universal space in which all were illuminated equally. Cicero, however, conceives of human life as existing in a clearing that light makes for our occupation. It shines in an 'economizing' fashion with respect to the space it illuminates, even though this clearing is a 'dazzling envelope . . . pure and absolute' (ibid.: 36). Darkness is beyond the clearing, a 'natural back ground zone'.

Cicero 'downsizes' the Ancient space that light illuminates from on high to one more in keeping with the finite spatial requirements within which an advanced (Roman) culture might take place. Not only are the good and the moral at the centre of this discrete clearing, which is visible and illuminated from above, but a second 'internalised' light begins to emanate from within, and takes the moral/aesthetic form of *virtu*. If the battle between light and dark suggested to Parmenides that nothing was self-evident, Cicero's repositioning of light begins an evolutionary process culminating in the notion that light illuminates being 'present to oneself'. The self starts to establish a moral claim on determining what might be true, and, with reference to issues examined in this chapter, manifests this claim in assertions linking the Enlightenment camera obscura, interior self and truth. With VR, an appeal is made to self-illumination to augment being present to oneself. Exterior light will enhance the truth of interior subjectivity, though this introduces the ironic risk of a remedievalisation or reexteriorisation of 'consciousness'.

When Christian thinking reworks Greco-Roman light metaphors, it therefore introduces a distinction between light a priori to earthly beings and *created* by God on the first day, and the multitude of earthly lights. In Exodus 3: 4, God appears to Moses in a burning bush. The Bible 'uses the element of light as the medium in which God becomes visible to man' (Jammer 1969: 36). The New Testament explicitly identifies God with light. In John 8: 12 it is written *Ego sum lux mundi*. God now becomes the reference source *behind* the light which emanates from His divine will. This distancing, which makes light a thing *and* a symbol, accords somewhat with a Neoplatonic positioning of seers as looking into a separate light, the source of which is withdrawn to 'on high'. The logical conflict between a Christian insistence on light versus evil and the earlier classically-inflected understanding is relaxed with the return of some of light's metaphysical powers to a God-as-origin.

Augustine's (354 BC–430 BC) conversion to Christianity was facilitated by his reading of Platonic philosophy. He redirects Christian theorisation of light back to the classical seeing *in* the light, yet he also conceives of an inner light that is 'behind' the self, a spatial move that returns the notion of origin or 'wherein' to light but also renders looking *into* it impossible (Blumenberg 1993: 50). Augustine posits a second differentiation between two kinds of light. *Lumen* is the objective, inexhaustible, intelligible and Divinely created radiance passing through and illuminating space. *Lux* is lumen's earthly, human reflection, our physiological experience of light and our capacity to receive it. Man becomes a light lit by light (ibid.: 43) and the connection between the eye and free will begins to be established.

Augustine's suggested relationship between the human eye and free will can be linked to his Platonic respect for a divinely inspired geometry whose reductive powers are made apparent via the eye acting as an agent for free will, as an intellectual mediator, *and* as a metaphor. Augustine's Platonism, however, also allows him to stress the primacy of geometry over perception. The eye may be central to geometry's ascendancy; however, Augustine distinguishes between vision and sight. In his words: 'reason advanced to the province of the eyes . . . It found . . . that nothing which the eyes beheld, could in any way be compared with what the mind discerned. These distinct and separate realities it also reduced to a branch of learning, and called it geometry.'[7]

Into the light

In contrast to Augustine, early Christian Neoplatonist mystics had conceived the finest access to truth as seeing *into* the light, as light was believed connected to the infinite and Heaven, and not of this earth. This effort at 'direct perception' reflects the suspicion of *logos*, and VR (often in the name of efficiency) indicates the ongoing search for pre-linguistic communication practices. This suspicion directed these Neoplatonists to give themselves over through direct perception to be dazzled by *lumen* in as unmediated a fashion as possible. Ironically, however, given belief that one looks *into* the light, the critical distance implied between seer and godhead, between receiver and Sender, demands a conduit for mediation, no matter how much its presence is decried by such mystics who are the culture-denying precursors of VR's promoters who advocate 'direct perception'.

Medieval Neoplatonism continues the tradition of an earlier Neoplatonism that had reversed the original Greek positioning of the seer *in* light, and within the 'wherein' of nature. The Dark Age seer looks *into* light in hope of entering its truth 'therein', or 'out there'. Yet medieval light cobbles this understanding to theories of light advanced by Cicero and Augustine. Medieval light is internalised to prevent 'the worldly dark from fully penetrating and disempowering the subject' (Blumenberg 1993: 51). The cavelike monastic cell becomes the bulwark of culture and the recess of memory (Carruthers 1990: 40). Something like a memory trace of Platonic light is carried within, while the barbarous reality of a natural world from which the light has been withdrawn is sealed off from view.[8]

Though the power of visual metaphor is somewhat diluted during the Middle Ages, Roger Bacon's *Opus Majus*, written during the 1260s, petitions papal authority to redirect Christian inquiry in accord with a *vision*ary perspective. Bacon, in placing vision directly on an axis of truth, follows Augustine in elevating geometry's status. Yet Bacon proposes geometry as an adjunct or enhancement to embodied vision. His *Opus Majus* reflects the thirteenth-century's interest in optics and mathematics that followed the renewed influence of Neoplatonist thought, and its conception of space as infinite and open (Jammer 1969: 39).

Victor Burgin's (1988) description of the Renaissance's synthesis of Euclidean geometry with the idea of a primary *perspective* suggests ways in which this synthesis fuelled the development of a parallel connection between an absolute light on high and the slowly emerging inner light of subjectivity. Burgin sets forth differences between two of Euclid's works, his *Elements of Geometry*, which codifies a number of earlier theorems which conflict with one another, and his *Optics*. It is in *Optics* that the 'cone of vision' is first theorised. In 1425 Brunelleschi theorised this cone to intersect with a plane surface, as part of devising his single point perspective programme. Although Euclidean geometry suggested an absolute and infinitely extensible 3-D space, the cone of vision helped to establish a somewhat contradictory belief that this infinite space had a centre. Imported into single point perspective technique, the cone suggested that the observer was at the centre of space (ibid.: 15), as is the case with the eighteenth-century *panorama*. Each modern observer is at the centre, in possession of a light that with practice can be directed outward or inward, forward or backward. This inner light illuminates an individual vision that extends outward along the infinite coordinates of a geometric and mental grid conceptually stamped on to the Earth's surface. After Brunelleschi and the cone of vision, this inner light traces the sight lines across an infinitely extensible space over which the individual eye may imaginatively voyage as if on high, as if light itself. The inner world of VR will create an imaginary space for further extending this interior voyage, suggesting that the interior is infinite if not eternal. With human agency placed at the centre of the dynamics of sight, the Platonic meaning of absolute light, more connected to vision and metaphor than sight and physiology, is inverted. The subject also is moving towards centre-stage and *into* the light. The physiology of sight is given an overlay of self-consciousness. Plato had argued the moral necessity for ascension into the Ideal light so that humans might possibly attain full being, metaphorically; ascended truth seekers would achieve the widened array of ideals and forms presented to them within the refined space of ideal vision. The Enlightenment does not so much change Plato's imperative as reconfigure the spatial metaphors by which this human duty is given direction. As suggested in the discussion of the camera obscura, seekers now must labour to find the light within, and the resulting inward orientation helps explain the primacy of modern subjectivity (Taylor 1994: 29).

There are intimations in Cicero, more fully articulated by Augustine, that free will and the eye have a role to play in the production of light or the Good. With the eighteenth-century Enlightenment belief that humans are endowed with a moral sense came the more developed understanding that this illuminating source of the Good also lies deep within us. Similar to how the early modern subject within a camera obscura produces meaning within an interior recess, the self comes to be seen as harbouring a separate luminary power from that residing 'on high'. This inner light is a metaphor for the Good, and the camera obscura and magic lantern are metaphors confirming different aspects of the belief that an interiorised light now shines from within. This difference can also be theorised by suggesting a parallel between the divine *lumen* and the camera obscura; the camera obscura is a technology of lumen. It truthfully reflects the objective world of exterior reality once wholly divinely given, though, for early modern subjects, also culturally authored. In contrast, the magic lantern, and its world of shadows and illusions, is a technology of *lux*, or *lumen*'s earthly, hence potentially more faulty, reflection. As the interiorised subject increasingly positions her or himself as the producer and judge of truth, the distinctions between *lumen* and *lux* become less hard-bounded.

Moreover, the early modern viewer, whether having recourse to a camera obscura or magic lantern or both, does not yet have the wherewithal or cultural need to imagine that he or she might sublate his or her identity to the light as a condition for imaginative entry to an immaterial virtual world constituted in illusion, luminosity, pixelation and information as data, and positioned imaginatively as more truthful than the exhausted 'real' of the natural world. Ironically, aspects of early modern scientific thinking nonetheless tend to support such imaginative twentieth-century sublation. With Issac Newton, the fundamental unity of matter and light is asserted (Koyré 1957: 207). Newton's unity can be read two ways. If light is matter then VR's metaphysical edge is potentially muted. However, it is equally possible to relocate materiality conceptually to an optical 'wherein'. Since Max Planck in 1900 and Albert Einstein in 1905, if light is a form of wave motion *and* a fast moving particle or 'packet', then it is possible to envision an optical immersive technology such as VR as rendering communication seemingly concrete even as VR dematerialises the physicality of the world it represents into a transcendent 'wherein'.

If it is accepted that modern thought retains a variety of subtle Neoplatonic influences, this distance between the self and the true light (of God) one seeks requires a conduit. The conduit metaphor of communications implies and demands uncontaminated passage of the message (from on high). Mediation across distance becomes the essence. When the metaphor of looking into the light is incorporated into virtual technologies, its earlier Neoplatonic moral imperative, which required purity of communication-at-a-distance from God to man, is updated and maintained by the assertion that technology is value free.

With respect to Neoplatonic mysticism and dazzlement, no one is able to accustom to the latter's absolute intensity, by which one is illuminated and blinded, has one's eyes fully open and resolutely shut. This mystical ambiguity was taken by early Neoplatonism to confirm God's illuminating and transcendent presence, which bypassed human communicatory and intellectual processes (ibid.: 45). To be dazzled is to be flooded by the universal light of God, a state of 'direct perception' achievable only by suspending the reflexivity and critical distance that normal cognition operating within a cultural milieu provides. Yet at the same time as this metaphysical directness-at-a-distance is being constituted as an axis for faith, Augustine also argues that one can open one's eyes in the dark or close them to the light, turning one's gaze inward. This free will in part depends upon the light that increasingly comes to be seen as 'shining within' as a reflection of God and therefore 'above' or 'before' culture. Sight-dependent subjectivity had been absent in classical thought, which, in its various metaphors of vision, had not accorded this degree of primacy to the eye. The interior self coming into being in Enlightenment thought is fertilised by philosophy's elevation of the eye's power, which is made to operate within the opening starting to develop between nature and culture in post-Hellenistic philosophy.

As if to anticipate individuated experiences within immersive virtual environments, Ciceronean and Neoplatonist direct perception attained by contemplation of pure light is an act of splendid isolation, and perhaps only conceivable within a sphere of dazzling *lux*ury[9] and culture – one which exerts a geographic or material influence on philosophical formation, the socio-political effects of which are often ignored. However, once one begins to communicate not only with God, but other people as well, the relationship between purity and the conduit metaphor of communication must evolve. If a (vertical) pure conduit was needed to transmit God's word in as uncorrupted a fashion as possible, thereby eliminating 'noise' from the heavenly transmission, when the conduit becomes 'horizontal', running between mundane, imperfect places, then the conduit's purity is implicitly available to purify the message being transmitted to 'imperfect' receptors. The reference is elevated above that to which it refers or from whom or where it was sent. The conduit or the technology then is believed to be not only potentially value free but is further privileged as morally superior to the message, sender and receiver. To communicate through a medium, therefore, is to have a sense that one's message might be touched by God. Enter metaphysics, and the more light-dependent the technology the more metaphysical the uses to which it might be put in seeking 'truth'. As if anticipating, for example, the pseudo-scientific New Age celebration of channelling's 'ability to resolve the technical problems of communication' (Ross 1991: 37), the purity of Neoplatonic light-as-conduit spiritualises information and the means of enlightenment and communication (see Davis 1993: 612).

In and of the light

I have made links to VR throughout the above history of light. I am now in a position to argue further continuities between earlier optical technologies, understandings of light and space and the contemporary virtual world. In *gestalt* terms, immersive virtual environments marry the 'modern ground' of an articulating Cartesian grid to a 'field' of the polyvalent 'identity formations' they situate. They are defined by light, the informing essence of sight. Given the spatial ambiguity that attends the Platonic 'localising' of truth-as-light in transcendency, *where* such a transcendent locality might be found would seem destined to remain a perpetual mystery. However, localising truth in transcendency via light implies that movement and, by extension, communication and its technologies become ironic sites of truth in and of themselves. In the West, such movement is often related to the emanating power of light. Truth becomes linked to movement and the pure, immaterial, and Ideal 'space' of communications. VR confirms the radical disconnection from real places with the modern practice of looking inside oneself for the light of truth. Real places are made to seem beside the point when truth is 'localised' to fibre-optically dependent transcendency, motion, and luminosity.

Stanley Cavell (1971: 102) notes that the Western 'condition' has accustomed itself to the naturalness of establishing a connection to the world by viewing it. He writes that 'we do not so much look at the world as look *out at it*, from behind the self'. With VR, users must first of all 'approach' the technology, a spatial move familiar to seekers entering the light *or* looking into it. In Neoplatonic fashion, users look into a virtual world composed of light. However, by then relocating a part of these individuals' sense of self to an icon located both in and of the light, VR collapses the Neoplatonic distance between light and self. This collapse is already underway with the stereoscope; however, by positioning the seer of and in the light, as both wherein and illuminated, VR goes beyond the stereoscope to suggest a transcendent doubling: both it and that part of the seer's iconised self 'within' the technology might now form a natural place. In the words of VR researchers Richard Held and Nathaniel Durlach: '[t]aking liberties with Shakespeare, we might say that "all the world's a display and all the individuals in it are operators in and on the display"' (1991: 232).

For Held and Durlach, it would seem as though users have become as one with the programme. Given the high status accorded interior subjectivity and self-identity, virtual environments may be seen as a relocation of the absolute light from on high to a place more convivial to this inward orientation, a hybrid of self-authored yet centrally authorised super-nature and peep-show rolled into one. The distinction between humans seeing the world around us and it showing itself to us is collapsed in VR, which organises a different binary in which users interact with images they can, in some programmes, alter or design, but only according to the preconditions designed into the technology. VR monitors users' body motility in part to reconfigure the images it presents. These images, however, are at least partially authored by the technology's designers, and subsequently translated into the code upon which VR relies.

Virtual environments, therefore, are also a pure interiorised space of culture, the virtual stage 'where' we are now expected to find and also be our own guiding lights. Recalling the contributions of the magic lantern towards thinking about virtual worlds, it is worthwhile to consider the emphasis on phantasm/fantasy accorded the magic lantern in contrast to the aura of science and truth that bathes the camera obscura. A nineteenth-century magic lantern experience took place in a cellar or darkened room and relied wholly on artificial light. The experience subverts the meaning of Plato's cave, a metaphor promoting separation of the faculty of sight from true knowledge. Though it is beyond the scope of the chapter to pursue, it is also worth considering that too much artificial light or *lux* deflects the quest for truth through the use of light into a pursuit of fantasy, today more commonly subsumed under the unquestioned rubric of 'pleasure'. In all of this the relationships between phantasm, utility, commerce and control that are established within virtual worlds remain underconsidered (Hillis forthcoming).

'Let's pretend': making communication replace existence

At the outset of this chapter, I mentioned Kevin Kelly's Hive Mind, identifying it as exhuming the Ancient metaphysical notion of World Soul.

Hive Mind is also the collective buzz of networks with/in which all bodies have become informational. If there is neither a natural location for the cosmic soul nor an otherworldly God available to Kelly today, there is the substitute possibility of fantasising the fibre-optics and light-dependent technologies of the Net and VR's spatial display as the immaterial, utopic embodiments of information as deity, a transcendent nowhere landscape 'where truth has gone'. The sublime netherworld of information permits the optical illusion that human bodies might merge with computers and the light 'within'. To quote the boy wonder computer whiz, Bryce Lynch, from the pilot episode for the short-lived TV series *Max Headroom*, 'You're looking at the future Mr. Grosman – people translated as data'.[10] To achieve such an incantatory state would be to ward off all the real-time viruses and other plagues of the flesh that bedevil contemporary supplicants who would gladly take leave of their 'impure' earthly form and their rootedness in the here and the now. This impurity, for those like Kelly, is constituted in a failure of the senses, belief in which can be traced back at least as far as Descartes. I think therefore I am is the opposite of the 'dumb terminal', a body or *automata* that communicates in a defective manner and therefore is in need of prosthetic devices to extend it towards enlightenment.[11] The purity of the conduit that Neoplatonic light demands to transmit its divine message from sender to receivers is conflated in Hive Mind into the network's ability to resolve the 'problems' of embodied, sensual communications. Held and Durlach's functionalist reduction of human experience to operators in and on the display shares the logic of Hive Mind, and reflects the ongoing wish that somehow communication technologies might both illuminate and stand in for embodiment and the ontological ground upon which (we think) we stand. Virtual reality: *as if*.

Notes

1 Mitchell's argument is cogent on this point; VR does offer the possibility for new subjectivities. However, *City of Bits* also contributes to the unproblematised hype surrounding VR. For example, in the passage just noted, Mitchell refers to 'immersion', as though the materiality of human bodies could reside in the immaterial 'wherein' of virtual environments. To suggest that the individual joins with the light of VR, as Mitchell implicitly does, is part of the historical process this paper outlines and discusses below.

2 The camera obscura, literally, a 'dark chamber', is '[a]n instrument consisting of a darkened chamber or box, into which light is admitted through a double convex lens, forming an image of external objects on a surface of paper, glass, etc., placed at the focus of the lens' (*Oxford English Dictionary*, s.v. 'camera obscura').

3 Brewster (1832) is disingenuous in this regard. He critiques the reliance upon smoke and mirror deceptions by reactionary rulers and despots seeking to maintain power through fear and illusion (pp. 56–57), yet he is also in awe of the contemporary smoke and mirrors technology. He not only describes how to construct it but is also favourable to its reception by a paying public (pp. 80–81).

4 The review of metaphors of light makes use of Blumenberg's 1957 history, 'Light as a Metaphor for Truth', published in English in 1993. For Blumenberg, the use of metaphor and narrative gives meaning to what would otherwise be a meaningless existence. Indeed, the philosophy and history of thought cannot be detached from metaphoric language.

5 The ancient notion of light as a 'wherein' that precedes the materiality to which it gives illumination and therefore spatial relations is not so very different from modern acknowledgements of the *basic* nature of light, and the modern theory of light which can only be stated in mathematical form (Brill 1980: 4). Brill notes that 'rather than trying to go further in "explaining" light, it is more useful to concentrate on its practical properties' (ibid.: 4), a statement in which epistemology swallows ontology and not unlike assertions by some geographers, who, in their fervour to study how 'space' is used, forget the politics that always freight its various conceptualisations and how they are then put into discourse.

6 Kitto (1964: 194) makes a similar observation to Taylor's: '[A]lthough Plato does not formally identify the Good with God, he speaks of its divine nature in such a way that formal identification would make but little difference'.

7 Augustine, *De ordine* 15: 42. Cited in Hofstadter and Kuhns 1976: 180.

8 This self-protective and insular medieval cultural move is not unlike that taken by many contemporary subjects choosing virtuality over reality.

9 The word *lux*ury contains a justification for the status quo of social inequalities and metropolitan privilege. *Lux*ury connotes the 'natural' stomping ground that is the due of those whose *lux* best reflects divine il*lumen*ation [*sic*].

10 The quote is from the pilot episode made by Chrysallis/Channel 4, *Blipverts*, 1985.

11 Enter Jaron Lanier's wish for a virtual world of 'post-symbolic communication' (Biocca and Lanier 1992: 161). Lanier's wish expresses an insecurity about how we operate as moral agents in the world. The cultural push towards all things virtual is a magical yearning that echoes Cabbalist assertions that a universal harmony is achievable through sound, shape and number (see Sack 1976: 321).

3 The telephone

Its social shaping and public negotiation in late nineteenth- and early twentieth-century London

Jeremy Stein

Introduction

In the light of recent advances in computing and telecommunications urban commentators have predicted the rise of 'infomational' and of 'networked' cities, places so interconnected by communications networks that cities are increasingly defined by their nodal position on the routes taken by global information flows, and by their ability to process and manage such information (Castells 1989, 1996; Graham and Marvin 1996). The consequences for the way cities are experienced and for the structure of time and space are profound. Yet these developments ought not to be viewed as historically new but as new phases of ongoing processes of change that began at least a century and a half ago with the construction of telegraph and telephone systems. By focusing on an earlier example of a 'virtual' or 'networked' city – the 'wiring' of London for its first telephone system in the late nineteenth and early twentieth century – this chapter aims to bring an historical perspective to contemporary technological change.

Scholars of the telephone have interpreted the technology from diverse perspectives, exploring many of the important questions usually asked of new communications technologies (see Pool 1977). Thus the majority of studies have considered the telephone's economic, social and geographic consequences: examining, for example, its impact on the speed and volume of information exchange, and on the operation of international financial markets (Garbade and Silber 1978; Thrift 1996; Michie 1997); its effects on community life and on traditional social networks based on face-to-face communication (Aronson 1971; Fischer 1992); and the dangers the telephone posed, because of its capacity for immediate communication, to Victorian and Edwardian societies and to their patterns of privacy, hierarchy, and strict social etiquette (Kern 1983). Further studies have explored the telephone's patterns of diffusion raising important questions about the diffusion process and about social access to technology (Robson 1973; Fischer 1988, 1992; Pike 1989; Martin 1991). The range of social responses to the telephone and to the emergence of electrification in Europe and North America is also well documented (Marvin 1988; Nye 1990). All of these studies in effect caution us against the uncritical acceptance of hyperbolic accounts of recent developments in information technology by demonstrating how at the end of the nineteenth century earlier 'new' and potentially threatening technologies generated a variety of utopian visions and cultural anxieties similar to that of our own age.

But my own interpretation of the telephone differs from these interpretations in three main respects: first, by focusing less on the telephone's consequences and more on the technology's social shaping; second, by embedding my discussion of technology in an urban context; and third, in my use of a social constructivist approach. Social constructivism, developed in recent years mainly by European historians and sociologists, sees technology as inherently social and thus as an object for sociological analysis. The approach has developed partly in response to criticisms of earlier approaches to the study of technology, mainly that of technological determinism. Social constructivism explores how different factors, agents and institutions shape the development of a technology. It rejects determinism and coherency in favour of contingency: that technology is continually shaped and reshaped, and in the process is subject to conflict, difference and resistance. It highlights technology's heterogeneous quality: the multiple influences on its development, and the constraints and possibilities these present; and emphasises that technology may have developed differently (Bijker *et al.* 1987; Bijker and Law 1992; Bijker 1995). Drawing on the insights of social constructivism I trace the development of London's early telephone system, and the public debates surrounding it. My approach raises a different set of social and geographic concerns to those interpretations of the telephone outlined above. I aim to show, for example, that the development of the telephone was literally 'grounded' in London's unique political and institutional geography, in its system of land structure, and in a range of political discourses. At the heart of these discourses were questions about the telephone's most appropriate uses, social access to it, its symbolic role for a capital and imperial city, and about the rights of the technology's institutional promoters to control the city's public and private spaces. These were issues of importance to London, but, as often, to cities in general thus raising important questions about the social shaping of technology in the environment of the modernising city. I argue that the telephone's development and geography was the outcome of a complex negotiation between a set of urban institutions and their ideologies, and a range of political discourses and public opinion. The means of debate – the use of rational argument – was an additional important element in the process of negotiation. The chapter is chronological and thematic in structure. Issues are discussed as they emerged historically.

The telephone, modernity and London, 'the world city'

The telephone was one of a set of technologies introduced to European and North American cities between 1870 and 1920. The telephone, invented and first patented in America in 1876, symbolised the growing significance in America of the electrical industry, the technologies associated with it, corporate business, professional associations and collaborative scientific research. Scholars agree that the telephone's inventors should be seen in the context of a large group of electrical men mostly working in the telegraph industry. As Platt observes, it was a general awareness of demands for better urban services among this fraternity that explains why several inventors in different countries would lay claim almost simultaneously to being first to perfect a telephone in the mid-1870s (Platt 1991: xvi). The telephone was thus closely associated with the second industrial revolution, a phrase used to denote the emergence between 1870 and 1920 of a set of new industries based on advances in electrical and chemical science, with associated changes in the structure and organisation of industry, including the advent of scientific management, mass advertising, mass consumption, and large-scale corporations (Hobsbawm 1968; Landes 1969; Chant 1989).

The telephone was one of several new space-binding technologies whose collective effect was to alter dramatically external relations of time and space between cities. The telephone directly contributed to processes of modernisation in the late nineteenth century and beyond. Its capacity to create instantaneous communication at a distance, and to destroy social and spatial barriers, was both a cause and effect of democratising tendencies in European societies, and extended an ongoing process of time-space convergence (Janelle 1968; Falk and Abler 1980; Kern 1983; Harvey 1989, 1990). This enhanced the growing sense of unity, a 'unity of disunity', among European urban populations, which Berman regards as an important element of the experience of modernity (Berman 1991: 15). During this time cities also experienced considerable internal changes. Networks of wires, pipes and cables provided a new range of urban services distributing water, power and information. Urban dwellers witnessed the construction of these networked cities and experienced the results (Tarr *et al.* 1987, Tarr and Dupuy 1988). In a British context little is known of how these wired cities came about or of how the inevitable problems in their construction – economic, social and administrative – were overcome.

Before embarking on an account of the construction of London's telephone network I want to make several initial points and to identify some of the main themes that emerge in my subsequent discussion. First, London's location at the centre of international trade and finance, and of the British Empire, ought not to imply that the city's communications systems arose largely out of pressures from abroad. Certainly London's position as a world city has significance, and makes it an obvious choice for the study of emerging telecommunications. However, the advent of the telephone, and the opening of telephone exchanges in London from 1880 onwards, was as much a consequence of internal urban pressures, of population growth and geographical expansion, as it was of external circumstances. The population of the county of London grew from 3 million to 4.5 million between the early 1860s and

1901, and the city's rate of growth between 1871 and 1901 was faster than the national average and faster than the provincial conurbations. As Asa Briggs describes it, London's growth was unique and 'seemed to obey no known laws' (Briggs 1968: 311–312). As with other expanding cities, London's spatial, demographic and commercial expansion stretched existing forms of communication, creating a need for such new responses as the telephone (Meier 1962). In colonial Melbourne, for example, as the city reached a critical threshold size there was a proliferation in systems of secondary communication: exchanges, agencies, trade journals, telegraph and telephone services, messengers and credit investigators; as distinct from 'primary' or face-to-face communication (Davison 1978: 131–133).

Second, London's existing institutional and political structures were important in shaping the way the telephone developed. In constructing their system, the private telephone companies faced a situation where much of the land upon which they wished to route their telephone wires was privately owned, or controlled by the city's public authorities. The latter in turn had highly complex and overlapping jurisdictions. Together these represented significant obstacles to the development of a telephone system. Third, the public debates surrounding the development of London's telephone system reveal much about the technology's social and symbolic value, and about the extent of social involvement in the shaping of technology by the city's diverse publics. The debates illustrate that in Victorian and Edwardian Britain it was predominantly the city's middle-class economic and political élites who were the major participants in the press and other public discussions on this issue. We should therefore interpret the telephone's development in light of these middle-class interests. We will see, for example, that the technology was in its early years primarily developed for its function as a business machine, to improve the efficiency of the city's trade and commerce, not for non-commercial purposes such as to enhance local community. Perhaps of equal importance to London's middle-class élites was the telephone's symbolic associations with modernity; for in their demands for an efficient telephone system was the desire that London should show itself to be a great capital and Imperial city by demonstrating that it was also technologically and commercially progressive.

Further issues arise when one considers the historical development of the telephone system and its geography. It must be stressed that in the short term the telephone was a relatively insignificant technology, with access to it effectively limited to wealthy individuals and to large businesses located in the central business districts of Britain's major cities. Rather its significance lay in the long-term creation of a telephone network, and in conjunction with other technologies, widened social access to various forms of non-face-to-face communication. Because of this, and because the telephone extended ongoing processes of social and spatial integration, the technology contributed to an extension of the public sphere (Habermas 1989). Over several decades the telephone system experienced considerable expansion. In London, for example, the number of telephone exchanges rose from eight to 30 between 1881 and 1893, and again from 47 to 62 between 1900 and 1912 (Baldwin 1925: 51–53, *Post 84*). By 1883 the volume of telephone messages in the city exceeded the number of postal telegrams, although the letter remained the most common means of non-face-to-face communication (*The Times*, 7 November 1883). Following the national pattern, whereby a diverse set of regional networks combined to form a basic national system by 1892, London's telephone network developed gradually over several decades. London was in fact relatively isolated by comparison with northern cities. Trunk lines connected London to Brighton in 1884 and to Croydon in 1888 but the line to Birmingham which connected London to Manchester, Liverpool and the industrial north, was only completed in 1890 (Robson 1973: 165–177). Although plans existed in 1888 to connect London with Bristol this western connection was still not made in 1892, though there was a fledgling international service (Robson 1973: 176; *The Times*, 5 July 1888). With the laying of the first telephone cable across the Channel in 1891 the Post Office opened a telephone service to Paris. That London was connected to Paris before Bristol indicates the telephone's economic and political significance in linking up foreign capitals and stock markets. Further cables were laid to France in 1897, and again in 1911 and 1912. A service to Brussels commenced in 1903. Both services showed steady increased use. The number of calls between England and France rose from 71,115 in 1909 to 96,806 in 1912, and in the same four-year period calls on the Anglo-Belgian service rose from 25,928 to 29,155 (*Post 86*). Prior to 1920, however, international telephony remained relatively undeveloped when compared to local, regional and national services. International calls were expensive, restricting their use mainly to government business and for a variety of commercial purposes to larger businesses and financial institutions.

Local social geography

The situation was highly problematic on the ground for within London private telephone companies faced a complex urban geography and administrative structure. The telephone's development exposed the patchwork quilt of British land structure and administrative political control. For example, when in 1880 the High Court declared the telephone to be a telegraph the private companies were forced to take licences from the Post Office, limiting them to tight areas of operation. In London, the United Telephone Company (UTC), was licensed to operate only within a radius of five miles from the General Post Office (Baldwin 1925: 51). In 1884 these early restrictions were relaxed, allowing private companies to construct long-distance trunk lines but failing to grant them sufficient powers to place telephone equipment on private property or on publicly controlled land (Perry 1992: 159).

Lacking statutory wayleave powers caused the private companies considerable problems. A wayleave was a legal document that declared the permission of a property owner to grant to another person or institution the right to use the property concerned for the purposes of routing their equipment on, over or under the ground. Parliament, having failed to grant private telephone companies monopoly rights of wayleave therefore forced companies to request from individual landowners and local authorities permission to locate telephone equipment on private and public land. Thus in constructing their system the telephone companies faced the political prejudice of London's public's bodies and the individualism of private property. In the case of individual or institutional landowners, many objected to private companies locating telephone equipment on their land because this interfered with their private property rights and because the equipment concerned was considered unsightly. Because of this the early private telephone companies paid employees on a commission basis to persuade people to allow them to erect telephone poles and wires on their roofs and gardens. In a city as complex as London this was no mean feat and had the effect of delaying the telephone systems expansion. The UTC, at its annual general meeting in 1887, admitted great difficulty in connecting the system to the suburbs due to problems of wayleave (*The Times*, 6 July 1887). In central London, where institutional landownership was considerable, it was necessary to gain the permission of large landowners. Institutional landowners could cause havoc for the telephone companies. For example, the refusal in April 1900 of the Bedford Estate to allow the National Telephone Company permission to place overhead wires on its land resulted in considerable inconvenience to the company and to businesses in surrounding districts, and was significant enough to be reported in the press. Similar troubles were experienced in Belgravia and Westminster (*The Times*, 18, 19, 21 April 1900).

Alternative development paths

The telephone was developed in Britain initially by private interests. Groups of financiers and merchants based in the City of London purchased from agents of the telephone's inventors rights to operate the telephone in Britain, and established two rival telephone companies, one in possession of Bell's, the other of Edison's patents. In 1880 these companies merged to form the United Telephone Company (UTC) (Baldwin 1925: 40). To circumvent the Post Office's restrictive licensing arrangements, the UTC established a set of subsidiary regional telephone companies, leasing to them the necessary telephone equipment but maintaining a controlling interest in these companies. Due to increased telephone business, the impending expiry of its patents, and the threat of competition, the UTC later amalgamated with its subsidiaries as the National Telephone Company Ltd (NTC). This process was largely completed by 1894, by which time the NTC controlled 90 per cent of the UK telephone market (Johannessen 1991: 158–159, 162).

From the outset the telephone was conceived by its promoters as essentially a business machine. Exchanges opened first in London's financial district, at

36 Coleman St and at 6 Lombard St, locating the telephone business close to potential sources of capital and to its most lucrative commercial market (Baldwin 1925: 51–53). Early advertising literature promoted the telephone's commercial advantages to the City of London's diverse commercial and trading community, stressing initially the telephone's capacity for improved *local* not long-distance communication. Alternative visions of the telephone's development did exist – for example, politicians, engineers and interested individuals debated in the press how to widen social access to the telephone, interventions that reveal aspects of the process whereby an élite technology is popularised. But, these arguments were largely about making the telephone available to a larger and more diverse commercial community, rather than about exploiting the telephone for non-commercial purposes, such as for social conversation or for enhancing the integration of local communities.

Direct involvement by the Post Office in the provision of a telephone exchange service was limited until the nationalisation of the NTC's local service on 31 December 1911. By July 1897, Post Office competition had resulted in only 37 telephone exchanges, none of which were in London, and 1,708 subscribers (compared with the NTC's 100,000); an average of 46 subscribers per exchange (*The Times*, 17 July 1897). Nevertheless by continuing to license telephone companies the Post Office continued to shape the environment in which the telephone business developed. At the municipal level, the telephone in Britain was not considered a public good as were water supply, gas, electricity and tramways, and thus developed differently from these essential urban services (Waller 1983: 302–306). Although the political ideology of mid- and late-Victorian municipalism did not exclude municipal ownership and operation of urban telecommunication systems, in practice this was unusual. Though municipal licences suggest that the creation of a national telephone monopoly run for private profit was not inevitable, of 1,334 local authorities only 13 applied for municipal licences and only six opened telephone exchanges (Perry 1992: 184).

Telephony in Britain remained a national rather than a municipal issue. This was because in the pre-1920 period the telephone was generally considered an item for the wealthy classes, not something that would become widely available. Believing the telephone's cost would severely limit its social use, municipalities were viewed as too small a social and technical unit to sustain a telephone system. Hence the telephone's development was largely shaped by national politics and ideologies. I show, for example, in subsequent sections how the institutions responsible for the telephone's management display the influence of entrepreneurial and professional ideals or ideologies common to mid- and late-Victorian Britain. The entrepreneurial ideal reached its zenith in the mid-Victorian period and was based upon the values of laissez-faire capitalism, on middle-class moral superiority, and on the superiority of active over passive capital. The professional ideal, which according to Perkin had largely undermined entrepreneurial capitalism by the end of the nineteenth century, was based on society's management by government, and by a cadre of technical and scientific experts (Perkin 1969; Perry 1977, 1992). The telephone was also appraised locally by a variety of institutions such as local authorities, chambers of commerce, and by individuals and public opinion. As with other technologies, to understand the telephone's social shaping in any particular place, one must consider the institutions, relevant social groups and individuals who appraise technology and combine the different geographical scales at which such appraisal takes place (Bijker 1995: 45–50).

The National Telephone Company and its critics

Until its nationalisation at the end of 1911 the NTC was subjected to intense criticism from politicians, local authorities, the press and from the public. The criticism reveals the various participants and their opinions in the public debate on the developing telephone system. The NTC, in defending itself, reveals the company's ideology and illustrates how modern organisations, increasingly subject to public scrutiny, had to defend their actions in the public sphere.

The NTC was mainly criticised for being an inefficient private monopoly that engaged in unfair monopolistic practices and charged excessive prices for an inferior service, which prevented the telephone becoming widely available. For example, in the 1890s, the Duke of Marlborough, the NTC's most ardent critic, made a series of public statements highly critical of the NTC and of the telephone's slow development in Britain. He thought matters were worse in London, where subscribers numbered only a few thousand, where the high cost of the telephone placed it beyond the means of the many (the NTC charged £20 p.a. to subscribers in London), and where the largely overhead single-wire system and primitive exchanges produced an inferior service (*The Times*, 29 August 1891). Convinced that a cheap and efficient telephone system could be developed in London, Marlborough established a rival telephone company, The New Telephone Company, which planned a twin-wire system of telephone exchanges to cover 23 square miles of central London and serve 25,000 subscribers with an annual tariff of £12 12s (Johannessen 1991: 183; *The Times*, 5 September 1891). The venture did not last, for in June 1892 the NTC purchased one third of the New Company's capital, and after the Duke of Marlborough's death, the entire company, which was finally wound up in December 1894 (*The Times*, 27 June 1892, 20 December 1894).

Marlborough's criticisms were significant because they were made publicly and because similar criticisms were made by politicians and newspaper editors. The NTC was forced to defend itself. In statements to Parliamentary Select Committees, to shareholders and subscribers, and in letters to the press, the NTC's representatives maintained that the company was denied rights that would make it capable of supplying an efficient national telephone system. For example, in 1892 J.S. Forbes, the NTC's Chairman, admitted that the company conducted a great deal of its business 'monstrously badly' but blamed Parliament for not granting the necessary powers. Forbes listed the obstacles in the company's way; they had to obtain licences from the Post Office and to pay the Government a 10 per cent royalty on gross receipts. They were subject to the unpredictability of Post Office telephone policy, to the threat of Post Office competition and to the threat that the Government might at any stage nationalise the telephone service. An important element of Forbes' argument was that the NTC lacked statutory wayleave powers: permission to bury or to erect poles and wires on privately owned or publicly administered land. The problem was manifested most in London where the telephone business was subject to the caprice of individuals and of landowners, and thus to constant disarrangement from the exercise of that caprice (Select Committee on the Telegraphs Bill 1892: 18–19).

The NTC's argument was a form of entrepreneurialism, in that if granted adequate legal powers, if given free rein to develop their system as a capitalistic enterprise, they would be able to construct an efficient telephone system (Perkin 1969: 221–230, 271–339). According to this view, lack of such an environment explained why the telephone system was not as efficient as it might be, and why comparisons with other countries and cities were false comparisons. In 1890, for example, the NTC's Chairman, F. R. Leyland, explained to shareholders the false comparison between the cost and efficiency of the telephone service in British and Scandinavian cities. Leyland explained that whereas in Britain lack of wayleave powers forced the NTC to choose circuitous routes for its wires, which added to costs, in Norway and Sweden the cost of transport of timber was less because poles were on the spot, and there was no 10 per cent royalty and no charge for wayleaves. London's relatively inefficient and expensive telephone service was blamed by Leyland on the high cost of wayleaves, double that of other British cities, and he claimed that nothing would change until the NTC was granted adequate powers (*The Times*, 12 July 1890). The Post Office at this time had limited statutory rights of wayleave granted to it under the Telegraph Acts of 1863 and 1878. These allowed it exclusive rights of wayleave on railway land and, with local authority approval, the right to bury wires and to erect poles on local authority land and, if refused, the right to appeal to a stipendiary magistrate. The NTC petitioned for similar powers. Both the NTC and its precursor the UTC sought to pass private Parliamentary Bills to achieve this end. All failed. The NTC also articulated its case publicly in several fora: in the press, before Parliamentary Select Committees, and to meetings of shareholders and of institutions convened to discuss aspects of the telephone question.

Other voices

In London the Post Office took little active involvement in telephone matters until the establishment of the Post Office London Telephone Service in 1901. Until then private telephone companies, engineers and London's public bodies, politicians, trade organisations, and telephone user groups debated how the telephone system should develop. All reported in the press, the issue was one of public concern and comment. The NTC's critics mostly espoused professionalism, putting the skills of engineers and technicians to public service, though the contrast between entrepreneurial and professional ideologies was not always clear. For example, the NTC's public statements after 1895 show increased concern to supply a telephone service in the public interest, whereas in 1908 the Government announced that the Post Office telephone service would henceforth be run along sound business principles (*The Times*, 2 April 1908). This shifting and blurring of ideologies should not obscure the fact that ideology mattered.

The most contentious issue during this period were the rates charged for the telephone service. Telephone rates were generally understood to be important because reductions in cost, as with the penny post and the sixpenny telegram, were known to widen social access to communication, which consequently extended the public sphere (Habermas 1989). The NTC was widely criticised for its excessive charges, as was the Post Office for its trunk line service and for its local service after nationalisation. In London, as elsewhere, several organisations campaigned for reductions in charges. Among these was the Association for the Protection of Telephone Subscribers, formed in April 1891, with 400 members, comprising 'the leading men and firms in every trade and profession in the metropolis' (*The Times*, 16 April 1891). The Association criticised the NTC for its high charges and inefficiency, petitioned for reductions in charges, and was successful in obtaining a reduction in private house telephone rentals, from £20 to £10 per annum (*The Pall Mall Gazette*, 20 January 1893; *The Times*, 20 March 1893).

The subject of telephone rates was also debated by London's Chamber of Commerce. At a special meeting of the Chamber in February 1892 held to discuss the state of the telephone service, the Chairman Sir A.K. Rollit spoke of the importance for commerce of improved transport and communication: despatch was an essential element of modern business, the saving of time being not the only source of profit, but a chief means of cheapening the cost of production and distribution. His opinion was that the telephone was not yet effective nor fully developed in Britain, and that other countries had greater advantages at less cost. After meeting with representatives of the National and New telephone companies, the Chamber's telephone committee concluded that to obtain an efficient telephone system for London, these companies required sufficient statutory powers. Following their advice the Chamber passed a resolution urging the Government to pass a General Powers Bill in Parliament to achieve this end, and to expand inter-trunk services throughout England to be run either by the state or by private companies (*The Times*, 23 February 1892).

Professional engineers were another important contributor to the debate about the telephone service. At times their technical expertise provided ammunition for other interested parties. The Duke of Marlborough, for example, based his scheme for a cheap and efficient London telephone service on the writings of William Preece, the Post Office's Chief Engineer, and on the calculations of another engineer who suggested that a telephone exchange system in London could be operated for an annual subscription of under £10 per subscriber. The most vociferous engineer was C.E. Webber. Webber, a Royal Engineer by training, was a former director of the UTC and a past President of the Institution of Electrical Engineers. He became an ardent critic of the NTC, and a great advocate of widened social access to the telephone. In 1892 he blamed the NTC for the retardation and inefficiency of the British telephone industry, arguing that if left in the hands of a private monopoly household use of the telephone would be limited to 200,000 subscribers (*The Times*, 24 August, 13 September 1892).

Considering they controlled much of London's public land, upon which the telephone companies wished to route their poles and wires, London's administrative bodies could not avoid involvement with the telephone's development. The power of local authorities over private companies was in fact considerable, demonstrated by the actions of the City of London Corporation which between 1891 and 1895 refused private companies permission to lay telephone wires beneath the City's public footpaths and carriageways. In so doing, the Corporation sought to use its power as the road authority in that district to obtain concessions from the private companies as to cost and effectiveness of the telephone service (*The Times*, 8 December 1893; Select Committee on the Telephone Service 1895: 111–117).

In its turn the London County Council (LCC) was actively involved in telephone matters. This was not surprising given that the vision of the Progressives, who dominated the LCC until 1907, was one of a municipality that fostered togetherness, in part through a co-ordinated transport system (Winter 1993: 191). Although the telephone was suggested as a possible remedy for relieving London's traffic congestion after the 1905 Royal Commission on Metropolitan Transport, the LCC's interest in telephones had more to do with curbing the excesses of a private monopoly than of fostering communal and civic spirit (Engineering 1905 80: 85, as cited in Winter 1993: 192–193). Thus in a deputation to the Postmaster-General in February 1895 the LCC's Deputy Chairman, Willoughby Dickinson, argued that London was in the hands of a 'gigantic monopoly' which gave a poor service and which charged higher rates for the telephone service than in other British or foreign cities. Dickinson believed that the telephone service could be provided in London at a cost of £10 p.a. The Postmaster-General disagreed, insisting that the telephone was a luxury whose cost could never be substantially reduced, thus preventing the telephone becoming available to a mass public (*The Times*, 5 February 1895). Sufficiently interested in the telephone service, the LCC commissioned two engineers' reports on the feasibility and cost of an LCC operated telephone service. The reports estimated the establishment costs of such a service at £350–400,000 with an annual maintenance cost of £75–90,000. This provided between 12 and 20 exchanges to serve over 10,000 subscribers. To reduce capital costs one plan involved using some of the Council's fire brigade stations as telephone exchanges with the advantage of assisting communication in the event of fires. Both reports estimated rates of subscription considerably below prevailing NTC rates, at between £8 and £10 per annum (Select Committee 1895, Appendices 4 and 5: 303–306). Although several years later the LCC considered applying for a municipal telephone licence, neither of these plans came to fruition. But they indicate alternative development trajectories, the seriousness with which the LCC treated the telephone question, and its efforts to cheapen, improve and to extend the service to a wider commercial public. Relations between the LCC and the NTC remained strained until 1901; the former seeking concessions from the NTC for the right to allow the company to place its wires beneath the city's streets, the latter arguing that the telephone service was impeded in London by unenlightened political leaders who denied them rights of wayleave with no right of appeal.

The telephone question

Public discussion about the telephone's development in London intersected with a debate about the telephone's development nationally. In 1892 the Government announced its intention to nationalise the trunk lines and to widen the powers of private companies in local areas, the aim being to extend telephone facilities to the public. Yet major issues remained unresolved, including whether the telephone service was best managed by the state, private enterprise or municipalities, the rates to be charged, and by implication how widely available socially the telephone ought to be. Another issue was the perceived backwardness of the United Kingdom telephone service relative to other countries. These issues were debated publicly under the heading 'The Telephone Question', a term first used by *The Times* in 1897 (*The Times*, 17 July 1897). Participants in this debate were national and local, including newspaper editors, news-reading publics, politicians, engineers, institutions such as the Post Office and national trade organisations, local authorities and individual landowners. The debate was mainly conducted through the press, demonstrating both the increased importance of national newspapers in the 1890s and 1900s and the greater social pressure on institutions to justify their actions publicly. The prinicipal means of debate was the use of rational arguments designed to appeal to the public's reason.

Developments in London contributed to and were influenced by the national debate on the telephone question. The issue of providing London with an adequate telephone system was the topic of seven conferences of the city's local authorities, held between 1898 and 1904, and hosted either by the LCC or the Corporation of London. The conferences dealt with various aspects of the London telephone service, including the NTC's Parliamentary Bills to enhance its powers, telephone rates and the possibility of London's local authorities combining to apply for a municipal telephone licence. In raising issues of concern to London's local authorities in connection with the development of a telephone service, the discussions reveal the telephone's social value to London's administrative bodies.

One issue in these debates was the right of private companies to control urban public space, in this case of telephone companies to place telephone equipment on publicly administered land. The issue was important and needed resolution before a telephone system could develop. The matter was expressed clearly by London's Lord Mayor in 1898 when speaking on behalf of the Corporation of London he declared that 'the roads and streets of the metropolis belonged to the people' and ought not to be given up to any trading company unless the rights of the public were protected. This comment by a senior public figure reveals how the development of the telephone network intersected with an ongoing negotiation and ideological struggle over the control of urban public space in the nineteenth century (Berman 1991; Atkins 1993; Goheen 1994). The 1898 conference in fact resolved to refuse the NTC permission to place its wires and pipes under London's streets until the public interest was secured by statute, as was the case for tramways and electric lighting companies (*The Times*, 18 March 1898). The issue was debated for several years, the NTC maintaining that it lacked appropriate powers to construct an efficient telephone system, and seeking to extend its powers through the passage of Private Parliamentary Bills. These Bills were vigorously opposed by local authorities in London and from across the country, many objecting to the principle of noncompliance with their ordinary rights.

A second issue was the state of the London telephone service, for the technology did not develop unproblematically or lead to automatic improvements in the speed and distance of communication. The early telephone system was in fact noted for its delays and inefficiency. In 1898 and 1899 London's local authorities passed resolutions condemning the NTC for its management of the telephone service, which it described as 'inadequate, inefficient and costly' (*The Times*, 18 March 1898). This was a persistent theme for the system was struck by delays, resulting from insufficient trunk line capacity, interference from neighbouring telegraph wires, operator error and occasional incompetence. Related to this was the perception that London's telephone service was deficient by comparison to other British and foreign cities. This issue was magnified in London, for at a time when it was generally assumed that technology was a measure of civilisation, and of Imperial greatness, commentators wondered how the Imperial capital and the world's largest city could be lacking in something so recognisably modern as a telephone service (Adas 1989; Headrick 1981). Hence in November 1897 a letter writer to *The Times* described as 'deplorable' how 'the greatest and richest city in the world should be at the tail of civilisation . . . instead of triumphantly at its head'; and in 1899 the engineer Charles Webber urged the Government to inaugurate in London a telephone service 'worthy of the metropolis of the Empire' (*The Times*, 3 November 1897, 23 January 1899).

A further issue was the telephone's cost and hence social access to the service. London's administrative bodies sought proposals to bring about a more efficient and socially accessible telephone service. LCC Chairman, Mr T McKinnon Wood, explained that there was 'almost unanimous agreement that use of the telephone under present conditions was grievously restricted; if cost could be materially lowered and the service improved in efficiency a great development of the system was to be expected'. McKinnon thought that London had a special grievance, having a costly and poorly developed telephone system relative to the city's size. Furthermore, only a small proportion of Londoners used the telephone, 'chiefly . . . the large commercial houses' while private use was limited (*The Times*, 15 June 1898). These comments typify a view common among London's politicians that the telephone had general public importance: to London's administration, to the efficiency of its business, and to the maintenance of its position at the centre of international trade and finance. The telephone's utility was not questioned. An efficient telephone system was something any city with pretensions for greatness ought to have. In a period when London was governed by a jigsaw of authorities with overlapping and unco-ordinated activity, and when Londoners were not known for their civic pride, such unity of interest was unusual. One can surmise that on a subject as important to London as the telephone question, London's different administrative institutions may not have been as far apart ideologically as their practical politics implied.

These issues were temporarily resolved in May 1899 when the Government announced its intention to allow local authorities to apply for telephone licences, and that the Post Office would soon open exchanges in London (*The Times*, 13 May 1899). Details of the Post Office system were announced in November 1901. A novelty of the system was the introduction of a party line and measured rate service in London, making it possible to subscribe to the telephone for as little as £5 10 p.a. This was in addition to fixed annual rates of subscription for unlimited use of the telephone. Thus the Post Office, responding to the recommendations of the 1898 Select Committee on the Telephone Service, sought to meet the needs of various classes of subscriber, including those who wished to make only moderate use of the telephone (Select Committee on Telephones 1898). An agreement between the NTC and the Post Office enabled inter-communication between the two systems and the two organisations agreed to provide identical services and to charge the same rates (*The Times*, 20 November 1901).

News of the system was greeted with acclaim by the editor of *The Times* who wrote that London was 'at last about to enjoy that improved and extended telephone service which has so long been promised' (*The Times*, 20 November 1901). Others were less enthusiastic. The LCC's Highways Committee welcomed intercommunication and the message rate service but felt that the aim of 'general, immediate and effective competition', recommended by the 1898 Select Committee, had not been attained (*The Times*, 25 November 1901). J.W. Benn, Chairman of the Highways Committee, argued that London's citizens and merchants had a right to a cheap and efficient telephone service but felt that the Government's proposals fell short of this ideal, being 50 to 70 per cent too high and, unless altered, likely to impede London's business (*The Times*, 26 November 1901). Others defended the Post Office, arguing that efficiency was as important as cost, and that the high charges were unique, resulting from the city's vast size, administrative complexity and from the cost and difficulty of obtaining wayleave agreements (*The Times*, 3, 5, 14 December 1901). London's public bodies were unconvinced. The Corporation of London's Streets Committee arranged a conference of London's local authorities to protest at the proposed scale of charges. The resulting conference, held in December 1901, criticised the Post Office for its high charges and for failing to provide by 'real and active' competition an efficient telephone service in London (*The Times*, 10, 23 December 1901).

The Post Office telephone system was designed to be technically up-to-date, and to avoid problems associated with earlier telephone systems. It was designed to serve a population of six million within an area of 640 square miles, whose boundaries were marked by Chipping Barnet and Enfield in the north, Bromley, Croydon and Redhill in the south, Romford in the east and Harrow and Hounslow in the west. The system was entirely underground, avoiding the problem of overhead wires, and constructed throughout by metallic circuits. Paper, a novel form of insulation, was used to prevent induction, instead of gutta percha, a vegetable product that was becoming scarce and expensive. Wires were wrapped with paper, twisted in pairs and then dried in ovens to remove moisture. These strands were twisted together and then covered with a leaden sheath to form cables. An additional advantage of using paper for insulation was that it had a lower electrostatic capacity, allowing speech to be carried four times further in length through a paper-covered cable than through one covered with gutta percha. Paper was also thinner, allowing five times as many wires to be packed into each subway conduit. Cast-iron pipes and earthenware ducts were used to carry the cables under the city's streets. Constructing subways for the wires and laying the cables was a colossal undertaking (see Plate 3.1). The Post Office's engineer-in-chief likened the work 'to the navigation of an unknown sea filled with shallows and rocks', because beneath London's streets there already existed an assortment of pipes and electrical cables, many of which were unmarked on the city's plans. Exchanges were equipped with multiple switchboards and the central battery system, the latest technical developments in telephony (*The Times*, 3 April 1902).

The inauguration of the Post Office telephone system did not lessen the criticism that London was, relatively at least, poorly and inefficiently served with telephones. Subscribers continued to complain of delays in connection, poor service and problems with intercommunication. Neither the Post Office nor the NTC were immune from criticism. In July 1903, the NTC's Chairman reported considerable expansion in the company's business but problems persisted, mainly the number of unexecuted orders, 10,563 in February 1904,

Plate 3.1 The networked city: constructing an underground subway for the Post Office's London Telephone Service, 1901
Source: E 320, courtesy of BT Archives.

up from 8,315 a year earlier. The Chairman blamed the additional capital expenditure necessary for each order, and the constant difficulty of obtaining wayleaves (*The Times*, 24 July 1903, 19 February 1904). In 1904 at a conference of London's local authorities convened to consider the Government's proposal to purchase the NTC's London system, the Lord Mayor stated that London was the prey of companies and trusts, while they waited for the Government to fulfil its promise to popularise the telephone. Resolutions were passed urging the Government not to purchase the NTC's undertaking until its licence expired, protesting at the Post Office telephone charges in London and at the 'inadequate and unsatisfactory trunk line service' (*The Times*, 13, 19 May 1904).

Central to the issue of telephone rates was who in fact the telephone was for. Many in the British telephone industry accepted that the potential market for telephones in Britain was huge, and only partially tapped, but how far was the market to extend socially? When the NTC sent its general manager and engineer-in-chief to America in 1904 they reported back that if free to develop their business the number of telephone users in Britain could be doubled in a short period (*The Times*, 19 February 1904). NTC officials blamed Britain's relative lack of progress in telephony on insufficient 'public education' and on political inaction and obstruction, referring to the Government's monopoly, the Treasury's unprogressive practices, and public conservatism – especially in the matter of wayleaves. Thus it was argued that, unlike in Britain, in North American cities hotels commonly had telephones in every room and municipalities allowed utilities to place all electrical wires underground, even insisting that this was done (*The Times*, 9 April 1904).

Discussion over telephone rates continued after the Post Office reached agreement in 1905 with the NTC to purchase its London plant in 1912, at the expiry of its licence. The agreement was welcomed because it guaranteed the telephone service's continuity and progressive expansion, though it remained the case that the telephone was largely used and conceived of as a business machine, not for wider public consumption (*The Times*, 10 August 1905). Yet there was also considerable agitation against the telephone being limited to the wealthy. When in 1907 the Government introduced into the House of Commons a Telegraph Bill, to raise £6 million to develop the telephone business over a four-year period, the main criticism was that it was a large sum of money for something that would benefit only the 'well-to-do' (*The Times*, 13 July 1907). Similar thinking prompted organisations in London and throughout the country to campaign for reductions in telephone rates. The Post Office itself sought to widen social access to the telephone by introducing measured rate services, designed to reduce the telephone's cost to small users, and by encouraging the telephone's use by marginal groups, for example in rural areas (*The Times*, 16 January 1906, 31 August, 18 October, 16 November 1907). Broader Post Office policy from 1908 was to conduct the telephone service on business-like principles thereby limiting significant reductions in charges. From about this time senior Post Office officials expressed the view that major changes in rates would have to await nationalisation of the telephone service. The main issue for them was to secure continuity of the telephone service (*The Times*, 17 July 1908, 5 March 1909). The onset of the First World War delayed significant moves towards popularisation which did not then resurface until the 1930s.

Throughout the discussion of telephone rates genuine attempts to widen social access to the telephone were balanced with considerations for the revenue, cost and efficiency of the telephone service. Despite attempts by some to make the telephone more popularly available, the telephone remained in the pre-1920 period a business instrument. The main issue was its cost, efficiency and systems of charging, mainly for the commercial community who used it. Such a conception underpinned NTC and Post Office policies which sought to make the telephone service pay. That the wider public might have need of a telephone and that this was a potentially profitable market had still to be learnt. Yet with the introduction of new services and reductions in cost the public was gradually granted access to telephonic communication, and admitted into a broadened public sphere.

Until its nationalisation in 1912, the NTC continued to criticise state and municipal management of the telephone service and to argue that the telephone was better managed by a private company. Attempts to extend its licence from 31 to 42 years were abandoned after 1892 when the Government announced its intention to purchase the NTC's trunk lines, as a first step in the nationalisation of the telephone service. After this date the NTC no longer challenged the decision to nationalise the telephone service. One might assume that by this time the NTC's senior personnel realised that public opinion, long critical of the NTC for its inefficiency, had shifted irretrievably in favour of the telephone's nationalisation. Another factor was the legal decision in 1880 which declared the telephone to be a 'telegraph', and which established early on the principle of state control of the telephone service. The NTC's officials must have realised from the outset that the prospect of nationalisation was really only a matter of time.

The issues associated with the 'telephone question' were resolved with the transfer of the NTC's remaining property and most of its staff to the Post Office on 31 December 1911. The resolution was only temporary. After the transition period telephone usage in Britain remained low by comparison with other countries. In 1921 there was one telephone for every 47 people in the United Kingdom, compared with one for every eight in the United States, and one for every ten in Canada. The onset of the First World War seriously impeded further development. Moreover, now entirely under state control there was still no unanimity between the Post Office and the Treasury on a policy of expanding the telephone service. Post Office administrators however now regarded efficiency, not lower rates, as their top priority (Perry 1992: 193–195). Earlier criticisms of the telephone service resurfaced under Post Office management. Thus subsequent Governments were criticised for their inconsistent telephone policy, for high telephone rates, for lack of investment, and for failing to develop a cheap, efficient and accessible telephone service comparable to that of other industrial countries (see for example Telephone Development Association 1930). These issues continued to be the subject of select committee inquiries (Select Committee on Telephone Charges 1920; Select Committee on the Telephone Service 1921, 1922).

Conclusions

The way the telephone system developed in Britain was not inevitable. In London the technology developed and was shaped by several parties. These included private telephone interests, the Post Office, London's numerous local authorities, engineers, the press and the public. These parties engaged in a public debate over the telephone question; a set of issues that concerned the cost, management and regulation of the telephone service. To make their case each party employed structured rational arguments designed to persuade educated urban publics and thus to influence public opinion. That arguments were expressed in public, either through the press or in venues where the press were certain to report on events, and that appeal was continually made to the public's rational mind are signs of the modern means by which the telephone question was debated and resolved.

The NTC, a private telephone monopoly, sought to provide the country's premier commercial market with an efficient telephone service but argued that it was continually obstructed in its task by limited statutory powers, and by fickle public authorities and private individuals. The NTC's entrepreneurial ideology was publicly articulated in the press and in a range of public venues. It sought to extend its powers through the passage of private Parliamentary Bills, and quickly repudiated public criticism of its actions. At times, the Post Office, subject to similar criticisms, used similar arguments. This demonstrates how private and public institutions were increasingly subject to public opinion and had ritually to defend their actions in the public sphere. In constructing their system the NTC faced the problem of a traditional land structure and the individualism of private property. London's local authorities argued collectively for an efficient and cheap telephone service, worthy of a national and Imperial capital city. These ideological arguments were played out physically over and beneath the city's streets, for instance when London's public bodies acted to deny the NTC access to the city's public space.

In the pre-1920 period the telephone was generally considered as a business machine for the wealthy classes. National trade organisations, telephone users and public bodies criticised the telephone service's inefficiency, and continually campaigned for reductions in charges. Except for a few solitary engineers and politicians, who made genuine arguments for the telephone's popularisation, the immediate concern was to improve and cheapen the telephone service for a self-interested commercial community and for middle class élites who were its prinicipal users. That the telephone could be generally useful, and profitably so, had still to be learnt. How this happened, and the processes of communication necessary for such a significant historical transformation, is an important subject for further research.

4 Consumers or workers?

Restructuring telecommunications in Aotearoa/New Zealand

Wendy Larner

Introduction

Much of the debate surrounding new communications technologies and social identities is polarised between celebratory and condemnatory accounts. Most visible are the discussions of new user identities, including those about net communities, cyber-cultures and cyborg selves. These accounts tend to portray either a 'brave new world' of fluid and multiple cyber-identities, or growing social inequality between electronic 'haves' and 'have-nots'. A parallel, but largely unrelated, literature on flexible work practices explores the relationships between new communications technologies and worker identities. In these analyses utopian visions of a post-industrial tele-cottaging world are played off against those of a panoptican associated with increased surveillance and control of workers.

Despite the hyperbole that accompanies these bi-polar debates, they serve to open important questions about the relationships between communications technologies, institutional forms and social identities. There can be no doubt that new communications technologies are associated with qualitatively new configurations of social and spatial power. The challenge, however, is to explore these issues without falling into the traps of technological determinism and/or epochal accounts of the 'postmodern' that assume unilinear correspondence between new communications technologies, institutions and identities.

Anti-essentialist accounts of identity formation, emerging out of feminist and poststructuralist theorising, allow us to address these questions in more nuanced ways. Rather than reading new identities as a direct consequence of broader technological and/or social processes, identity formation is understood to have its own dynamic involving multiple and heterogeneous social locales (Dean 1994: 165). Because these accounts do not assume any necessary relationship between communications technologies and social relations, attention is shifted to the complex systems of meaning and social practice through which institutions and identities are established in particular forms in specific historical conjunctures. The most obvious advantage of such approaches is that they allow us to move beyond formulations based on singular and overarching conceptual frameworks.

Drawing on these anti-essentialist theorisations, this chapter discusses relationships between new communications technologies, institutional forms and social identities in the telecommunications industry in Aotearoa/New Zealand. While there is general agreement that telecommunications are now an advanced producer service industry at the core of a new 'information economy', there has been little empirical research on the implications of the shift from national infrastructure to internationalised industry. Moreover, the few existing social scientific studies tend to assume pre-given identities and focus on the 'impacts' of restructuring, usually in terms of access and/or employment. Thus the analysis presented, which develops an account of changes in the identities 'consumer' and 'worker', is an intervention into both theoretical and substantive literatures on new communications technologies and social identities.

The 'New Zealand Experiment'

Once understood primarily as national infrastructure, telecommunications have been transformed into a new growth industry in which both the quantity and quality of products and services are burgeoning. Multinational telecommunications companies, particularly those from North America, now dominate the industry through both direct foreign investment and strategic alliances. While the future global structure of the industry remains unclear, the economic significance of the industry is not. As one prominent scholar observed, 'It has been evident, since the earliest moves, communications was destined to be a, if not the dominant industry in the twenty-first century (Schiller 1989: 113). In this context, an analysis of the telecommunications industry is a particularly apt lens through which to investigate the new configurations of social and spatial power.

The New Zealand telecommunications industry is unparalleled in both the speed and extent to which policy makers moved towards deregulation and privatisation. In less than a decade the provision of telecommunications services in New Zealand has shifted from being the responsibility of a government department with a mandate to provide a universal public service, to that of an internationalised producer service industry largely owned and controlled by multinational telecommunications companies. International comparisons suggest that New Zealand now has one of the most liberalised telecommunications sector of all OECD countries (see, for examples, *The Economist* 1991, *Wired* 1995, World Economic Forum 1996).

New Zealand also has a unique regulatory framework. Rather than a separate telecommunications regulatory authority such as the Canadian Radio-television and Telecommunications Commission (CRTC) or the British Office of Telecommunications (Oftel), general competition laws and 'self-regulation' govern the conduct of firms in the sector. 'Light handed regulation', as this regulatory framework is known, is based on the assumption that it is preferable to create incentives for market participants to negotiate their own solutions, resorting to the legal system if necessary, than it is for a regulatory body to intervene directly (Belgrave 1993: 2). Correspondingly, the official policy for the sector is that the Government will confine its role to that of ensuring the operation of these markets. This new form of governance is a variant of neo-liberalism that I refer to as 'market governance' (see Larner 1997a, 1997b).

The explicit rationale for market governance is to improve the provision of telecommunications to consumers. A recent speech by Hunter Donaldson, General Manager of the Communications Division of the Ministry of Commerce, outlined the characteristics of the new form of governance in the following terms:

> The underlying approach to the Government's telecommunications policy is that competition is the best regulator of the market. An open and competitive market place is most likely to produce an efficient and internationally competitive industry in New Zealand. Particular objectives are to ensure that New Zealand consumers of telecommunications services enjoy the best possible service offerings at the lowest possible cost.
>
> (Donaldson 1994: 2)

New Zealanders, as consumers, are encouraged to make active and informed choices about their use of telecommunications goods and services, thereby ensuring the efficacy of market governance.

My claim is that the introduction of market governance is best understood not as a measurable improvement in objective economic conditions (as proclaimed by its advocates), nor as a successful ideological bluff by foreign multinationals and their supporters (as argued by many opponents of recent changes), but rather as a qualitative shift in the source of social meaning from the sphere of production to consumption. In other words, market governance has not involved simply abandoning the claims of a national citizenry in order to attract international capital. Rather it has involved reconstituting those claims as those of consumers. In turn, labour has been recommodified and workers are no longer treated as a social collectivity with legitimate political claims on employers and the state.

This chapter shows how the shift from state governance to market governance was integrally associated with this redefinition of social identities. The analysis proceeds as follows. Following a brief theoretical discussion, I examine the emergence of the consumer as the hegemonic identity category in the telecommunications industry. I trace major shifts in the use of this term and relate these back to the distinct phases in the transformation from state governance to market governance. I show that in the context of market governance the identity of the consumer has inherited the legacy of the social. The second part of the chapter explores the implications of this shift for workers, showing how the social content of this identity was emptied out. In the final section, I discuss the implications of the argument made herein for political strategy. My claim is that anti-essentialist understandings of identity allow us to emphasise that consumer and worker are not necessarily mutually exclusive and antagonistic identities. In turn, this opens new political possibilities.

Theorising identity

The association between contemporary forms of restructuring and new social identities – most notably individualistic identities such as those of consumer, client and taxpayer – has been widely acknowledged. What has not been as widely developed is discussion of how new identity categories are constituted and come to be hegemonic. Indeed, often the assumption appears to be that the power of these new identities comes simply and directly from their mobilisation in the rhetoric of dominant political and economic groups.

Identity formation, or more specifically attempts to shape and reshape forms of subjectivity, is not simply the top-down effect of, or a response to, socio-structural shifts. Contemporary feminist and post-structuralist accounts of identity formation have shown that identities do not exist prior to, or outside of, specific social formations. Rather they are constituted in and through social formations, emerging in relation to shifting contexts made up of economic and social conditions, cultural and political institutions and ideologies (Alcoff 1988: 433). Thus, rather than attempting to establish the authenticity of any particular identity, it is more useful to investigate how identities emerge from a multiplicity of subject positions, showing that they develop in relation to each other, to endogenous formations and external influences, and accepting that they are likely to be multiple, unstable and historically discontinuous.

Of particular relevance to this chapter are those accounts of identity formation identified with neo-Foucauldian discussions of 'governmentality'. As Miller and Rose (1995: 428) explain, 'transformation in identity should not be studied just at the level of culture, nor solely in terms of the history of ideas about the self. A genealogy of identity must address the practices that act on human beings and human conduct in specific domains of existence, and the systems of thought that underpin these practices and are embodied within them'. In this literature it is argued that careful attention should be paid to not only the discourses through which identities are framed, but also to political technologies – the social practices through which these identities are constituted and consolidated. In this way it becomes possible to explore in more detail how certain forms of identity become aligned with particular political projects without portraying them as a direct consequence of broader social transformations.

Beginning from these theoretical premises has major implications for the way in which discussions about restructuring and social identities are framed. More traditional analyses would begin by identifying the location of specific actors in the production process, or by their membership of a particular political entity such as the nation-state, and then consider how new forms of identity develop in relation to changes in the structural location of these actors. Instead, this analysis begins from the assumption that the invocation of specific social identities is a central component of the processes associated with restructuring. Thus by examining shifts in identity categories, and their implications for social relationships, it is possible to gain a better understanding of how the changes associated with the consolidation of market governance in the New Zealand telecommunications sector were able to happen.

The consumer

At first glance the centrality of the consumer – sometimes known as the customer or user – to the new form of governance is both explicit and unambiguous. The aim of market governance is to improve service and reduce prices for consumers. Closer inspection, however, reveals major shifts in the use of the term 'consumer', associated with particular historical moments in the transformation of the sector. With an examination of the identity shifts associated with the four phases of restructuring in the telecommunications sector – corporatisation, network deregulation, privatisation and competition – I make visible both the power relations involved in the formation of this particular identity, and the other forms of identity against which it has been constituted. This allows me to 'denaturalise' the constitution of the consumer as the dominant form of social identity in the New Zealand telecommunications industry.

A discourse about the needs of consumers first emerged early in the corporatisation process. This discursive formation was linked to the demand for greater efficiency within, and better service from, the Post Office. Thus while the Annual Report of the Post Office had long identified its constituency as a universal 'public' or 'the people of New Zealand', the Mason–Morris report on the reform of the Post Office explicitly identified its mandate as being to ensure 'that the Post Office carries out its functions in the most' efficient and beneficial way for consumers and users of its services (Mason and Morris 1986: 6). At this time, however, explicit mobilisation of this identity was largely confined to business and large commercial users who were actively agitating for liberalisation of the telecommunications sector.

Consequently in the political debates around corporatisation – the process by which the Post Office was transformed into three State Owned Enterprises – claims invoking the identity of the consumer played very much a secondary role to those of the taxpayer and the worker. The Government framed the benefits of corporatisation primarily in terms of the better use of taxpayer's money (Larner 1997b). Considerable emphasis was also placed on the role that corporatisation would play in improving employment conditions within the public service – providing a framework within which 'public sector employees can contribute effectively and creatively to the economy' (Hansard 1986: 4725).

In these debates the identity of the consumer was primarily invoked by those contesting corporatisation. The National Party, in their capacity as the official Opposition, claimed that the corporatisation process was designed to 'rip off the consumer' (Hansard 1987: 9328). The basis for their argument was that SOEs could exploit their monopoly position by increasing prices, and that New Zealanders would end up paying twice for telecommunications – once as taxpayers and once as consumers. The other organisation to mobilise the identity of the consumer, via that of 'the ordinary customer', was the Post Office Union. They argued that the Government's plans would damage the ability of the Post Office to provide services to all New Zealanders at the lowest cost possible and would undermine public ownership.

Thus the debates over corporatisation were, in part, a contestation over the relative significance of political claims made on the basis of three different forms of identity: namely the taxpayer, the worker and the consumer. The political rationale for corporatisation was premised on the identity of the taxpayer although, presumably at least in part because of the Labour government's historical links with organised labour, the benefits of corporatisation for workers were also explicitly highlighted. The identity of the consumer, on the other hand, entered into these debates only indirectly – mobilised by those contesting corporatisation, often taking the form of customers or users, and in response to widely divergent political agendas.

That said, the appearance of the consumer in the corporatisation debates marked a significant moment in the reconstitution of social identities in the telecommunications sector. First, these claims mark the beginnings of a shift towards the universalising of the consumer as an identity category. Both National and the Post Office Union discursively constituted New Zealanders as consumers. Following the naming of the consumer in these debates, political claims made on the basis of this identity could be legitimately inserted into the domain of policy formation (Yeatman 1990: 154). Given the existence of other discourses premised on the consumer, including those of large business users and the Treasury (who had mobilised the identities of users and customers in their arguments for state sector reform), it is not surprising that political claims based on the identity of the consumer began to take on greater significance – albeit in diverse and contradictory forms.

It was in the debates over network deregulation that the consumer began to consolidate as the dominant form of social identity. Network deregulation involved the redefinition of the responsibility of the state as that of facilitating competition, as firms in a newly created 'market' took over responsibility for the provision of telecommunications. Unlike the corporatisation process, in which Labour politicians discursively constituted New Zealanders as taxpayers, the benefits of network deregulation were framed primarily in terms of the benefits to New Zealanders as consumers. The political rationale for network deregulation was that the introduction of competition would improve efficiency, and consequently bring down the costs of telecommunications goods and services for both business and domestic consumers.

The most obvious explanation for the new dominance of the consumer over the taxpayer and worker is that it reflected the co-option of the political process by large commercial users. Certainly business groups including TUANZ, the ITANZ and the New Zealand Business Roundtable were lobbying hard for more competition in order stimulate economic efficiency and drive down the costs of telecommunications for business and industry consumers. Yet public demand for change in the telecommunications sector was also widespread, particularly in Auckland where there had been a major telephone crash in 1987 (Joseph 1993: 39).

Moreover, the only major organisation actively to oppose network deregulation, the New Zealand Post Office Union, also framed their campaign primarily in terms of the consumer. In the absence of a domestic consumer's association in the sector, it fell to the Union to point out that while it was claimed that network deregulation was in the interests of the consumer, it was actually business consumers that were likely to be the primary beneficiaries of deregulation. In their campaign the Post Office Union emphasised the importance of retaining universal service, arguing that any move away from universality would reduce the quality of service to residential and rural customers. As part of their strategy the Union argued (unsuccessfully) for consumer representation on Telecom's Board, the setting up of a watchdog consumer group, and the establishment of a Department of Communications.

While the Post Office Union's arguments for consumer representation were not taken up directly, they did have the indirect consequence of both centring the identity of the consumer and influencing the form of the political technologies associated with market governance. Labour's response to public concern about the risks of network deregulation was to encourage Telecom NZ to make public three undertakings designed to explicitly protect the interests of domestic customers. These included promises to maintain a free-calling option for all residential telephone customers; to ensure that the rate of increase in residential telephone rentals would not increase in real terms relative to the Consumers Price Index; and that line rentals for residential users in rural areas would be no higher than standard residential rentals. These protections for domestic consumers were written into the company's Articles of Association and, following privatisation became enshrined as 'the Kiwi Share'. Rather than just a publicity exercise, or even as a neutral regulatory mechanism, these undertakings can be understood as a political technology that contributed towards the constitution of the consumer as the dominant form of identity in this sector.

Thus the emergence of the identity category of the consumer did not occur in a direct top down fashion. Rather it reflected the coalescing of a number of diverse discursive formulations, with different political origins, that came together in the debates around network deregulation, and resulted in an understanding that the new form of governance within the telecommunications sector should be based on the consumer. Moreover, the centrality of this identity category was reinforced by the political technologies put into place in the telecommunications sector.

The hegemony of the consumer continued to be consolidated with the shift towards market governance. Discussion of the pros and cons of privatisation – which involved the sale of Telecom NZ to a consortium dominated by Ameritech and Bell Atlantic – was almost entirely focused on the costs and benefits to the consumer. Indeed, it could be argued that it was because of the centrality of the identity of the consumer to the privatisation debates that concern about foreign ownership of a key infrastructural industry could be marginalised. The Parliamentary debate, for example, was not so much about the merits of privatisation as a concept, but rather revolved around how effectively Kiwi Share provisions would protect New Zealand consumers following the sale. Consequently National's argument for a cap on the level of foreign ownership was not framed as an argument for national ownership *per se*, but rather reflected their doubts about the Kiwi Share. Placing a limit on the level of overseas ownership was presented as the means by which they would ensure that the telephone system remained under the 'effective and real control' of New Zealanders (Hansard 1990: 806).

However it was only following the emergence of competition, as marked by the establishment of Clear Communications in 1990, that the consumer became the hegemonic identity category. As Clear's CEO explained, 'When we started to build this business back in 1990, everyone at Clear knew that the only lasting competitive advantage we could bring to the market was a quality based company focused not on satisfying our customer's needs but exceeding them. We aimed from the outset to offer a customer experience that was seamless, reliable, efficient and friendly' (Makin 1994: 6). The success of Clear Communications was to exceed everyone's expectations, including their own. By 1995 they accounted for 22 per cent of the national tolls markets and 23 per cent of the international tolls market (Ministry of Commerce 1995). As Telecom NZ found itself rapidly losing market share, the two companies became engaged in a heated battle for customers. In the course of this battle prices dropped dramatically, new services were introduced and there was considerable improvement in the overall quality of telecommunications services. It was with these developments that the discourse about the benefits of competition for the consumer was consolidated.

The consolidation of the consumer as the hegemonic identity category in the telecommunications sector marks the emergence of a new social order. In this new order corporate social responsibilities are constituted as involving consumers. Consequently Telecom NZ emphasises its involvement in social programmes including the 111 emergency service and special needs programmes, together with major sponsorship programmes such as the Telecom Regional Arts Awards, New Zealand Ballet, and support for school education. As the company explains, 'Our aim of helping to strengthen the social fabric of the country and improve the quality of life of its people is all part of being a company that is open and receptive to the needs of its customers: the community' (Telecom NZ 1991: 54). Significantly, in this new understanding the social is no longer territorially bounded, nor is it premised on a single locus of solidarity. Rather the social is being defined in terms of targeted services and programmes to address the needs of specific groups. Thus this new version of the social profoundly challenges forms of social solidarity premised on national coherence.

In short, what I am arguing is that the emergence of market governance was integrally associated with the entrenching of the consumer as the hegemonic form of social identity in the New Zealand telecommunications industry. Not only was this identity socially created, but it also meant different things at different times to different groups. It was only with the consolidation of market governance that the multiple and often contradictory meanings of the consumer were to condense into a hegemonic identity category that was able to form the basis for a new social order.

The worker

It is not surprising to discover that the consolidation of the consumer as the dominant form of social identity in the New Zealand telecommunications sector had major implications for the identity category of worker. Under state governance it was the identity of the worker that linked the economic and the social in the context of a nationally bounded territory (see, for example, McDowell 1991). Under market governance the social content of this identity has been emptied out. Instead the worker is constituted as an economic identity that comprises the crucial link between individual and enterprise in the pursuit of international competitiveness. Again, however, the transformation in this identity category was not straightforward, but was the outcome of a complex and contradictory process involving both discursive shifts and political technologies.

While the corporatisation debates were characterised by the mobilisation of three different identities, it is clear that all participants in these debates recognised the validity of the political claims made on the basis of the identity 'worker'. Indeed, mobilisation of this identity was crucial in convincing New Zealanders that corporatisation, despite its unprecedented nature, was an appropriate policy for a Labour government to be advocating. In overall terms it was argued that job growth would result as demand for the provision of telecommunications goods and services increased. More specifically, it was claimed that corporatisation and the deregulation of customer premises equipment would provide greater opportunities for Post Office workers.

The beginnings of the reconstitution of this identity category can be found in the debates around network deregulation. In these debates there were two crucial shifts in the way in which workers were constituted in relation to consumers. First, in contrast to the corporatisation debates, the possibility that network deregulation might involve job loss in Telecom NZ was publicly acknowledged. However the political claims of workers within Telecom NZ were juxtaposed against the overall benefits to both workers and consumers of a rapidly growing industry. Second, there emerged a discussion about the need to improve labour productivity within the industry in order to reduce costs for consumers. It is with these shifts that the basis for the marginalisation of workers was established.

That said, the legitimacy of political claims made by workers continued to be widely recognised during the period between network deregulation and privatisation. During this period Telecom NZ was involved in a major restructuring exercise. One component of this restructuring involved what were then considered to be fairly large scale redundancies, primarily associated with the introduction of new technologies and the contracting out of 'non-core' activities. Telecom NZ also dramatically reduced training activities in the lead-up to privatisation. Yet despite shifts towards the centring of the identity of the consumer, and the corresponding de-centring of workers, there was still widespread recognition of the rights of workers, with Telecom's management working closely with the Post Office Union to manage both technological change and redundancies.

It was only with the consolidation of competition that the identity of worker lost effectiveness as the basis for political claims. The usual explanation for the new hard-line approach to workers is to argue that it was the result of privatisation, and that it was associated with managerial changes within Telecom NZ. Certainly, following the sale of the company, there was the appointment of a more market driven management and board of directors who brought a more confrontational North American style of negotiating to industrial relations issues (Anderson 1992). In and of itself, however, this explanation is not sufficient. Even the Post Office Union was not unhappy with the outcome of the sale, observing that the North American buyers of the company had both the experience and track record to suggest they would look after their workers. Indeed, it is interesting to note that it was the Union's perception that the US directors had a much clearer understanding of the importance of employees than did their NZ counterparts (Anderson 1992). The significance of privatisation for the argument I am making lies in the further reconstitution of the identity of worker. During this period there were two further significant discursive shifts. First, whereas in the debates around network deregulation the interests of workers had been played off against consumers, in the privatisation debates workers were largely invisible. Indeed one of the few times that workers are mentioned in this debate is in the context of the observation that unemployed people may not be able to buy shares (Hansard 1990: 828). Second, in the literature promoting Telecom NZ to overseas investors, labour productivity, measured as access lines per employee, was widely used as a means of demonstrating the potential for future 'productivity enhancements'. It was in this context that the redundancies associated with network deregulation begin to be reconstituted as efficiency gains and the discursive framework for the re-commodification of labour was consolidated.

In the early 1990s a number of different developments crystallised to consolidate the new approach to workers. The privatisation of Telecom NZ had been rushed through Parliament before the election of 1990. One of the first steps taken by the new National government was the introduction of the Employment Contracts Act. This new industrial relations legislation was designed to deregulate the labour market. During the first term of this government immigration legislation was also liberalised, and training costs shifted to the individual through new industry training organisations. Moreover, these shifts in the form of labour relations occurred in a context where internationalisation of the domestic economy had become the dominant political rationality. The combined effect of these changes was not only to create a context in which the cost of labour was seen as determined by market demand and supply, but also the labour market began to operate in an internationalised frame of reference (Larner 1996, 1997c).

It was in 1993, following the establishment of Clear Communications and in this new political-economic environment, that Telecom NZ announced an 'extensive restructuring programme . . . designed to improve customer service, efficiency and reduce operating costs' (Telecom NZ 1993: 1). There was considerable speculation that the company's new strategy would involve more redundancies. This speculation was to prove well founded. On 16 February 1993 Telecom NZ announced it would lay off nearly 40 per cent of remaining staff. These figures represented a 70 per cent decline in the total workforce of Telecom NZ since corporatisation. This announcement was made in tandem with that of a record operating profit, and amid predictions that the planned job cuts would add another $100 million a year to future profits (*New Zealand Herald*, 17 February 1993).

The significance of the 1993 announcement, above and beyond the pain of those immediately affected by the job cuts, was another marked shift in the discourse about workers. This shift involved the rejection of the social aspects of this identity. The most obvious indicator of this shift was that whereas previously workers had been referred to as employees or staff, in the new discourse workers were more often referred to as an operating cost. Moreover, the frame of reference for assessing the appropriate level of such costs was international. As Peter Shirtcliffe, Chair of the Telecom NZ Board, explained to the PTTI following a formal complaint about the level of redundancies, 'The marketplace is now so open it must be assumed that this will be the standard for the future. This assessment process indicates that Telecom would need to substantially reduce its costs and its number of employees over the next years if it is to remain competitive' (Shirtcliffe 1993).

In this discursive formation new political technologies took centre stage. Once measures of customer service and labour productivity were quantified they could be used not only as a measure of internal performance, but also as a point of comparison against other telecommunications companies both nationally and internationally. With efficiency understood in terms of labour productivity, measured as employees per line, it was inevitable that increasing efficiency would be understood as involving fewer workers doing more work. Even the business media seemed startled by the extent to which the company seemed prepared to go. The National Business Review, for example, observed that 'Telecom is set to be amongst the highest in the world at 220 [employees per line], it might be sacrificing future earnings by damaging service, stressing out employees, losing trained personnel, constraining growth and producing a time bomb of inefficiency' (NBR 19 February 1993).

The announcement of the new round of layoffs was accompanied by a new hard-line approach with the Communication and Energy Workers Union, the successor union to the Post Office Union. From management's point of view the new approach was an attempt to address the effects of competition in a context where the consumer had become the hegemonic social identity. Consumer demand was mobilised to justify the changes; customers were portrayed as demanding round the clock service and the company argued it could not afford to provide such services at current pay rates. The CEWU, on the other hand, saw the new approach as an attempt to break its power, which, in turn, would allow serious erosion of wages and conditions for telecommunications workers. Significantly, given the argument made above, media reaction to the ensuing industrial action was to accuse both management and the Union of compromising consumer interests.

In late 1995 the CEWU went into liquidation. The decision to voluntarily liquidate the Union was triggered by the combined effects of a cash crisis and a continuing fall in membership. Despite efforts by the Engineers Union to sign up those workers previously covered by the CEWU, less than a year later only 30 per cent of Telecom employees remained in collective contracts. In this regard, claims that Telecom NZ took advantage of the collapse of the CEWU to actively de-unionise its workforce appear warranted. Certainly, any understanding of the Union as being in a social partnership with the company had been fundamentally undermined. In its place was a new understanding that the cost of labour, like any other resource, should be determined by the market.

My point is not that Telecom NZ treats its employees badly. Rather it is to demonstrate that the consolidation of the consumer as the hegemonic social identity had the consequence of marginalising the political claims of workers. The following quote encapsulates the new approach:

> One public criticism of Telecom is its seemingly endless shedding of staff. . . . Early in 1993 when the company announced plans to reduce staff by another 5000, people began to wonder about the merits of privatisation in a company like Telecom. Was it a monstrous multinational sacrificing people for profits? But it should be remembered who is the ultimate winner in such a company's drive for profits: the consumer. Profits are not possible unless the company is providing goods and

> services at a standard and price the customer wants. People will not make toll calls through the company that hires the most people. They will make them through the company that gives the best deal, that is, charges the least.
>
> (Coddington 1993: 112)

On one level, in tracing the emergence of this new version of the worker, this section has told a relatively straightforward story about the re-commodification of labour. As the identity of the consumer inherited the legacy of the social, so too was the social content of the worker emptied out. More generally, however, the introduction of market relations into the sphere of employment has seen a redefinition of this identity. In the context of market governance the worker is constituted as an economic identity that comprises the crucial link between individual and enterprise in the pursuit for international competitiveness (see Larner 1997c for more detail). It is through the enhancing of the capacities of individual workers that the success of firms, and correspondingly macroeconomic success, is ensured.

Political perils and possibilities

In the first section of this chapter I showed that market governance in the telecommunications sector has been associated with the emergence of a new social identity – the consumer. In the second section, I explored the implications of this shift for the identity of worker, showing that the social content of this identity has been emptied out. On one level, the specific claims made in this chapter support more general assertions, in that it is widely acknowledged that the categories of both consumer and worker play fundamentally different roles in the new political-economic environment. However, the originality of this chapter lies in the way in which it theorises these shifts. Rather than assuming a general and/or determining relationship between new communications technologies, institutional arrangements and social identity, I demonstrated how new versions of social identities emerge out of multiple processes and come to be seen as natural and normal.

Most immediately, this analysis makes it possible to understand how it was that the new form of governance could be constituted as being in the best interests of New Zealanders. In particular, large-scale redundancies could be constituted as politically acceptable because of the redefinition of the individual citizen as a consumer. Hence statements such as the following:

> In 1987 Telecom had an astonishing 25,000 on the payroll. It's a point that needs to be hammered home in the context of politicians' efforts to make job creation an economic priority. In fact the health of any business, and therefore the wider economy, should be measured by the opposite. It's far better for the country to have a slimline Telecom and modern

> technology than be faced with the Kiwi version of the agonies faced by IBM or GM.
>
> (NBR 19 February 1993)

In this editorial 'the country' is clearly not composed of workers, but rather consumers who benefit from access to communications technologies provided at the lowest possible cost.

More generally, the analysis makes explicit the point that consumer and worker are not necessarily mutually exclusive and antagonistic identities, despite the best efforts of neo-liberal politicians and business leaders to deny their coexistence. Moreover, with the recognition that consumers are also workers, and vice versa, it then becomes possible to look for the moments when the contradictions of market governance are made apparent. In this context it is useful to recall the attempts of the Post Office Union to mobilise the identity of the consumer in their attempt to preserve social cohesion. Another starting point for such an analysis might be found in the fact that one of the few indications of an active public response to the 1,993 layoffs involved reports from Clear that they had been inundated with new customers.

In this regard, it should be clear that the argument presented in this chapter has relevance beyond the small (but significant) New Zealand case. This analysis allows us to contest the neo-liberal vision of a world comprised of individualistic consumers intent on maximising self-interest through their choices in the market place. As Daniel Miller (1995: 17) explains, this consumer is 'the fictive consumer of economic models, the aggregate of desocialised, individual, rational choice-makers, the source of whose demands and desires is understood as entirely irrelevant to politics as it already was to economics'. Such a vision is obviously a 'god trick' (Haraway 1991a) premised on an essentialist understanding of the consumer, and denying multiple forms of social inequality.

At the same time, however, we are able to avoid the trap of the 'two sides of the same coin' in developing this critique. The danger is that we also constitute consumers and workers as mutually exclusive identities that can be played off against each other. All too often arguments against neo-liberalism are framed in terms of the innate individualism and inequality of consumer choice, and emphasise the consequences of these qualities for social integration and moral cohesion. In contrast, the analysis developed in this chapter suggests that instead of dismissing the consumer as an ideological ploy of the New Right, there are important questions to be asked about the possibilities for emancipatory politics premised on this identity.

Daniel Miller (1995) reiterates this point in his review of new studies of consumption. His claim is that 'the contradictions of dialectical development are increasingly manifested within individuals...in their dual existence as labour and consumer' (p. 49). He calls for research that transcends simplistic images of the good and bad consumer, and focuses instead on the complex and contradictory forms that consumption can take. In this way, he suggests, we might begin to understand more about how the actions we take as consumers have consequence for our status as workers; 'we acknowledge the massive influence of consumption upon the political economy, while acknowledging the political economy inscribed in the historical projects given to people as consumers' (p. 54).

To support these claims does not necessarily imply that sociologies of consumption are now more relevant than sociologies of production, nor is it to idealise consumer based social movements. Rather it is to suggest that anti-essentialist understandings of identity may help us to identify progressive versions of the consumer, without forcing us to claim that consumption based identities are inherently emancipatory or otherwise. Moreover, as we investigate how these identities come to be constituted in particular ways, through both discourse and social practice, it is likely that multiple and contradictory forms of these identities will become visible. In turn, exploring these heterogeneous forms, and the political claims they enable, may well serve to re-politicise political-economic identities in new ways.

In sum, this chapter problematises the possibility of deriving the political orientation of a social subject from a fixed identity (Daly 1991: 88). Instead I have emphasised the importance of theorising social identities as open sites of potential contest and, correspondingly, as both provisional and contingent (Butler 1990). This requires careful consideration of 'the ways in which these identities are under permanent threat of subversion by different articulations, other discourses, and are continually having to be redefined and renegotiated' (Daly 1991: 93). With such an approach, I would argue, it becomes possible to move past the polarised debates that have accompanied discussions of new communications technologies, institutional forms and social identities.

Conclusion

In this chapter I have shown how the shift to market governance in the New Zealand telecommunications sector has been integrally linked to the emergence of the consumer as the dominant form of social identity. Rather than seeing shifts in identity as a response to more fundamental technological and/or social transformations, this chapter demonstrates that identity formation, or more specifically attempts to shape and reshape forms of subjectivity, has its own dynamic. Moreover, I have stressed that it is not simply ideological categories that situate members of a social group in a particular process, and prescribe certain identities for them (Barry *et al.* 1996: 10–11). Rather I have developed a more active account of new identities emerging from a constellation of social, political, and economic forces. In doing so I have demonstrated that market governance is transparently invested in particular economic agendas, political ends and social formations, but in a way that avoids epochal formulations and centres changes in social identities.

Acknowledgements

Thanks to Wallace Clement, Rianne Mahon and Janet Siltanen for their support. The editors of this book and the participants of the Centre for Labour and Community Research Workshop: 'Labour and Technology – Being Changed, Making Change', held 21 March 1996 at Carleton University, Ottawa, also made useful comments.

5 Transnationalism, technoscience and difference

The analysis of material–semiotic practices

Laura Chernaik

This chapter addresses the relation between technoscience and transnationalism, two of the material–semiotic discourses and practices which interact to produce complex objects and representations. My main concerns, here, are metatheoretical and methodological. I argue that an analysis which focuses on processes of transnationalism and globalisation/localisation alone would be incomplete. So would an analysis that focused solely on technoscience. Our objects of analysis are produced by a multiplex interaction that features transnationalism and technoscience. A successful account has, therefore, to involve both political economy and 'New Science Studies'. We should address both local and global aspects of these discursive practices, such that questions of region too become significant. It is equally important, as I argue in my conclusion, to examine the ways in which in the contemporary period the material–semiotic practices of race, gender, and sexuality act through, and on, technoscience and transnationalism.

My approach to the study of transnationalism is based on neo-marxist political economy, social-history and social and historical geography.[1] I draw heavily on the work of the neo-marxist geographer, David Harvey. Harvey used Regulation School economics as the basis for an extremely useful concept, the 'spatial fix'. The Regulation School is concerned with the social reproduction of capitalism; it analyses the way in which production and consumption are regulated in a regime of accumulation. The modes of regulation include social and ideological formations, institutions and actions. Harvey argues that transnationalism is capital's 'spatial fix' of accumulation's crisis; an addictive, repetitive attempt to use displacement in space to keep the system going. That is, capitalism makes use of spatial displacement – opening up new markets, appropriating raw materials, locating production and design in whatever region or country is cheapest – in order to deal with its recurring crises. It then needs more: another fix or a larger fix.

The 'spatial fix' is a particularly good theoretical concept because it is able to account for several of the characteristics of transnationalism. First, it can account for the way that production is separated into different stages, with design and assembly taking place in different countries. If we take a computer as an example, the circuit boards might be made in South East Asian *maquiladoras*, and the software in silicon valley. It can also account for the transition in the developed countries to so-called service-based economies, with shrinking core employment and increasing peripheral employment. There is some question about the key date for this latter transition of the developed countries from manufacturing-based economies to service-sector based economies. Harvey dates it as 'around 1973', that is, when the United States went off the gold standard. That standard had been set in the Bretton Woods agreement at the end of the Second World War. It is therefore possible to argue that the key date for our argument falls much earlier than 1973. In addition, the shift between manufacturing and the service sector is only part of the historical process. Transnationalism and its globalisation/localisation dynamic are conditioned by the histories of colonisation, decolonisation and postcolonialism.

If my approach to transnationalism is based on neo-marxist political economy inflected, as I discuss later, through feminist and postcolonial studies such as Grewal and Kaplan (1994), my approach to the study of technoscience is based on the New Science Studies. There are two major strands to the New Science Studies, a more social-historical approach centred on the work of Donna Haraway and a more anthropological approach centred on the work of Bruno Latour. Both approaches address questions of difference. Where she deals with difference, Haraway's work is concerned with race and gender, as well as class. Latour, however, is more interested in difference as an abstract, almost mathematical concept, than in, for example, gender or sexuality. Haraway and Latour's approach to difference is in turn derived from the work of Gilles Deleuze and Felix Guattari.

The most original, and useful, element in Deleuze and Guattari's philosophy is their development of 'non-particularistic' difference. This notion is in opposition to the way that difference is sometimes conceptualised; in particular, in much political economy. A good example of the 'particularistic' difference that Deleuze and Guattari oppose can be found in a relatively early text by David Harvey (1982), in the introduction to which he makes it clear that he *contrasts* difference to universalisation and the other generalising, abstracting moves that are part of theorisation, as if difference is a counterweight to the 'abstract conceptions' of theory. Harvey's most recent work (Harvey, D. 1996), however, rejects this position, arguing, instead, that difference can be at the heart of theorisation. *A Thousand Plateaus* is the second volume of *Capitalism and Schizophrenia*, both volumes of which were co-written by the philosopher Deleuze and the psychoanalyst Guattari, as were the monographs on Proust and Kafka. Deleuze also wrote, separately, a number of philosophical monographs on Spinoza, Nietzsche, Bergson and other thinkers; however much they might demand of the reader, it is in these philosophical monographs that Deleuze remains closest to traditional philosophical modes of argumentation and logic. Philosophy, like many other disciplines, has dominant traditions – canons of centrally important texts – with their attendant schools, and competing, subaltern traditions. The two dominant traditions are Anglo-American philosophy and Continental philosophy, each of which breaks down further; thus, Continental philosophy includes, for example, Phenomenology, Kantian and Hegelian philosophies. What makes Deleuze particularly difficult, is that he situates himself outside of all of the major philosophical traditions, as a Spinozan. He even suggests, with some justification, that his work and that of the other philosophers he writes about and by whom he is influenced form an affinity group rather than a philosophical school: they make up a movement rather than a filiation. Or, in Deleuzean terms, philosophical traditions are striated (they are structured, binaristic and metric – that is, measurable) and molar (they are totalities with form and function). As Elizabeth Grosz points out (in Boundas and Olkowski 1994), for Deleuze and Guattari the molar identity is particularistic and difference is universal, is, – or reaches – a plane of consistency. Thus, in *Capitalism and Schizophrenia*, for example, Deleuze and Guattari suggest that classical marxist theory is 'molar' (about totalities) and based on identity (class). So far, the argument is familiar; and, indeed, Deleuze and Guattari move from this point, as one would expect, to a critique of Hegelianism, and its influence on marxism. Many recent critiques of classical marxism focus on its mistaken subsumption of gender, sexuality, and race. It is argued that these analytical concepts are, generically, 'specificities' which classical marxism's logic overlooks, or overcomes. However, Deleuze and Guattari do not oppose gender, race, and sexuality, as 'specificities', to class. They argue, instead, that class-based marxism is about identity; it is therefore particularistic. Class-based theories can therefore *not* be opposed, logically, to theories of specificity. Instead, they claim, there is a way to conceptualise difference as universal – universal in the way that molar theory tries – and fails – to be. This non-particularistic difference or 'plane of consistency' is not transcendent – it is not outside the world. It is found, or more precisely, can be analysed, in the world, in nature. So, molar practices are – or constitute – identities: class, race, gender, sexuality. That is, subjects – psychological subjects (personal identity) and political subjects (political identity). Molar political identities lead to political infighting and fragmentation. So, what alternative do we have? According to Deleuze and Guattari, the alternative to fragmentation is difference and the plane of consistency. Difference and the plane of consistency are not just a matter for subjects: they have to do with, they act in and through, both subjects and objects. For Deleuze and Guattari, rather than 'subjects' perceiving and analysing 'objects', both object and subject are joined in 'affects, subjectless individuations that constitute collective assemblages'. It is this argument about subject and object being joined, not opposed, that is picked up by, and further developed by, the New Science Studies. This new way of thinking about subject and object is useful for anthropological or social-historical studies of science; and, if the scholars concerned wish to use their anthropological or historical approach in a politically motivated critique, the new way of conceiving subject and object provides a sound methodological foundation for the attempt.

However, rather than making explicit the extent to which these reconstituted concepts enable a radically different philosophy of science, Deleuze and Guattari keep their arguments allusive and indirect. They shift back and forth between language, arguments, examples and claims that seem to belong to the philosophy of history, and language, claims, examples and arguments that seem more part of the philosophy of science. This shifting and combination of arguments and discourses is, I grant, frustrating for a reader, but it is, at the same time, this which enables Deleuze and Guattari to make one of their most interesting claims; to have inaugurated a new kind of philosophy, a philosophy of externality. This suggestion is followed up by Elisabeth Grosz in her recent work (1994, 1995a). However, for now, it is enough simply to trace the logic of the argument behind the claim: Both the philosophy of science, and, strictly speaking, the philosophy of history, are, in phenomenological terms, 'intentional', in relation to an object. They are philosophies of externality, rather than of internality. Thus, in their terms, Deleuze and Guattari are philosophers of externality, unlike Hegel, who, in this reading, becomes the philosopher of the State, and in a peculiar way, of interiority (consciousness, thought).

Many of the plateaus, or chapters, of *A Thousand Plateaus* take as their object research in the physical, biological, and social sciences. In each case, the argument is political. Each plateau is used to produce a phenomenological intentionality that is, in Sandoval's terms, an 'oppositional consciousness' (1991), a kind of 'situated knowledge' (Haraway 1991b, 1997), or, in Deleuzean terms, a 'becoming-minoritarian'. As I have said, Deleuze and Guattari's highly allusive style, and tendency to shift, without warning, from one field of study to another, makes it quite difficult to keep track of the argument in each plateau, or to disentangle the political, historical claims from the substantive assertions about scientific facts and biological and geological processes. That is, one may grant that scientific knowledge is produced historically, as what Foucault called 'power/knowledge', but one must still be precise, and distinguish the scope of one's claim. Is it a claim about the production of scientific knowledge; a claim about the production of historical knowledge; or a claim about the production of both kinds of knowledge? In the latter case, one must also decide whether one will argue that the two kinds of knowledge are bound together either ontologically or epistemologically, or are simply linked heuristically in one's argument. Grosz (1994) investigates some of Deleuze and Guattari's scientific claims at length. She shows that 'minority science' or marginal research is not epistemologically different from mainstream work. Whatever one's field of study, whatever one's object of analysis, the knowledge one produces is 'situated', rather than 'free-floating' or 'objective'. This, as Deleuze and Guattari acknowledge, and as the New Science Studies stresses, is what good social and natural scientists do in practice, as well as what many of them argue as theory. Situated knowledges or becomings-minoritarian are not in opposition to universalisation.

For Deleuze and Guattari, as I have explained, difference universalises. It does so, however, not by repeating its identity infinite numbers of times but by 'proceeding to the plane of consistency'. The detailed arguments that provide the groundwork for this idea of non-particularistic, universalising difference belong to yet another philosophical field: that concerned with language and action. Just like Judith Butler (1990, 1993), whose notion of normativeness/deviance is much like Deleuze and Guattari's odd, technical use of the term 'minority', Deleuze and Guattari draw on a theory of performativeness. Deleuze and Guattari use the word 'majority' in a technical, non-numerical sense; a majority, for them, is an identity, a 'state or standard' in relation to which others 'deviate' (Deleuze and Guattari 1987: 291–292). Butler derives her notion of the 'performative' from the philosopher of language, J.L. Austin (1955), who argued that the 'speech-act' was a crucial and neglected aspect of language. Deleuze also draws on Austin, making unexpected connections between his work and that of a much earlier philosopher, the seventeenth-century Spinoza. Deleuze argues that the two most influential contemporary philosophies of language, the first of which analyses language as 'information', and the second of which discusses it in terms of 'connection' or intersubjectivity, are both wrong. Language should, instead, be understood in terms of 'expression', and 'expression' understood as a speech-act, a performative. Rather than content pre-existing, outside expression, expression is a performative, and one, which like Žižek's analysis of retroactive perfomatives (1993), can produce what it seems to represent. Žižek argues that the process which Lacanians call 'Lack' is retroactively constitutive; the very small child begins to get a sense of itself as a person, a subject, separate from its mother or other primary caregiver, and this 'act' retroactively constitutes a past imaginary unity with the mother, now lost. This is Lack. The thing which Lack loses is the Real: that which hurts, that which one never had, that which only exists in the past, constituted retroactively by oneself, by one's loss of it. What makes Žižek's work so original, and so moving, is that he uses his insight into Lacanianism to analyse one's most painful loves and losses; he shows how Lack becomes the basis not just of love but of hate: nationalism and ethnic hatred, in particular.

The second volume of Deleuze and Guattari's *Capitalism and Schizophrenia*, which is my primary concern here, deals with retroactive constitution, but does not address Lack, *per se* which is dealt with in the first volume. Deleuze and Guattari are greatly troubled by the implications and consequences of the way that we separate and oppose mind and body. The first volume of *Capitalism and Schizophrenia* focuses on mind, addressing Hegelianism and its influence on marxism, and the second volume, which is my concern in this chapter, focuses on body, addressing both scientific and historical discourses and practices. Deleuze is greatly influenced by Spinoza, (this edition, 1955) who came up with a surprising solution to the mind–body problem: there is one substance, infinite and unitary, which has two modes of expression, extension and thought. As Spinoza believed in full space, extension also means that which is extended: matter. There is no mind–body split, because they are different modes of the same substance. Extension and thought are, in a sense, each one (so that Spinoza verges on a kind of pantheism); they are also multiple, with an infinity of attributes, each creature a mode of these attributes or a modification of substance. Deleuze's notion of performativeness is taken from a key concept of Spinoza, the notion of 'affectio': affections or affects. As Deleuze explains in *Spinoza: Practical Philosophy* (1988b), 'affects' are, first, the modes of substance or its attributes. They then take on, or can be considered at, a second level – as 'that which happens to the mode, the modifications of the mode, the effects of other modes on it.' And third, 'affect' means 'states', 'variations', 'degrees of perfection': transitions from one state, image, or idea to another. The ordinary English meaning of affect, emotion, is very different from Deleuze (or Spinoza's) usage. Deleuze is not writing about interiority at all. Brian Massumi (1992) argues that 'affect', as the term is used in Deleuze's work, is (the body's) 'capacity to affect and be affected, to act and to perceive' (Massumi 1992: 100): closest, I would suggest, to the second meaning of affect Deleuze identifies in Spinoza. It is this second meaning of 'affectio', in Spinoza's work, which Deleuze draws on for his performative theory of language. The third meaning becomes deterritorialised in Deleuze's work, moving into mathematical fields, as 'degrees of freedom', and into physical and musical fields, as 'frequency' and 'resonance', notions drawn on by the New Science Studies. This Spinozan approach to theorisation is also the context from which Deleuze developed his notion of assemblages, or 'abstract machines'.

Deleuze argues that becomings proceed to the 'plane of consistency', to a 'body without organs', by way of 'diagrams' (abstractions, in phenomenological terms), and abstract machines. Deleuze's notion of the abstract machine is reached by means of a critique of the notion of the 'organism', and its associated concept, 'organ'. For Deleuze, 'organism', as a concept, is used to theorise a body as a totality. An organ, similarly, is a discrete part, out of which organisms are made. These are rather old fashioned biological notions; as Deleuze points out, and as Donna Haraway has argued in greater detail, contemporary scientists are more apt to analyse homeostatic mechanisms than organs, and populations than organisms. Deleuze uses the term machine as a move away from these organicist connotations. A machine, for Deleuze, is not just something inorganic. He uses the term in an unusual sense: for Deleuze, a 'machine' is an assemblage of heterogeneous entities: a 'nomad war machine', for example. Haraway, and the scholars influenced by her, call these 'cyborgs': for example, a pilot with a head-mounted display, or a technofeminist scholar.

There are several differences between Deleuzean machines and Harawayan cyborgs. A Deleuzean machine is a *collectivity* made from many entities; the separate entities are from different categories and, in interaction, construct a new way of being and thinking. Their example is a 'nomad war machine': horses, tents, people, collectively a threat to 'sedentary civilisation', to social stability and stultifying respectability. A contemporary example, for them, would be New Age Travellers: perceived as a threat by the bastions of social stability, living in their vehicles, rather than using them instrumentally. Deleuze and Guattari sympathise, politically, with anarchism, and underestimate the constraints of *Gemeinschafts* (communities) in their rejection of *Gesellschafts* (hierarchical societies). For Haraway, in contrast, the boundary-crossing that makes an *individual* a cyborg means that her identity is 'multiple', that is, non-unitary.

For Stone (1995), the internet is a particularly useful place to explore the multiplicity of identity. Typing at a keyboard, one interacts with someone else typing at a keyboard. Since neither is face-to-face, either can pretend, can take on any 'identity', or can explore his or her own identity in new ways. For Stone, cyberspace becomes an occasion for self-fashioning, and an instance of the performativity that is always at the heart of identity. Play in cyberspace can have real effects. The cyberspace produced by computer scientists and made available for other agents via the internet, and by means of the other cybernetic tools that in our use of them almost become prostheses, is the medium for virtual actions and virtual speech-acts. Deleuze used the word 'virtual' to name the 'unconscious' aspects of language, the nine-tenths of the iceberg; the 'actualised' statements we speak are the tenth part, above the waterline. Deleuze used this trope to suggest that we may be limiting ourselves by dividing ontological states into possible and actual. Actual and virtual might be a better pair of notions. Stone deterritorialises Deleuze's questions about existence and agency, moving them into a new scientific domain, information technology. She is thus able to argue that the virtuality of action and entities, in the internet, is only an extreme case of performativeness.

Stone draws on an aspect of Deleuze and Guattari's work, the theory of performance, expression, and affect, that is closely related to current debates in poststructuralist theory and Continental philosophy.[2] Donna Haraway and Bruno Latour, however, focus less on the implications of their arguments for Continental philosophy than on developing the aspects of Deleuze and Guattari's arguments that go furthest towards a reformulation, or in some respects, a refoundation, of the study of science. It is for this reason that the approaches are collectively known as New Science Studies. Both focus on questions of difference. They each argue, following Deleuze and Guattari, that difference is not the same as the particular. Latour develops a counter-intuitive but very productive argument about the degree to which the Enlightenment project and modernity have not so much failed as never been properly inaugurated. Haraway draws on Deleuze and Guattari's arguments about performance and performativity, using them as the basis of a radically new way of understanding technoscience and technoscientific practice in terms of action. She goes far beyond her earlier work on feminist cyborgs, and, instead of just concentrating on boundary crossing machine–human hybrids, looks at a much wider range of 'human and non-human actors'.

Both Donna Haraway and Bruno Latour have argued, speaking of two periods when there was a transition between different economic and cultural formations – that is, the seventeenth century and our present period – that there was an interesting and important contradiction between what was said and what was done. In the seventeenth century, as many recent historians of science have argued, there was a lot of rhetoric about the separation of science and politics. The scientific revolution, it was argued, had brought about, or ought to bring about, this separation. Science, and later, social science, would be value-free and the political part of politics could be separated out from a scientific part of politics. This last was eventually known as management, or social policy. As Haraway and Latour both argue, this may have been what people said, but it was certainly not what people did. Rather than these separations, it was actually hybrids that were produced. The hybrids they consider are various – hybrid objects, hybrid subjects and hybrid discourses. Influential New Science Studies texts include, for example Shapin and Schaffer's *Leviathan and the Airpump*, Latour and Woolgar's *Laboratory Life*, Haraway's *Primate Visions* and, as I discuss in this chapter, *Modest-Witness@ New-Millennium*. Under hybrid discourse we ought also to include the boundary crossing terminology of so much poststructuralism and postmodernism. However, some of the most troubling criticisms of deconstruction, poststructuralism and postmodernism are targeted at the political implications of antihumanism or, generally speaking, of anti-enlightenment arguments. Many critics have objected that, whatever the strength of a particular deconstruction of an individual Enlightenment notion or text, there is still a lot that is useful, for politically motivated people – and scholars – in humanism and the Enlightenment. Thus the importance of Haraway and Latour's argument. They argue that, rather than having to be postmodern in a Lyotardian or Baudrillardian way – and therefore having to reject such good as the Enlightenment stood for, along with the bad – we can be amodern. We can argue that the modernity that was called for never really took place, never really arrived.

Thus Latour (1987, 1993), for example, argues that whilst modernity and 'postmodernity' have set up oppositions between the material and the semiotic (for example, science vs. politics, history vs. ideology, the natural and social sciences vs. the humanities) and created large numbers of hybrid discourses and cyborg subjects and formations we have in fact 'never been modern'. The opposition between the material and the semiotic, science and politics defines modernity, and yet, since, in the process, it is hybrids rather than separated discourses and entities that are produced, modernity has, in a way, not yet happened. Modernity has both created 'science' and 'progress', and yet not been modern. It is for this reason that Latour concludes that a thoroughgoing Science Studies would have to call itself 'amodern', rather than 'modern' or 'postmodern'.

Donna Haraway takes this claim further, developing a more detailed historical argument. She also brings in the questions about race, gender and sexuality which Latour neglects. (Both Latour and Haraway discuss class, although, as I argue in my conclusion, they each move away from the familiar terms of political economy). Discourses, as Donna Haraway (1997) explains, 'are not just "words"; they are material–semiotic practices through which objects of attention and knowing subjects are both constituted' (Haraway 1997: 218). The typography of the phrase, 'material–semiotic', is significant: a dash, not a slash; a relation, rather than a binary opposition. Rather than the 'modern', 'scientific' opposition between subject and object – between, on the one hand, meaning, intentionality, and, on the other hand, history, objectivity, measurability, and universalisability – Haraway draws on Deleuze and Guattari and represents object and subject linked together and transformed, actants in 'material–semiotic practices'.

Haraway moves on from these general points about modernity to a specific claim about the Enlightenment. She suggests that the Enlightenment and Scientific Revolution were far less effective in ending religious master narratives than familiar narratives of progress and science maintain. The material – historical contingency – has shaped what seems most semiotic – and most universalisable – in science and technology. The semiotic – for example, Judeo-Christian figurations, the 'Second Millennium' of Haraway's e-mail address title – has been an irreducible part of modern science. The anthropological studies of scientific labs (Traweek 1988; Latour and Woolgar 1979; Latour 1987, etc.), the social-histories of science and technology (Haraway 1989, 1997; Shapin and Schaffer 1985), both major strands in the New Science Studies, do more than just provide a historical context for truths, the epistemological validity of which depends on a separability from that historical context. Instead, Haraway claims, following Sandra Harding (1992), the epistemological claims (valid, pragmatic, sometimes even life or death claims) of science rest on 'situated knowledges' and 'strong objectivity'.

What interact in 'material–semiotic practices' are not 'subject' and 'object' but 'sticky threads' (Haraway 1997: 68) that can be teased apart by the critic and historian to show 'heterogeneous and continual construction through historically located practice, where the actors are not all human' (Haraway 1997: 68). 'Subject' and 'object' are what used to be called 'reifications', but what now, in her most psychoanalytic argument to date, Haraway calls 'fetishes', 'models' that 'obscure the constitutive tropic nature of themselves and of worlds' (Haraway 1997: 136).

Haraway is drawing here on the work of an extremely influential American philosopher of history and specialist in the early modern period, Hayden White. White argues that, as historians, we need to pay attention to tropes – figures of speech. The American literary critic Kenneth Burke said that there were four tropes: metaphor, metonymy, synecdoche and irony. Metaphors are used to make claims about identity (my love is a rose, Americans are pragmatic); metonymy has to do with relations of part to part (the logic of the example, brand names used as generic terms, etc.); synecdoche figures the relation of part to whole; and irony is a method of using language in a way that highlights the meanings most different from the literal. Modernity makes subject and object seem things in themselves, non-tropic, non-figurative, self-identical. However, Haraway argues, both subjects and objects really are tropic, figurative. It actually takes a reification to conceptualise something as an object.

So, instead, we can make use of a tropic model, and (re-)conceptualise an 'object' as a non-human actor. Rather than a scientist (a subject) studying an object (a virus), a scientist (a human actor) studies a virus (a non-human actor). Why does this matter? What verb do you use to describe the actions of a virus? Does a virus 'invade' the body, or would another trope be better? If a virus 'invades', a virus is like an enemy soldier, or even worse, like the disguised 'enemy within'. The trope affects how viruses and people with viruses are perceived, which affects funding, which impacts on the production of knowledge. Or, shifting focus in our example, a person 'invaded' by a virus is a 'victim'; a person 'with' a virus is just a person with a virus. The opposition is between, on the one hand, being a 'victim' of AIDS, and, on the other hand, being a person Living With AIDS (a PWA).

We would also focus on 'practices'. Rather than arguing that science 'discovers' knowledge like a 'new-found land', and, often, theorising this process by means of sexual, racial and religious tropes, we would argue that human and non-human actors engage in material–semiotic practices, theorizable in terms of strong objectivity and situated knowledges. Our analysis would be pragmatic and material, and would respect the 'meaningfulness' of the scientific discipline: good science and bad science. It may seem counter-intuitive to think of Boyle's airpump (Haraway 1997; Shapin and Schaffer 1985; Potter, in Haraway 1997), a particle accelerator (Traweek 1988), an oncomouse, that is, a mouse with human breast-cancer genes (Haraway 1997), or a 'DNA labeling kit' (Haraway 1997: 157) as an actor, or actant, but it proves very useful. It means that the practitioners of the New Science Studies can argue a non-relativist position, making strong epistemological and ethical claims. Like the queer theorists who draw on Foucault (Halperin 1995; Bersani 1995), Haraway is able to link together arguments about agency, and arguments about discourse. Even if, as Halperin (1995) argues, the claim[3] that a strictly Foucauldian position creates problems for agency is overstated, it is still refreshing to find arguments about agency, and arguments about discourses, combined in such a thoroughgoing way.

Haraway's book begins with an extraordinary first chapter on Robert Boyle's 'modest witness'. She brilliantly analyses the way this trope was used, at the time, to figure, and thus to enable, the grounding of a modern, classed, raced and gendered social order on the set of concepts which became known, and valued, as 'scientific rationality'. Haraway then uses this argument as the basis for a number of original, and explicitly political, detailed studies of contemporary technoscience. She clearly draws upon Deleuzean oppositions, 'machines', and flows:

> Nature and Society, animal and man, machine and organism: The terms collapse into each other (Haraway 1997: 120); 'the operating mechanisms, called pragmatics. How do critical theoretical practices deal with the materialised semiotic fields that are technoscientific bodies?'
>
> (Haraway 1997: 121)

Haraway has a choice of ways in which to follow Deleuze and Guattari. She can emphasise one aspect of their work, and trace the collapsing terms further and further, into 'becomings' and abstract, pure 'difference'. Or, on the other hand, she can link her analyses of technoscience to the pragmatics[4] and to the performative theory Deleuze and Guattari touched on in *A Thousand Plateaus*. *Modest-Witness* adopts the second option. In her first chapter on Boyle, Haraway builds on *Leviathan and the Airpump* (Shapin and Schaffer 1985) and the more recent feminist articles on Boyle (Potter, in Haraway 1997; Heath, in Haraway, 1997), developing a theory of actors 'not all of whom are human'. She emphasises the political issues most crucial to the human actors: the oppositions between public and private, and science and politics or ideology. She focuses on the epistemological issues most pertinent to the interactions of human and non-human actors: strong objectivity and democratic technoscience. She thus focuses on difference, but not on 'becomings'. In sympathy with Grosz and Braidotti's critiques of Deleuze and Guattari (both in Boundas and Olkowski 1994), she takes great care to avoid the kind of arguments about pure, abstract difference to which Deleuze and Guattari were led by their deployment of such concepts as 'becoming-woman' and 'becoming-animal'.

The focus on actors and actants, however, links up with something curiously American – myth production. Each chapter is both a detailed, historically based, analysis of a 'material–semiotic object' and a presentation of a myth – an empowering or disempowering narrative about an exemplary figure. Some figures are human actants, ranging from famous scientists to the Jane and John Does of careers guidance booklets. Some are non-human actants, ranging from oncomice – mice with human breast cancer genes – to DNA labeling kits. If White is correct – and I think he is – about narrative and rhetoric in history then Haraway's myth production is just a way of telling different stories – minority histories – of the seventeenth century and the present. She is attempting to produce an alternative genealogy for the present, in order to bring about a more just future. Haraway's rejection of synecdoche – of false universalisation, subsumption, and the hypostatisation of a monolithic structure – leads her to adapt a metonymic approach, 'myth production'. However, I am not convinced that myth production is the right answer. It is true that if the only two alternatives are synecdoche and metonymy, then, if we reject synecdoche (subsumptive versions of neomarxism or feminism) we must chose metonymy (one of many myths, none of them master narratives). But, even if Burke is right that there are four tropes, and White is right about the significance of these tropes to historical analysis, and even if subsumption is a form of synecdoche and myth production a form of metonymy, the statement 'we must choose myth production, if we reject subsumption' is *not* the only valid conclusion. The parallels break down. If culture is not unitary, we are not restricted to only two choices: master narrative or a myth. We have a third alternative.

As Grewal and Kaplan (1994) have argued, the binary opposition between general and particular is not necessarily the most useful way to approach the problems of theorisation. Instead, concepts such as 'situated knowledges' and 'scattered hegemonies' can be more productive. Drawing on Appadurai (1990) and Hannerz (1987, 1989, 1992, 1996), they argue that the theorists who analyse 'cultural flows' are often limited by the way that they also rely on notions of 'cultural homogenisation' (Grewal and Kaplan 1994: 13). Instead, they suggest that like Hannerz, we should acknowledge that most cultures are 'creolised', not 'homogeneous' (Grewal and Kaplan 1994: 14). As Grewal's concept of 'scattered hegemonies' suggests, we may have more choices than just 'synecdoche' and 'metonymy'. Shifting the ground from postmodern*ism* to postmodern*ity*, we can analyse transnational 'cultural flows' and 'travelling theory' without having to posit a monolithic, unitary culture, or a monolithic, unitary ideological discourse (the 'whole' of the part-to-whole synecdochal figure), while, at the same time, avoiding the excessively metonymic 'logic of examples' of isolated case studies.

Haraway is afraid we only have two alternatives: metonymic, case-studies-like texts, or synecdochal, subsumptive texts. However, if we follow Grewal and Kaplan, and Hannerz, and conceptualise culture as non-monolithic and creolised, we have a third alternative to the two possibilities Haraway considers. Culture-formation is, indeed, comparable to the way that two languages interact to produce a creole. The trope is useful: it can help us conceptualise the way that technoscience and transnationalism interact. Hannerz's 'creole' figure enables us to recognise what is at stake. First, culture is not unitary, and, second, and even more importantly, cultures generally do not have unitary origins. However, the trope only takes us so far. Culture isn't a language. It is, or is made up of material–semiotic practices: the material and the semiotic intertwine and are not ultimately separable.

So, if we keep both these material–semiotic discourses, transnationalism and technoscience, in mind, we might begin to see how, for example, the collapse of Barings Bank, or the turmoil in the Japanese financial system could be analysed. In both cases, we are dealing with complex local/global transnational dynamics, interacting with technoscience. An analysis would have to engage with both the material and the semiotic: both the technology which enables derivatives trading, the interdependence of the various stock markets, and the production processes used in the 'little tiger' and Japanese factories, and the 'expertise' aura that makes oversight, and public accountability, so rare.

For a contrasting example, focusing on technoscientific empowerment, rather than on exclusion, let us look at the reasons for the wide choice of alternative and mainstream local newspapers in San Francisco. In the midst of such tragedy, by the very actions taken in response to the tragedy, the gay community's response to AIDS has produced a community-wide techno-scientific empowerment that interacts with, and produces, changes in a transnational economy. The great number of scientifically well-informed, politically active people Living With Aids has led to a situation in which, induced by readership demand, both the gay papers and the other local San Francisco papers have the most up-to-date and sophisticated coverage of treatment and research. The newspapers and other media provide a good source of information, and keep people informed, prepared to demand the latest treatments, and eager to join the activist groups.

It is especially clear, given these examples, that contemporary society is best analysed in terms of performance and pragmatics. Many new avenues of research can be opened up; many new investigations of the contemporary interactions of technoscience, transnationalism, race, gender, and sexuality can be based on this paradigm, 'material–semiotic practices': the interaction of language and action, pragmatics and praxis.

Notes

1 The following paragraphs are taken, with minor adaptations, from Chernaik, Laura (1996), 'Spatial Displacements: transnationalism and the new social movements', *Gender, Place and Culture* 3: 3, pp. 251–275, in which I discuss, at length, the neo-marxist analysis of transnationalism.

2 See, for example, the body of work by writers as diverse as Butler and Žižek.

3 See Hoy (1986) for a number of well-argued examples of this claim.

4 'Pragmatics': theory of language and action. 'Pragmatism': an American philosophical movement, based on the work of William James (1907), that concentrates on effects and use, rather than on meaning.

6 The convergence of virtual and actual in the Global Matrix

Artificial life, geo-economics and psychogeography

Otto Imken

Introduction: cyberspace and the Matrix

I think it is best to drop all trepidation and begin to treat the global telecommunications *Matrix* like a new artificial-life form: not a mere organism (which is still trapped by its limited functionalities and restrictive stratifications) but a non-linear, asymmetrical, chaotically-assembled functionality with much more potential freedom than that of an entity encased in skin or limited to being an agglomeration of discrete organs. A new being made up of widely distributed hardware, software, and pulses of electricity coursing through its nervous system is now stretching its exoskeleton across the planet, into the upper atmosphere crowded with satellites, and even out to incorporate data from sensors on the Galileo space probe currently orbiting Jupiter.

The global Matrix is made up of multiple, complex processes with near-infinite levels of fractal detail and intensity, all built from an enormous number of constituent bits and bytes, a number which is growing exponentially every day. The Matrix as a whole has not stopped growing since the first electronic computer was activated. The information-age trend is in one direction only: toward greater complexity, greater speed of transmission of electronic messages, greater throughput of data, and greater range of functionality.

Moore's Law – first proposed in 1965 by Gordon Moore, one of the founders of the semiconductor manufacturer Intel – states that the computing power of a microchip doubles every 18 months, and for 30 years this has remained true to provide us today with PCs more powerful than almost any computer mainframe extant 20 years ago.[1] Increased power is no doubt wasted by underdemanding users and overextended software packages, but the increasing availability of computing power continues to open new doors for applications that cannot help but permanently change our lives and culture. Today video phones and language-translation programs are finally becoming reliable while desktop publishing and photo-manipulation programs designed for million-dollar glossy magazines are available to every aspiring writer or editor with a PC and a blank disk; tomorrow instant connection with and information up/download from any other node on the Matrix, in whatever form desired will become possible: for a price, of course.

I will show how this growth and complexification process is effecting dramatic change not only on the Internet, but also in the perhaps surprisingly interlinked fields of artificial life, global economics and political philosophy. This chapter will attempt to describe the new phenomenon of the Matrix as an emerging artificial life form fully differentiated from the much more limited construct that is cyberspace. The complexity of the Matrix necessitates rethinking basic notions such as autonomy, competition and life itself, filtering these notions through the framework of not only evolutionary biology but also the insights of the non-linear economics of technological growth, as well as the situational politics of event-oriented analysis and action. In the end, the Matrix is an evolutionary, spatio-temporal process of connection and intertwining, not a virtual geography. The Matrix is the creative, functional reality of connectivity: the dynamic result of linking this and that and thereby gaining access to the power of the shared knowledge of these two nodes and all of the others in their network.

First, a clear distinction must be made between the Matrix and its more widely-known graphical subset called cyberspace. The Matrix is the asymmetrical grouping of heterogeneous and combinatorial virtual and actual communication spaces which fortuitously cling together to form the global Matrix out of our linked telecommunications and computer networks (i.e. as a meta-phase-space, graphing the interactions of all of its component spaces; or alternately, as the set of machinic functionalities). It is not made up simply of all e-mail messages, or all web-pages, or all local-area network sites. The Matrix is not merely the realm of personal computers: it transmits every telephone call, even those made from a human to a fax machine or made directly from computer to computer. Additionally, there exists a very secure network, not much spoken of because of its ubiquity, which directly connects hundreds of thousands of cash machines globally, linking them to bank mainframes, to stockbrokers, to credit card companies and to every shop that takes credit cards or electronically approves cheques. Satellites constantly monitor and report the positions of people and objects on the planet's surface, while transmitting television, telephone and military signals. If you start adding together all these artificially-generated spaces and their real-life components, it soon becomes clear that the Matrix encompasses the heterogeneous electronic zones of machinic functionality, both virtual and actual, associated with every electrified appliance from toaster to mobile phone to supercomputer, all linked through the power grid which underlies each of these component spaces. Cyberspace is just a small subset of this virtual/actual space.

From the viewpoint of the computer programmer, cyberspace is the interface between the computer and the human, between the virtual and actual, digital and analog, and the programmer's job is to make that interface as seamless and intuitively functional a space as possible so that humans can get the most productivity or entertainment possible out of their machines. In order to increase productivity and consumer enjoyment, the interface must be friendly and comfortable for the user, disguising the raw chaotic flux of digital bits with multicoloured tree-structures, imaginary desktops, and self-descriptive graphics, which allow one to access the connections of the Matrix without needing to understand its machine code of 0's and 1's, electronic handshakes or transmission protocols.

In many ways, the most convincing representation of this new dataspace, and certainly more convincing than the better known and originary vision of cyberspace outlined in Gibson's *Neuromancer*, is Neal Stephenson's Metaverse; a graphical dataspace posited in his novel *Snow Crash* (Gibson 1984; Stephenson 1993). In Stephenson's Metaverse, the most visible and accessible part is a consumer mall, a re-creation of Walt Disney's re-creation of Main Street USA. Here, while its undeveloped regions remain a black desert of empty memory space dotted with secret caches, online visitors' virtual bodies are displayed as identical Ken and Barbie clones. As Stephenson has recognised, people are drawn in masses to the things that seem familiar, and the design of cyberspace will be influenced greatly by this money-making imperative: Placate the humans. Give them what they think they want. And yes, the mall in cyberspace will have every store, they will sell everything, and the people will go.[2]

But this same psychological trait which draws many human consumers toward what they easily recognise works at the same time to inhibit their perception of the heterogeneity of the Matrix. People see and enjoy the virtual storefronts but do not access the full extent of the paradigm change wrought by the global event of heterogeneous interconnectivity. This lacuna encourages the replication of past development and marketing strategies on the Matrix without substantially attempting to understand or even exploit the radical changes that exponentially growing interconnectivity brings. This imperative to re-create the past is being applied mainly to what I am differentiating as cyberspace. The Matrix, on the other hand, is being left unthought, undifferentiated, and conceived negatively as the chaos of a new market waiting to be capitalised as soon as it equalises itself into something recognisable like cyberspace.

In *VRML*, a book about the Virtual Reality Modelling Language (the main computer language currently used to create the graphical interface of 3-D cyberspace), Mark Pesce makes the imperative to be user-friendly into a fundamental law of the new universe he is helping to bring to life. Pesce's vision of cyberspace (which is not all that different from Stephenson's Metaverse) is worth quoting at length – though one needs to keep in mind that his vision is not only that of a talented programmer writing for other programmers, but also of a programmer working in the middle of a cutthroat, multi-billion-dollar, consumer service industry. Discussing the limitations of the World Wide Web and its static, relatively isolated spaces, Pesce says:

> The Web, with its roots in the hyperspaces of island universes, will need to transcend them in order to provide the three spatial principles required for a human-navigable environment: ubiquity, uniformity and unity. There must be a single, infinitely large cyberspace – even if there are others – which is everywhere, continuous and regular. When such a cyberspace exists, we can knit together our islands into a continuous whole. It doesn't mean that my house will be next to yours, but that if I travel from my house to yours, I travel through all of the intervening space. This is very important for the users of cyberspace – with a unified layout, people can remember where they are and what's around them. Without this, people will find cyberspace rather disorienting and discontinuous – something the real world is not. In a unified cyberspace, you can make maps, or stop somewhere and ask directions.
>
> (Pesce 1995: 317)

As one who finds the real spatial world 'disorienting and discontinuous' every day, and who believes that the world is produced precisely by non-linear flux and chaotic intensity, I must immediately note that Pesce's 'unified cyberspace' is still necessarily a subset of the global Matrix: a specific graphical interface, created so that most of us do not have to learn to program in order to make use of what our interconnected machines are now capable of doing. Moreover, the principles of uniformity, ubiquity, and unity are not inherent to the Matrix. Rather, they are overcoded by Pesce and others in order to create cyberspace as the humane, consumer-friendly face of the Matrix. These three principles are not necessary for life, or even for understanding or navigation, but are instead remnants of an anthropocentric, rationalising worldview which has been successful in the past at consolidating the power of centralising forces. Cyberspace is being created as a closed system with definable, familiar limits while the Matrix is opening in every direction such that it can even be thought of as abstract extension. Homogenising principles such as Pesce's will have a profound effect on the shape of cyberspace, but the Matrix is bigger than such humanistic and consumerist concerns and will continue to evolve on its own, creating its own form according to its functionalities, a form unrelated to any centralising, hierarchical structures and unresponsive to our perceptions of it. The Matrix is a dynamic set of functions, not representations.

Instead of settling on the flat plane of cyberspace, which is too susceptible to centralisation, in attempting to describe the Matrix we must therefore acknowledge multiple, non-interchangeable dimensions. These multiple dimensions cannot be easily mapped or traversed and require alternative means of navigation and construction, ones rooted in an event-oriented situationist approach. One fairly all-encompassing attempt at a definition of the Matrix with fewer philosophical prejudices built in (but which the author still refers to as cyberspace) is given by architect Marcus Novak, who integrates samples of Bruce Sterling, Wendy Kellog, William Gibson and others. He summarises:

> Cyberspace is a completely spatialized visualization of all information in global information processing systems, along pathways provided by present and future communications networks, enabling full copresence and interaction of multiple users, allowing input and output from and to the full human sensorium, permitting simulations of real and virtual realities, remote data collection and control through telepresence, and total integration and inter-communication with a full range of intelligent products and environments in real space.
>
> (Novak 1991: 225)

Novak sees cyberspace navigation not as traversing an homogeneous field, but as synthesizing different kinds of information into a self-coherent but temporary image.[3] While cyberspace remains overly spatialised, this definition retains the vital heterogeneity of what I will call the Matrix without imposing an abundance of human overcodings and preconceptions. Vital because if we want to understand anything about the Matrix, we must see it on its own terms, and perhaps even transform ourselves to become more like it; more heterogeneous. In order to survive and prosper along with the ever-expanding Matrix, we must destratify, de-homogenise ourselves and our thinking ever further. The global telecommunications Matrix is a being both evolving and mutating beyond the control of any individual organism, organisation or state, and it will continue to take on forms and functions unplanned and unimagined by today's developers of cyberspace.

Artificial life

Some researchers in biology and computer programming have already recognised the need to 'forsake rational design . . . for techniques based on the blind forces of biological evolution' (Coveney and Highfield 1995: 238). The science of Artificial Life studies virtual (computer-generated) populations and how they undergo processes such as reproduction, competition and evolution, the simulation of which often causes unpredictable and complex life-like behaviour to emerge. When programmed with the ability to compete, evolve and reproduce, populations of machine code begin to move in chaotically rhythmic patterns, growing in complexity, interconnectedness and functionality. In the following quote, which name-checks many of the founders of the discipline of Artificial Life and their basic theories, Steven Levy pieces together a useful definition of life:

> If [John] von Neumann established that life existed as an emergent information process; if [Stuart] Kauffman was among those who told us that through self-organization life wanted to happen; if [Christopher]

> Langton, [Jim] Crutchfield, and [J. Doyne] Farmer informed us that among life's properties was a preference for locating itself just this side of chaos; then [Daniel] Hillis, in ratifying computationally the work of biologists such as [William] Hamilton, hinted that life was a symbiotic process that virtually required the company of deadly rivals. Equilibrium was an illusion; order finds itself from a relentlessly troubled sea.
>
> (Levy 1992: 203)

Life does not occur at a state of equilibrium, but has been shown to be a chaotic, self-organising process emerging out of the increasing complexity of a given population. Life occurs in multiplicities not in individuals, who could never exist for long as the only one of their species. Complexity arises when increasing connectivity creates dynamic new possibilities amongst previously isolated components: new processes such as competition, reproduction, mutation and especially evolution. 'And indeed evolution was something based on simple rules that yielded wondrously complicated results' (Levy 1992: 196).

The commercially sponsored evolution of Artificial Life is being accomplished back on the Matrix by corporations such as British Telecom, which currently routes phone conversations using mobile, ant-like software programs that swarm through its phone system and follow a few simple rules of thumb to route calls where there is the least traffic at any one moment (Coveney and Highfield 1995: 251). Semi-intelligent software agents (bots, spiders and daemons) search the internet and look for specific types of activity or information, gather it and bring it back to their master's console. Semi-autonomous computer viruses arise from Bulgaria to Beijing and can finally be transmitted by e-mail. This long-mythical threat became real in 1995 and will become a graver problem with the prevalence of Java and other object-oriented programming languages that are capable of exchanging discrete (possibly infected) software objects over the internet. Virus-objects will be able to disguise themselves, be downloaded by unsuspecting net-surfers, and enact complex operations on their new host computers, without the approval and outside the control of the host or its user. For example, the Wahoo virus of 1996 seeks out all Microsoft Word documents on its host computer and changes every other word in the document to 'Wahoo', while ignoring everything else on the hard drive. Once launched, viruses are outside the control of their creators, making them semi-autonomous agents seeking out nothing more than partial autonomy within whatever host they can latch on to. The autonomy of a self-reproducing information virus is not the same as that which we attribute to ourselves, but it has many of the same characteristics (e.g. self-reflexivity, reproductive capacities, self-propulsion, and context-based decision-making) often considered necessary components of autonomous existence.

Unleashed into the electronic environs of the Matrix, viruses are left to their own (initially-limited) devices to survive, compete and reproduce. But as they have become more complex and capable of mutation, they have discovered a possible route to long-term survival. Mark Ludwig, a leading researcher of software viruses, has a strong position on the future evolution of computer viruses:

> Given our current understanding of evolution, the question isn't 'what if' at all. It's merely a question of when. When will a self-reproducing program in the right location in gene space find itself in the right environment and begin the whole amazing chain of electronic life? It's merely a question of when the equivalent of the Cambrian explosion will take place.
>
> (Ludwig 1996: 242)

Particular viruses or intelligent agents are less important than the environment in which they breed and compete, the connectivity of the Matrix: 'the interesting thing in natural selection is not the evolution of a single species, but things like the coevolution of hosts and parasites' says Daniel Hillis (Levy 1992: 203; cf. DeLanda 1994). Teenage hackers of the 1980s are reborn as security consultants in the 1990s, using the knowledge they gained illegally to combat computer crime and to earn a comfortable living. These hacker/consultants are using the same tactics and techniques as before, but now for different purposes. Their role in the ecology of the Matrix has evolved from chaotic attractor (infectious wound) into a stabilising force. At the same time, virus developers have evolved much more complex subroutines which are capable of mutating around even the strongest security algorithms. This evolutionary spiral only helps to populate the bestiary of the Matrix with more viable, more autonomous artificial lifeforms. With the ability to act (semi-) autonomously being distributed more widely across platforms and programs connected to the Matrix, the possibility of centralising power along the mainframe/dumb terminal model lessens. Info-power is not suddenly rationed out more fairly, but it does become much more difficult to control directly in mass amounts or for extended periods of time. One must remember that the seemingly iron grip of the Microsoft/Intel monopoly on personal computers has been in place for less than twenty years and yet provokes the same Big Brother rhetoric which surrounded IBM in the 1960s and 1970s when they once dominated the computer industry. A single species needs interactions with others in order to thrive.

The global Matrix exemplifies a smooth space which effectuates complex, non-linear interaction between the virtual and actual, thereby creating new and unexpected possibilities. The most distinctive feature of the Matrix is undoubtedly its distribution of control and communications, which are dispersed throughout a meshwork web of interconnected but heterogeneous multiplicities. Multiplicities exhibit emergent properties that cannot be deduced from an individual part, properties that will not emerge until the process is actually run through. A myriad of collective possibility opens up to a million ants that are not available to (or deducible from) one ant. Flocks of birds actually exhibit faster reaction times in their movements than a single bird can achieve alone (Waldrop 1992: 241–243). More is qualitatively different. Therefore creativity (or originality) is a distributed, collective process of multiplicities. Assembled machinically, a complex system creatively produces attractors around which the system's energies and components orbit, and these new attractors mark points of reference or information nodes which did not exist before (Coveney and Highfield 1995: 166–167).

The Matrix exhibits all of the qualities of a complex, dynamical system: multiple, heterogeneous components interconnected at varying strengths to form an open, dissipative structure which both inputs and outputs energy and data; the abilities to mutate, evolve and reproduce; and a semi-awareness, or self-referentiality with regard to its own actions and environment. What keeps the Matrix from being a true life form, of course, is the role that human operators continue to play, performing the activities which the Matrix cannot yet do on its own – especially the reproduction of hardware and software (Levy 1992: 26–46). But the Matrix as we have defined it – a global network of both virtual and actual environments – will always include humans as part of the loop. Whether the relationship will remain one of master-to-servant is an open question.

Economic complexity

The virtual/actual environment which the Matrix provides is complexifying on more than just an ecological level. In the 1980s, when the major world stock markets were first linked directly by satellites, the amount of capital existing on the markets grew immediately by 5 per cent (Kelly 1994b: 226). Networking capital markets globally and in real time did not just make it quicker and easier for money to flow back and forth – it actually created more money on the capital markets than there was a moment before. Capital appeared which previously had been denied access to the markets by inefficiencies in the previous data infrastructure and its methods of communication. This new level of interconnectivity and complexity, which marked a qualitative shift (or a singularity) of the geo-economy, was described already by Karl Marx in the 1850s, when he showed how speeding up the circulation of capital will actually create more capital, along with a new form of capitalism: not only an accelerated one, but an economic process less attached to the physical unit value of labour input.[4] This is the emergence of qualitatively new methods of capitalisation in the wired economy, one of the most important transformations due to the speed and number of agents interconnected with the global Matrix.

This geo-economy, like the Matrix, is made up of multiple congruent and conflicting processes, and it is very easy (and often profitable) to over-emphasise some tendencies while downplaying others. The increasing accessibility of the data libraries of the internet marks a growing and widely noted decentralisation of information and power in computerised regions. At the same time, 1995 and 1996 were both record years for corporate mergers and acquisitions with multiple worldwide industries centralising all of their productivity and capital in smaller numbers of larger corporations. In the USA, the world's largest producer and consumer, oligopolies were either created or strengthened in: pharmaceuticals, television broadcasting, microchip production, retail sales, insurance, auto manufacturing, financial services and the various branches of telecommunications technology and services.

While consumers' choice of available services and products are growing daily, the choice between providers of those services is shrinking just as quickly until a few companies in numerous vital sectors now own oligopolistic market shares in their specialities. This has strong distortional effects on the emerging Matrix whose functionality underlies the geo-economy. Global corporations are centralising power in the consumer economy in conflict with the decentralising tendencies of the Matrix. Open competition is the most robust producer of creative evolution and mutation within complex systems and markets, so oligopolistic centralisation tends to produce distortions, inefficiencies and a weakening of the overall creative potential of a system.[5]

Whatever the short-term reasons behind these recent mergers, they display the competing machinic processes of the *market* and the *anti-market*, which denote respectively the free exchange of ideas and goods horizontally vs. the controlled regulation and distribution of ideas and goods vertically from the top down (DeLanda 1996). This is not to say that oligopolisation is incompatible with decentralisation since some of the most profitable corporations have 'reengineered' their organisational structures into horizontal networks of partners who distribute power away from the centre of their corporation. They mimic the new functionalities of the Matrix by creating networks of production, distribution and consumption; by rerouting around barriers to trade and communication; and by creating unforeseeable virtual spaces, all done with an eye on boosting profits.

With ideas and goods travelling around the world at the speed of light and of a Boeing 777, respectively, formerly closed, local markets are now in direct contact and competition with each other. A large corporation obviously has the upper hand at the beginning of the full integration of the global economy, since it takes a lot of money to operate in multiple markets worldwide. But the technologies of connection that allow globalisation to occur in the first place bring with them the decentralising tendency of the Matrix, which can open the door for niche services, global demand for a local product, or a million-dollar bidding war over a single good idea. A fully distributed network allows for interaction in any direction.

Within such a system, both parts of the bottom-up/top-down binary can be idealised. The market has been fetishised by equilibrium-seeking, neoclassical economists as the magical balance sheet which automatically equalises supply and demand, while the anti-market has been simplified as the authoritarian power of the monarch or élites dictating orders. Instead, the market and anti-market should be seen as multiplicities, as complex, non-linear dynamics in which real processes interact and are shaped, coerced and enabled by their either smooth or striated environment; together composing an environment which is never fully one type or the other, which is always changing itself. The striating and smoothing, converging and diverging tendencies of market and anti-market therefore come to resemble the co-evolution of hosts and parasites discussed by Hillis. Decentralisation towards a truly open marketplace will not solve any global problems, but it does help to make certain types of autocratic solutions less functional and therefore less plausible, while activating ever more functions of the Matrix.

Instant connections

Looking at thirteenth-century Europe, one can map fairly precisely the spread of ideas from town to town, along market routes and the lines of advance of invading armies, and by tracking the date and place of the transcription of certain books. In modern times, how long did it take before every human with access to a television or radio had heard the words 'cold war' or 'personal computer'? 10 years, 30 years? And communication moves faster every day. Instead of memes (i.e. easily transmitted ideas) travelling in a continuous line from market town to town, today there is the sudden appearance of a product or popular idea out of nowhere, followed quickly by complete lateral distribution and saturation: those who like the meme take it up and infect others with it, while others ignore it as best they can. But the key here is that any meme is instantly accessible to almost anyone who is connected to our global media distribution net. The Matrix is beginning to look less like Pesce's uniformly distributed space and more like a global computation device for producing, storing and transferring data, memes and goods.

Two quotes to expand this idea of global computation and instantaneous contact:

> In computation there is no analogue of distance. One memory location is as easily influenced as another.
>
> (Daniel Hillis in Kelly 1994a: 73)

> With geographic contraction, the reduction of distances, territory has lost its significance in favour of the projectile.
>
> (Virilio 1986: 133)

All manner of filters, switches and translators no doubt work to alter memes and other data along the way to consumers, but this is a property of any form of communication: the much-discussed hermeneutic circle that is at the rotted root of postmodernism. The hermeneutic circle is an engineering problem of tracking differential values of data filtration and translation, and not an ontological problem of finding truth. All data must be interpreted and filtered to be useful, and the problem today is not the modern problem of gathering or creating the data, but the quintessential information-age problem of processing the sheer quantity of data available, which clogs all the filters and overheats the capacitors.

With the exponential increase of both data creation and distribution, the baseline-ambient environment of our saturated society is now noise. We are forced to burrow through the accumulating layers of conflicting and overlapping information to pick out connections to what we might find useful. The feedback cycles of our culture are already too quick and unrelenting to make room for much that is slow or sentimentalist. Now that the clearly defined moralities of the past have fallen away dramatically in influence, we have entered into another age of materialism and pragmatism where you either recognise the effective reality on your own life of the Ebola virus or 100 million Chinese migrating to their coastal cities to escape poverty and isolation, or else you live in a theological or deconstructive house of cards.[6] The global communications net, the Matrix, links us not only to auto dealers and the Louvre but to flesh-eating bacteria and cultural landslides. With the ubiquity of jet travel Ebola and other new viral entities are global creatures, while the recent addition in China of 100 million people to the global, urban workforce in the space of a decade has profound repercussions on, for instance, the price of Nike shoes in Chicago, the profitability of processing food for export in Marseilles, and the projected demand for silicon chips in Asia over the next ten years. Urbanisation has been a hallmark of the twentieth-century but a shift of the speed and magnitude of China's will have unprecedented effects worldwide. Even those who attempt to disconnect from, or are not yet connected to, the global grid still live in its glow.

Maps and controls

In order to understand how complexifying systems of communications and exchange function, study of the Matrix must be grounded in a materialist pragmatics. It is easy to make long-term predictions and spew cyperbole that has no relation to reality, whether virtual or actual, but pragmatic knowledge is required to understand this massive shift in functionality linked to the Matrix. Analogously, a new field of psychogeographical study is opening up to map the Matrix and this new field will no doubt one day be as established as those other fields – biology or sociology – that investigate the habits and peculiarities of life forms and their organisation. Only this field will investigate the habits and peculiarities of several phyla of new life forms, each constructed from a combination of virtual, artificial and organic components.

Guy DeBord showed quite forcefully how the control nodes of information-age capitalism (what he called the Society of the Spectacle in the 1960s, the

Integrated Spectacle in the early 1990s) are inherently secret nodes which connect and route flows of capital, money and raw cultural intensity into pre-chosen outlets (Debord 1977, 1991). The results are always made visible by accumulations of physical wealth and power, but the connections, switches and controls which link the input and output of the black box that is anti-market capitalism always remain hidden, distorted or partially blocked from view.[7]

With incomplete knowledge a given, situational control is always the art of experimentation, of reaching certain desired levels of interactivity, stimulation or complexity to see what will happen: to approach the unknown, not to reiterate what is already known nor to pile up carbon copies of past pleasurable experiences. One must deal with the secrecy, disinformation and noise in the system in order to avoid recuperative traps set up to restratify control. Regardless of the expertise of the situationist at internalising psychogeographical maps of his or her local attractors which show the best routes to desired situational bifurcation states, it remains the case that any attractor worth being interested in is in fact chaotic by definition, unpredictable.[8] With their strong dependence upon hard-to-measure initial conditions, these dynamic patterns of energy which flow around multi-dimensional (chaotic) attractors will never be fully compressed into internal models (schemata). Rather, they must be played out in real-time to see the actual outcome of a given, directional situation. Psychogeographic maps, akin to flow charts, must therefore be of this spatio-temporal process and not of any virtual or actual terrain.

A key to gaining situational control is learning to map the psychogeographical phase space. This involves: 1) knowing the environmental possibilities for energy flow in an area; 2) knowing the tendencies and directionalities of the energy available in a given situation; and 3) knowing the grammars and languages of the different programs running or entities interacting. With a well-wrought internal model or computer-generated phase-space map, one can in fact tilt a situation into various desired basins of attraction, and then, with a higher-level (almost instinctual) knowledge of the attractor's neighbourhood, its quirks, probabilities, and back-alley pathways, the expert situationist can reach desired results. The 'goal,' if there is any, is 'creative evolution' or directed bifurcation into a new level of possibilities wherein something original can emerge or be created. As Georges Bataille pointed out, our ignorance of our own material context 'causes us to *undergo* what we could *bring about* in our own way, if we understood' (Bataille 1988: 23).

Situational control entails not being merely a means, but setting up a positive feedback loop via viable connections (healthy and willing) so that one is also the end as well as the means. Situational control, or autonomy, emerges from collective, complex interaction; it is something that can never happen in isolation (or in a closed system). A software virus, for example, becomes autonomous only in the context of its hosts. We have so much information (and know how to look at the information in sufficiently complex, self-conscious ways) that control can never be individual. Situationists attempted to create event-oriented maps, intensive maps which were tied not merely to the terrain, but which attempted an understanding of geography's relation to society, to urban blight, to architecture, to human habit, to transportation planning, to political history, to economic activity and to dreams. The political failure of situationism was precisely the belief that individuals could control or create the political situations they were involved in. Being a means to others' ends cannot be avoided, and has been shown to be the peak meditation achieved through the materialist trend in philosophy (cf. the later Nietzsche; Sade's ritualism and rigidity as his personalised form of situational control; Bataille's thoughts on community).

We must take these hard-sought results of past experiments and use them to de-isolate, *to connect*. As Deleuze and Guattari say in *Anti-Oedipus*, we need to accelerate the decoding and deterritorialisation which capitalism is always trying to keep under its own profit-producing control (Deleuze and Guattari 1984: 321). Controlled, safe deterritorialisation is the same as pre-capitalised recuperation. The only way to escape the trap of recuperation is to give up the desire for control: not to allow yourself to be controlled by the powers-that-be, to be out of the control circuit. Learning to recognise near-random (or far from equilibrium) situations and nudging them in one direction or another at a crucial node, to help or to allow some new conglomeration to emerge will become a highly-developed and sought-after skill: opening a temporary, syncretic assemblage within the Matrix (Fishman 1995; Gibson 1996; Stephenson 1994).

Understanding the material world has complexified to such an extent that almost no human today can know clearly both how the surface of a silicon chip works and how the evolution of geological strata progresses. Following financial derivatives markets requires a career of analysis and creativity, leaving no human time for a thorough comprehension of even a related financial sector of functions, such as venture capitalisation, which requires its own experts to function efficiently. Autonomy must be dehumanised and decentralised from the notion of an independent judicial process of exerting free will, to become a process of occupying a functional role: of telephone psychic, DJ, entrepreneur, welfare recipient, of teacher or student. Autonomy activates when choosing the functionality available which increases local field intensity. The Matrix is complexifying by creating new functionalities, a process entirely different from global expansion, accumulation of wealth, or upgrading the old.

With the proper build-up (either ritualised or spontaneous) to the edge of chaos, a skilled cartographer of the psychogeographical dynamics around a libidinal space of attractors can navigate amongst the pulls and pushes (like a bit of code running through the back streets of its motherboard, effortlessly recalling every nook and cranny of the circuit architecture) to reach singularities, crucial nodes of peak interaction and productivity per labour input. At these points (or spaces or curves), a seemingly minimal input (an incantation, a sacrifice, a dozen electrons, a kiss) can spontaneously alter the structure and even content of libidinal space, the input being hyper-magnified, in a butterfly-effect, because of the positioning of the fluxes right on the edge of chaos, at the point of self-criticality where complexity and innovation emerge.

Most interesting is the middle level of functionally-effective virtual spaces: data interchanges, satellite positioning grids, graphs of television-rating monitors bulging and bursting in rhythm to political scandals and toilet breaks during commercials. This is where virtual and actual space open on to each other and become undecidable; this is where autonomy either appears or slips away, in the synthesis of human and machine. Philosophy and criticism in general need to return to practical knowledge in order to understand how things function and evolve both on the Matrix and off. Acknowledging the reality of Artificial Life and the influence of the geo-economy on society and thought, as well as implementing a more event-oriented analysis of the material world are all necessary steps forward which I hope to have begun to explain here in terms of the Matrix.

Finally, many might interpret a convergence of networked machines and humans as collectivism, or a globalised electronic life form as a threat. But we do not really know what it will mean. The notion of convergence on to the Matrix marks the possible existence of a point of 'self-criticality' (Waldrop 1992: 304–306) or a future singularity arising in the global communications network. A singularity denotes a specific time-space in the life span of a given system where a system which is far from equilibrium reaches a level of high complexity and suddenly, qualitatively changes its pattern of organisation. The Matrix complexifies every hour and at some point in the near future, the sheer volume and complexity of interactions, both virtual and actual, will push the Matrix to a threshold of self-criticality where something qualitatively new and as yet unimaginable will arrive. Keep your eyes open.[9]

Notes

1 'Today's $2,000 lap-top computer is many times more powerful than a $10m main-frame computer was in the mid-1970s' ('The Hitchhiker's Guide to Cybernomics: A Survey of the World Economy' in *The Economist*, 28 September, 1996: 3).

2 An approximation of Metaverse is in fact being constructed as I write by both Microsoft programmers and networked video-game players and creators. See Laidlaw 1996.

3 'There are no objects in cyberspace, only collections of attributes given names by travellers, and thus assembled for temporary use, only to be automatically dismantled again within a short time-span' (Novak 1991: 229; 235).

4 Marx's idea in the *Grundrisse* that 'the velocity of turnover' can substitute for 'the *volume* of capital' (Marx's emphasis) leads him to speculate: 'Does not a moment of value-determination enter in independently of labour, not arising directly from it, but originating in circulation itself?' (Marx 1993: 519).

5 Although falling communications, computing and transaction costs allow more small firms than ever before to compete both locally and globally, 'in a growing number of industries there is a natural tendency for the market leader to get further ahead, causing a monopolistic concentration of business'. 'Increasing returns' kick in for particular market leaders when they are in a position to combine: 1) high fixed costs (such as Research and Development); 2) beneficial network externalities (consumers wanting to use a product because of its compatibility with other useful products); and 3) customer lock-in (when the complexity of learning how to use new products makes consumers slow to change) ('The Hitchhiker's Guide to Cybernomics: A Survey of the World Economy', *op. cit.*: 35–36); see also Arthur 1994b, 1996).

6 One hundred million is in fact a very conservative estimate of the growth in the urban population of mainland China since the 1980s. *China News Digest* reported the State Statistics Bureau's announcement that 'by the end of 1996 there were 515.1 million people living in urban areas in China, which accounts for about 43 per cent of its 1.22 billion population, AFP reported, quoting Thursday's *China Daily*. In 1994, only 20 per cent of the population were considered urban residents' (Weijun Liu and Ray Zhang, 'Urban Population in China Boosts to 43 per cent,' *China News Digest*, 13 October 1997).

7 Buzzing with intensity, the Matrix is a curved, Riemannian (non-Euclidean) time-space connecting multiplicities in complex ways, giving nearly instantaneous data communication around the world through a web of tangled connections weaving itself into existence at every moment. Being non-linear and chaotic, the Matrix is filled with instances of what the psychogeographers DeBord and Ivan Chtcheglov called 'the Northwest Passage' or 'the reversible connecting factor': non-linear shortcuts from the top of one curve to another. Greil Marcus translates this situationist idea as the instant route to total change, but maybe they are just bifurcations: points of qualitative change where intensity peaks or falls in space, where systems begin to self-organise (Marcus 1989: 385–390). This function is taken up by *hyperlinks* which allow instantaneous access to any page on the Web from any other: they do not follow an arboreal organisational structure but instead crosscut hierarchies to land directly at their related site.

8 See Waldrop 1992: 225–226 for the differences between point, periodic and chaotic attractors.

9 With a few specific exceptions, the reader may substitute the Deleuzo-Guattarian term 'the Machinic Phylum' for 'the Matrix' throughout this article. A full explication of the connections, however, requires a full article devoted to the Machinic Phylum. In the interim, see Deleuze and Guattari 1987 and DeLanda 1991.

Part II

Cyberscapes

7 From city space to cyberspace

Jennifer S. Light

Legend has it that Senator Albert Gore Jr conceived the notion of the 'information superhighway', an infrastructure system analogous to the physical highways his father, Senator Albert Gore Sr, helped to create in the 1950s. While at first many suspicious commentators questioned the superhigh-way analogy, these high-speed communications networks are now often portrayed like the physical highways of the mid-twentieth century, tearing holes through communities. Such portrayals of media and cities share one overarching theme. It is a consistently pessimistic vision of how the rapidly growing use of information technologies will further exacerbate a decline of civic life and community. This perspective views public life in decline as accelerated by a culture of simulation. Physical spaces are replaced by digital ones – first television and film and now the Internet.[1]

In this essay, I explore and ultimately question such grim portraits of a bleak future. The essay has three sections. The first examines several arguments about the decline of civic public life – that there is a decline in public space, that urban environments are increasingly commodified, and that architectural authenticity is in decline. A concern with simulations of urbanity emerges as a central theme. The second section introduces recent suggestions that new electronic technologies must invariably have negative effects on public life and community. This section notes remarkable similarities between arguments made by scholars who focus on architectural changes in cities as an explanation for declining civic and community engagement, and more recent arguments by cyberpessimists who anticipate developments in communication technologies as precipitating a similar fate. The third section draws from several examples in the history of technology, and current developments in cyberspace, to reach a more optimistic conclusion. While no one can predict the full range of implications of the Internet, it is nevertheless clear that arguments about its relationship to cultural 'decline' are based on questionable assumptions. This essay presents a more optimistic view – that the Internet presents exciting opportunities to revitalise civic engagement in new ways.

Narratives of decline

> Why do they build special Hollywood towns? One hardly knows where the real city stops and the fantasy city begins. Did I not see a church yesterday and believe it belonged to a studio – only to find out it was a real church? What is real here and what is unreal? Do people live in Los Angeles or are they only playing at life?[2]
>
> (Moeschin 1931: 98)

> A real city is full of life, with ever-changing moods and patterns: the morning mood, the bustling day, the softness of evening and the mysteries of night, the city on workdays so different from the city on Sundays and holidays. In contrast are America's 'downtown' areas, those part-time ghost towns, spawned by one-sided and one-track development of many of our city cores, which are busy eight hours a day on weekdays and deadly silent and unpopulated in the evening, during the night, on Saturdays and Sundays.
>
> (Gruen 1964: 27)

> Something unusual is happening today in the relation between the real and the imaginary, reality and its representations.
>
> (Soja 1996: 242)

Three comments, from three observers of urban life, spanning three generations. From the 1930s, to the 1960s, to the 1990s the names change but the theme remains constant. The authenticity of American cities is in decline. In recent decades, a rich and textured literature has pressed this theme (Gruen 1957; Keyes 1973; Oldenburg 1989; Whyte 1988; Hiss 1990; Sorkin 1992b; Beauregard 1993). Much of it focuses on the decline of informal public life and community, both in cities and small towns. Observers link this loss to eroding concepts of civic space, to postwar suburbanisation, to inner city crime, and to the phenomenal growth of privatised public spaces including malls, festival marketplaces, common-interest developments (CIDs), business-improvement districts (BIDs) and gated communities (Deutsche 1996; Goss 1996; Sorkin 1992; McKenzie 1994; Huxtable 1997). Taken together, these narratives portray a bleak outlook for vibrant civic life, and for any sense of urban community.[3]

Charges about the decline of cities, popular both on the left and right of the political spectrum, are linked both to the physical realities of cities and to the imagined cities that represent the public sphere. In this formulation, 'public space' represents both physical public space and the idea of democracy. To a number of urban theorists, parks, streets and squares are physical embodiments of the principles of democracy and community. Changes to the physical fabric of a city thus throw these principles into question. A continuing erosion of the public sphere goes hand in hand with the privatisation of city streets and other 'public' spaces. Specifically, the physical artifacts representing the idea of a public sphere for democratic debate are pushed aside and replaced by 'simulations'.[4] The common theme for these writers is a critique of the illusion that these spaces are heralded as revitalising public life when in fact they are privatised, exclusionary spaces. As the number of such simulations grows, they are described as signalling not only a loss of authenticity, but also a loss of 'reality'.

This essay examines three specific narratives in depth. The three are worth examining because they represent perspectives about what has gone wrong with cities, and what can be done to repair the damage. I bear in mind the work of Robert Beauregard, who:

> assumes the discourse on urban decline to be more than the objective reporting of an uncontestable reality and pursues [instead] an interpretation . . . that considers how the discourse functions ideologically to shape our attention, provide reasons for how we should act in response, and convey a comprehensible, compelling, and reassuring story of the fate of the twentieth-century city in the United States.
>
> (1993: xi)

In examining these narratives I want to show that, just as there has never in fact been a simple and open public space, neither are those other motifs evident in such narratives entirely new. Rather, concerns over both the commodification and increasing inauthenticity of architecture, and city life more generally, have a long history. Together with a more nuanced account of the changing nature of public space, such a history leads to a rather different interpretation of recent trends.

Streets and the privatisation of space

Begin with the city street, as described by Boddy:

> Over the past decade, new extensions to the city have appeared in downtowns across the continent. In cities as various as Minneapolis, Dallas, Montreal, and Charlotte, raised pedestrian bridges connect dispersed new towers into a linked system; mazes of tunnels lead from public transit to workplace without recourse to conventional streets; people-mover transit systems glide above the scuffling passions of streetbound cities. Grafted on to the living tissue of existing downtowns, these new urban prosthetics seem benign at first, artificial arms and plastic tubes needed to maintain essential civic functions. Promoted as devices to beat the environmental extremes of heat, cold, or humidity that make conventional streets unbearable, they seem mere tools, value-free extensions of the existing urban realm.
>
> They are anything but that. These pedestrian routes and their attached towers, shopping centres, food fairs and cultural complexes provide a

> filtered version of the experience of cities, a simulation of urbanity. By eliminating the most fundamental of urban activities – people walking along streets – the new pedestrian systems underground and overhead are changing the nature of the North American city.
>
> (Boddy 1992: 124)

Boddy contrasts exclusionary skyways and underground streets to regular city streets. He asserts these privatised alternatives to traditional downtown streets provide a thoroughly inadequate substitute. Skyways are mere 'simulations of urbanity'.[5] Boddy argues that downtown streets are emblematic of public space where different sectors of society can mix and mingle. 'Their replacement by the sealed realm overhead and underground has enormous implications for all aspects of political life' (Boddy 1992: 125). Similarly, Ray Oldenburg, lamenting the course of American urban development, writes that 'the grass roots of our democracy are correspondingly weaker than in the past, and our individual lives are not as rich' (Oldenburg 1989: xii). Framing new streetscapes as inauthentic is a position adopted to distinguish them from more real environments that existed in the past.

Boddy's focus is on disruptions and dislocations caused by new urban skyways. Yet his vision of the street is itself a simulation, an ideal. For example, through the nineteenth and twentieth centuries, cities were seen not as positive spaces, but often as in decline – beset with festering problems of dirt, disease and sanitation (Beauregard 1993; Griscom 1845; Wright 1980). Both Boddy and Oldenburg suggest a lost wholeness of civic interconnection. Yet this narrative of wholeness is contradicted by urban studies focusing on cities as sites of social atomisation and danger (Robins 1995; Lofland 1973). Boddy, Sorkin and other authors emphasise exclusions from the 'theme park city', but we should not forget how exclusions proliferated in past city forms – often based on race, class, and gender (Lofland 1973; Wilson 1995; Morford 1867).

Indeed, narratives of decline can benefit from other perspectives. For example, in *La Tribune* of 18 October 1868, Emile Zola described the streets of Paris, 'I know that M. Haussmann does not like *les fêtes populaires*. He has banned almost all those that took place in the old days in the recently annexed districts; he is pitiless in his campaign against hawkers and peddlers. In his dreams, he must see Paris as a gigantic checkerboard, possessed of a geometrical symmetry' (Clark 1985: 280 n115). Just because streets were ostensibly public did not mean everyone was welcome there.[6] In America, the streetcorner soda shop, fondly described in Ray Oldenburg's *The Great Good Place* (1989), was for a long time an all-male establishment (Smith 1992: 360).

Today as well – outside the simulated streets Boddy describes – such exclusions remain, from the mostly white centres of cities like historic Charleston, South Carolina, surrounded by rings of African-American neighbourhoods, to streets in many American cities where few women feel safe to stroll. This fuller picture points to a central simplification in the narrative of the decline of public space and its replacement by privatisation: the drawing of black and white opposition between privatised spaces and the abstract ideal of a public space open to all (Deutsche 1996; Mitchell 1995).

Proposals have been offered to counteract a perceived decline in public life. Supporters of the New Urbanism, such as Andres Duany and Elizabeth Plater-Zyberk, have designed new towns such as Seaside, Florida, categorised as a TND – traditional neighbourhood design, with idealised streetscapes and housing designs of a bygone era (Al-Hindi and Staddon 1997). In a similar spirit, the Disney Corporation built Celebration, a planned Florida town which opened in late 1996. Ironically, such planned communities – built in the spirit of reviving community in physical space – are precisely those privatised public spaces critiqued by Sorkin and fellow theorists such as Trevor Boddy, Mike Davis, and Evan McKenzie (Davis 1992 ; Boddy 1992; McKenzie 1994; Flanagan 1996). Equating open spaces with the democratic public sphere is a leap of faith, too easily made (Mitchell 1995). While these places may offer human-scaled planning, the ownership of land by a single developer, the private sector's assumption of public services, the rise of homeowner associations and residential private government, and private security forces do not promise a public sphere. In Seaside:

> The town has posted signs discouraging vehicular intruders, reminding them that the streets are private property. Fear of vandalism has prompted owners to suggest the use of street entry-gates and has led the development corporation to hire a night security guard.
>
> (Audirac and Shermyen 1994: 168)

Some have gone so far as to ban political signs altogether (McKenzie 1994).[7]

Don Mitchell has addressed the contested literal and metaphorical meanings of 'public space'. In one vision, public space is a physical manifestation of the public sphere – a politicised space for different voices to come together to constitute 'the public'. In another vision, public space is an ordered, safe environment where people can see and be seen – a courtyard, a city park, a beachside pier. While Boddy envisions streets as public spaces in the first meaning, the streets he describes are examples of the second. For similar reasons, communities designed in the spirit of the New Urbanism have come under scholarly fire (Al Hindi and Staddon 1997; McKenzie 1994). These disagreements are not unlike the different visions that led to conflict over the uses of People's Park as described by Don Mitchell (Mitchell 1995).[8]

Shopping malls and the commodification of space

A second theme of urban decline points to the commodification of space. Shopping malls are featured prominently in contemporary criticism. They have come to symbolise inauthenticity, simulation, homogeneity, consumption and surveillance. Margaret Crawford describes the West Edmonton Mall: 'Confusion proliferates at every level; past and future collapse meaninglessly into the present; barriers between real and fake, near and far, dissolve as history, nature, technology, are indifferently processed by the mall's fantasy machine' (Crawford 1992: 4). In this formulation, malls emblematise the 'culture of consumption'. Their interiors simulate the shopfronts and squares of a pedestrian town – yet their focus is profit per square foot. In a related development, main streets are themselves purchased by developers and turned into outdoor malls.[9] Sites of politics are thus replaced by sites of consumption that offer little opportunity for politics.

While some authors have critiqued malls and the ways space is commodified and depoliticised as a uniquely contemporary (and largely North American) phenomenon (Huxtable 1997), concerns about the commodification of space are in fact much older. In nineteenth-century Paris, for example, changes to the urban fabric made working-class protests more difficult while they increased the circulation of goods through the city and Crawford recognises nineteenth-century Paris, and in particular its department stores, as one analogue to shopping malls. Indeed, her description of the West Edmonton Mall recalls characterisations of nineteenth-century Paris where by the 1860s, newspapers routinely used the words 'parade, phantasmagoria, dream, dumbshow, mirage and masquerade' to describe the city (Clark 1985: 66). While Crawford draws upon the nineteenth century to offer an engaging history of consumption and the Paris department stores, she does not comment on the parallel between climate-controlled malls, which simulate both nature and city spaces, and the many other artificial environments of nineteenth-century Paris. All-weather enclosed exteriors and virtual travel experiences have been around for some time, from arcades to wintergardens to the Expositions Universelles (Friedberg 1993). Like the Mall of America with its aim to encapsulate the entire world within its walls, the nineteenth-century Expositions Universelles did the same. Indeed, Anne Friedburg and Vanessa Schwartz have contested the positions of Jean Baudrillard, Umberto Eco and Fredric Jameson who assert that hyperreality is a postmodern American cultural form (Schwartz 1994; Friedberg 1993). They relate hyperreality to transformations in Paris under Haussmann and Napoleon III. For example, Victor Hugo called this Paris a 'copy'; Benjamin an 'artificial city' (Hugo 1963 (1862): 508; Buck-Morss 1989: 90).

As Crawford suggests, not only did architectural and planning changes increase the circulation of goods, but the standardisation of prices turned shopping into an activity where looking replaced the need to buy. Her description stops short of suggesting how streets and spaces of the city were commodified. Yet one observer described the city itself as a department store, 'The crowd is the veil through which the familiar city lures the *flâneur* like a phantasmagoria. In it the city is now a landscape, now a room. Both, then, constitute the department store that puts even *flânerie* to use for commodity circulation' (Benjamin 1978: 156). In Benjamin's formulation, the crowd and the city were commodified as they became part of the spectacle enlisted to sell merchandise. The city became a department store, with flâneurs part of the crowd luring consumers inside.

One might go so far as to say that Paris, like American festival marketplaces today, became a commodity that was itself visually consumed. In the months before the 1867 Exposition Universelle, which brought millions of tourists to Paris, concerns for the appearance of Paris increased. An American visitor to Paris in 1867 described how 'superficially, under the Emperor's command, Paris washed its face and put on its Sunday raiment, very early in the event . . . the dust commanded to lie still, and not offend the eyes and nostrils; the trees to leaf (not *leave*) at their very earliest' (Morford 1867: 40). Benjamin has described how in this Paris, streets were like works of art. This representation ranged from the perspectival views they offered, to the fact that 'streets, before their completion, were draped in canvas and unveiled like monuments' (Benjamin 1978: 159). Paris itself became an object to be visually consumed and collected; a later Exposition was the birthplace of the picture postcard that advertised the city.

The argument that malls are the ruin of democracy – that they are essentially commercial spaces unsuccessfully simulating public spaces – is in principle compelling. For example, Frieden and Sagalyn summarise states' different legal rulings on what is and is not permitted in shopping malls (Frieden and Sagalyn 1989). Yet the argument is weakened because it ignores other versions of the past in which many of the sites of 'public life' were in fact commercial spaces – including, for example, department stores and cafes. Further, it assumes that citizens have few if any alternatives to shopping malls (Gregson 1995).

This critique of malls also misses some alternative interpretations. Crawford's description views malls as places of consumer domination.[10] Yet this position on 'the mall's fantasy machine' is architecturally determinist, viewing consumers as passive rather than active and even imaginative users of space and products. Crawford did not interview shoppers, but rather looked at how the mall trade categorises them. This methodology makes it very difficult to find creative uses of mallspace. Other scholars have tracked mall users differently, finding they are not duped into consumption, but rather that they use spaces within malls, without buying, for other purposes – for example to socialise or to keep cool on a warm day (Fiske 1989; Gregson 1995). Bernard Frieden and Lynne Sagalyn suggest much criticism of malls is too academic and élitist, that it misunderstands 'suburban folks' and overlooks the communal function of shopping malls (Frieden and Sagalyn 1989: 72). New methodologies such as focus groups reveal how what happens in seemingly homogeneous spaces like malls is actually quite complex (Jackson and Holbrook 1996).[11] What is called 'the mall' is far more differentiated than the literature suggests. While indeed there are examples of a few developers who have numerous malls to their credit, not all malls are alike – and this is largely due to the ways people bring themselves into the shopping experience (Holbrook and Jackson 1995; De Certeau 1984; Francaviglia 1996; Gregson 1995).

Inauthentic architectures and simulated history

A third theme of urban decline criticises contemporary architectural styles as inauthentic:

> This new realm is a city of simulations, television city, the city as theme park. This is nowhere more visible than in its architecture, in buildings that rely for their authority on images drawn from history, from a spuriously appropriated past that substitutes for a more exigent and examined present.
>
> (Sorkin 1992b: xiv)

This frequent critique of 'postmodern' architecture is often directed at shopping malls and commercial architectures, with their pastiche of styles from different historical periods and geographical contexts. As Umberto Eco writes:

> The Wall Street area in New York is composed of skyscrapers, neo-Gothic cathedrals, neoclassical Parthenons, and primary cubelike structures. Its builders were no less daring than the Hearsts and the Ringlings, and you can also find here a Palazzo Strozzi, property of the Federal Reserve Bank of New York, complete with rustication and all. Built in 1924 of 'Indiana limestone and Ohio sandstone,' it ceases its Renaissance imitation at the third floor, rightly, and continues with eight more stories of its own invention, then displays Guelph battlements, then continues as a skyscraper. But there is nothing to object to here, because lower Manhattan is a masterpiece of living architecture, crooked like the lower line of Cowboy Kate's teeth; skyscrapers and Gothic cathedral compose what has been called a jam session in stone, certainly the greatest in the history of mankind. Here, moreover, the Gothic and the neoclassical do not seem the effect of cold reasoning; they illustrate the revivalist awareness of the period when they were built, and so they aren't fakes, at least no more than the Madeleine is, in Paris, and they are not incredible, any more than the Victor Emmanuel monument is, in Rome.
>
> (1986b (1975): 28)

Eco contrasts a reverential review of New York revivalist architecture with a derisory account of falsehoods in more recent American architectural attempts to revive the past. He differentiates between nineteenth-century pastiche and contemporary postmodern style, between European cities and American ones. In both cases, he argues, the former is more authentic.

Yet do contemporary, glass-covered mall atria differ so much from nineteenth-century railway stations where 'steam engines and the machines of mass production were invariably housed in structures that looked like botched versions of Greek temples or medieval cathedrals' (Relph 1987: 26–27)? Whether Rome's borrowings from Egypt and Greece, Romanesque and Renaissance borrowings of classical architectural models, or the neo-Gothic and Orientalist architectures of the nineteenth century, a long history exists of looking to the past and to other geographic areas for architectural authenticity. Mall developers' adaptation of architecture to connote elegance is perhaps not so different from nineteenth-century adaptations of gothic styles to inspire museums as cathedrals of learning (Lenoir and Ross 1996). Indeed, critics of contemporary commercial architecture might examine the work of Edward Relph, who writes that 'By 1880 almost every sub-style of architecture had been revived, modified, and combined with all the others' (Relph 1987: 26–27).

Nineteenth-century architectural theorists, such as Eugène Emmanuel Viollet-le-Duc and John Ruskin, concerned with preservation and restoration, also debated architectural imitation versus originality. Simulated architectures were evident in Victorian Britain where the Cambridge Camden Society practised a kind of restoration which 'could involve comprehensive demolition and rebuilding to a historically speculative design', since their 'mission in restoration was to recover the "original" appearance of a building whether from existing evidence or from supposition' (Chitty 1995: 107).[12] In some schools of Victorian thought, 'an architecturally correct copy of an ancient architectural feature was considered to be as good as, or even better than, the real thing' (Chitty 1995: 107). In a different twist on the decline of authenticity, architect Albert Simons critiqued architectural generalities which defied local styles. Expressing concern about architectural homogenisation, for example, in his 1932 comments on a fence in Charleston, South Carolina, Simons noted that: 'I do not particularly like their idea of cribbing a Long Island Fence from *House and Garden*. . . . *House and Garden* and other publications of the same sort have done more to reduce American taste to a monotonous standard with a rank new England flavor than anything else I know of' (Datel 1990: 210).

Criticisms of a decline in architectural authenticity seem to romanticise the past in making a distinction between the revivals of nineteenth-century architecture and those of today. They draw the kind of high art/low art distinction that has been the subject of much criticism from students of vernacular architecture (Rubin 1979). Eco's commentary differs from the critiques of inauthenticity focusing on urban forms which simulate public or democratic space. His argument is ostensibly about reality versus virtuality – or in Eco's blunt terminology, 'fakes'. Yet it is equally concerned about what reality should be – an unmediated, non-commercial experience. The debate about reality versus virtuality thus seems to function as a language for other arguments that remain unarticulated.

In some critiques of inauthentic practices, then, what is counterposed as more authentic represents a distinct class bias. Tying these architectures to a widespread sense of urban decline is a set of assumptions about what makes urban culture great (Rubin 1979). As Barbara Rubin points out:

> The diversity of an urban population, or the cosmopolitan range of goods and services exchanged, is rarely taken as an index of urban success by students of urban culture. Instead, urban success is found in a catalogue of a city's noncommercial, nonindustrial institutions: a philharmonic orchestra, art museums, parks, religions and historical shrines, theaters, fine-arts architecture, and unified, monumentalising plans.
>
> (1979: 341)

These assumptions about what constitutes urban success are distinctly middle-class and upper middle-class, practically guaranteeing findings of failure in the architectures of commerce. While some commentators critique commercial architectures for zoning out 'the public', the spaces are nevertheless being used by a wide range of people. Rubin dates concern about consumption implying culture in decline to the Chicago Exposition, where a walled-off area was set aside for commercial activities: 'The emerging ideology of urban aesthetics could not admit commerce, since by definition commerce could not be aesthetic' (Rubin 1979: 345). This line of attack continues in critiques of festival marketplaces, which discuss how at places like Faneuil Hall Marketplace or South Street Seaport, historic buildings are enlisted to sell merchandise (Huxtable 1997; Boyer 1992). Architectural revival is thus valued differently depending upon its audience.

> Here is urban renewal with a sinister twist, an architecture of deception which, in its happy-face familiarity, constantly distances itself from the most fundamental realities.
>
> (Sorkin 1992b: xiv)

The fundamental rationale for each critique – of skyways, malls, and certain architectural combinations – is that 'traditional' architectural forms are somehow more real and true. As Sorkin writes in the introduction to his collection:

> 'City air makes people free' goes a medieval maxim.[13] The cautionary essays collected here describe an ill wind blowing through our cities, an atmosphere that has the potential to irretrievably alter the character of cities as the preeminent sites of democracy and pleasure. The familiar sites of traditional cities, the streets and squares, courtyards and parks, are our great scenes of the civic, visible and accessible, our binding agents. By describing the alternative, this book pleads for a return to a more authentic urbanity, a city based on physical proximity and free movement and a sense that the city is our best expression of a desire for collectivity. As spatiality ebbs, so does intimacy. The privatized city of bits is a lie,

> simulating its connections, obliterating the power of its citizens either to act alone or to act together . . . In the 'public' spaces of the theme park or the shopping mall, speech itself is restricted: there are no demonstrations in Disneyland. The effort to reclaim the city is the struggle of Democracy itself.
>
> (Sorkin 1992b: xv)

This statement sets up a black and white distinction between the unreality of privatised Disney and the reality of the outside world. Yet there *are* demonstrations in Disneyland, and there were demonstrations long before Sorkin's article went to press. The *San Francisco Foghorn* described 'several strikes that have plagued the park including one in 1984 that lasted for 22 days' (http://foghorn.usfca.edu/archives/fall.95/f09/entertainment/mouse.html). The notion of Disneyland as a space beyond the potential conflicts of public space has been exaggerated.

In summary, the 'decline and fall' narrative suggests that new strategies for urban revitalisation are not resulting in improvements, and that true democracy can only be found through more traditional architectural forms. The criticisms of new forms for zoning out some of the public are partially justified, since malls and skyways do not, in fact, equally serve all citizens. Yet we must ask whether a nostalgia for traditional urban forms is perhaps misplaced, since in their time these stood for a far more limited definition of 'the public'. Such nostalgia paired with critiques of decline can be seen as continuing a trend to find authenticity in the architectural past. Yet one wonders, when both conservative and liberal writers turn backward to affirm a similar ideal, how different are the versions of the past they keep in mind?

Linking city space to cyberspace

From Adorno to the Lynds to Habermas to Oldenburg to Sorkin, writers characterise the rise of home-delivery media as a central factor in the decline of cities as civic spaces (Adorno 1990 (1934); Lynd and Lynd 1956; Habermas 1992; Oldenburg 1989; Putnam 1995). For example, Robert and Helen Lynd's classic study of Middletown found that with the use of telephones, visiting decreased (Lynd and Lynd 1956). Ray Oldenburg and Neil Postman tie telephones, televisions, VCRs and other home entertainment to the closing of neighbourhood establishments (Postman 1992; Oldenburg 1989). Such observations about American culture are extensions of the critique of American individualism levelled by Alexis de Tocqueville, who wrote that American individualism might be the downfall of public life. 'Individualism is a calm and considered feeling which disposes each citizen to isolate himself from the mass of his fellows and withdraw into the circle of family and friends; with this little society formed to his taste, he gladly leaves the greater society to look after itself' (Bellah 1985: 37). While he admired Americans for their commitment to participation in public life, de Tocqueville simultaneously felt that Americans' predilection for individual freedom could result in isolation and thereby interfere with the future of their freedom. Now, computer technologies are coming under fire as time spent at the computer translates into time not spent on the street, removing much-needed attention from city problems (Zimmerman 1986). Dolores Hayden has described this disappearing community and public life as part of the uniquely American drive to create an ideal home, rather than an ideal city (Hayden 1984).[14]

Like cyberpunk stories of escape into virtuality, academic narratives of the decline of cities and of public life tell of a decline of reality, and of its replacement by simulations. Linked to commentaries on the rise of virtual spaces, then, are criticisms of the growing hyperreality of physical spaces – critiques that sometimes encompass entire cities. Criticisms of hyperreality have generally focused on a few cities and city types. They include Los Angeles (Soja 1996; Davis 1992), Las Vegas (Eco 1986b (1975); Kunstler 1993) and planned communities such as Disney's Celebration, Florida (McKenzie 1994).[15] For example, Umberto Eco has advanced the notion that Americans prefer fakes, reconstructions and simulations, a practice somehow tied to a need to fabricate history in a historyless country (Eco 1986b (1975)). As he put it, if Europe's cities have amusement parks, in America entire cities are entertainment. Describing California and Florida as 'artificial regions' (Eco 1986b (1975): 26) he writes, 'The fact is that the United States is filled with cities that imitate a city' (Eco 1986b (1975): 40). Soja offers a similar report on Los Angeles: 'The simpler worlds of the artificial theme park are no longer the only places where the disappearance of the real is revealingly concealed' (Soja 1996: 278).

Not only are urban forms described as sites of simulation standing on the spaces of older more authentic cities, but commentators describe the decline of cities as a 'fall' into mediation, drawing analogies between cities and media. From Jean Baudrillard's remark that the shopping mall resembles 'a giant montage factory' (Baudrillard 1995a (1981): 76), to Michael Sorkin's observation that 'the structure of this city is a lot like television' (Sorkin 1992b: xi), to Edward Soja's comments on 'real-reel life' (Soja 1996: 238) the mediated city is a common motif. This motif criticises the effects of media on cities, suggesting that 'television harbingers a totally recast public realm, a city remade in the image of TV' (Sorkin 1992a: 71). To update such an academic narrative for the information age, Christine Boyer has drawn 'an analogy between the computer matrix and the space of the city' (Boyer 1996: 9), a parallel popularised in cyberpunk science fiction (Gibson 1984; Stephenson 1992).

A common theme in these portrayals is not only that cities have become more physically dispersed in recent decades (a fact few would dispute), but that simultaneously they are becoming dematerialised in people's imaginations. As Sorkin puts it, 'recent years have seen the emergence of a wholly new kind of city, a city without a place attached to it' (Sorkin 1992b: xi).

Obviously a city needs a physical place; Sorkin's point is that cities today have little connection to their local geographies.

Like Eco and Soja, James Kunstler deems several American cities Capitals of Unreality: 'Atlantic City, unlike Disney World, was once a genuine town [from the 1880s to 1929], but has evolved unhappily into a place of the most extravagant unreality' (Kunstler 1993: 217). In his commentary on the most 'virtual' American city of the late twentieth century, Baudrillard echoes Eco's stance to make a similar point about Los Angeles. Baudrillard describes how Disneyland, with its Main Street USA, allows Americans to pretend that the rest of our landscape is real:

> Disneyland exists in order to hide that it is the 'real' country, all of 'real' America that *is* Disneyland . . . Disneyland is presented as imaginary in order to make us believe that the rest is real, whereas all of Los Angeles and the America that surrounds it are no longer real, but belong to the hyperreal order and to the order of simulation.
>
> (Baudrillard 1995a (1981): 12)

With such an assumption behind many commentaries, popular comparisons between dematerialised cities and cyberspaces as mediated places without physical structure begin to make some sense.[16] This is the charge of works like Boyer's *Cybercities* (1996) and Mark Slouka's *War of the Worlds: Cyberspace and the High-Tech Assault on Reality* (1996). With cities ever more like media and public spaces disappearing, these writers foresee a circular pattern. People's withdrawal will lead to further disintegration of the physical environment, which will increase the pull of virtual worlds. Characterising the 'invisible or disappearing city' (Boyer 1996: 19), Boyer writes:

> I tend to agree with William Gibson, who decided even before writing *Neuromancer* that what was happening in the space behind the video screen was more interesting than what was happening in the space in front of it – in other words, that cyberspace pulls the user into the receding space of the electronic matrix in total withdrawal of the world.
>
> (Boyer 1996: 11)

Boyer notes that 'the decline of reality as a serious referent' is an object of her study, and she stresses this as an issue that must be addressed (Boyer: 7). Mark Slouka extends this argument.[17] In his chapter 'The Road to Unreality', Slouka criticises a:

> trend so pervasive as to be almost invisible: our growing separation from reality . . . Let me state my case as simply as possible: I believe it is possible to see, in a number of technologies spawned by recent developments in the computer world, an attack on reality as human beings have always known it.
>
> (1996: 1–4)

The emphasis here is that media are disorienting. In Sorkin's formulation, virtual reality is electronic deception and unreality – 'the electronic construction of images that are indistinguishable from the nominal realities they purport to represent' (Sorkin 1992b: 71).[18] And Soja writes that:

> as our ability to tell the difference between what is real and what is imagined weakens, another kind of reality – a hypperreality – flourishes and increasingly flows into everyday life . . . Baudrillard has not been alone in drawing our attention to the fact that reality is no longer what it used to be.
>
> (1996: 239–240)

It is not only cities that are disappearing, but 'reality' – an idea here which stands in for authenticity, community, public space and an idealised non-commercial realm.

While cities have changed physically across this century – for example through the building of highways and shopping malls, and a general trend towards decentralisation – drastic changes across the past two decades are not so easily identified. Peter Larkham and Elizabeth Wilson point to how one of the tensions in thinking about cities is the challenge of stasis versus change, of conservation versus restoration (Wilson 1997; Larkham 1996). As an ideal, 'the city' is invented and reinvented, and cannot be treated as a static referent. Writing about the strengths of traditional cities therefore becomes complex because 'the city' – both its physical forms and the ways of conceptualising it – has evolved over many centuries.[19] Previous commentators also mourned the passing of an authentic city. Today's critique thus ends up, ironically, locating authenticity in the past when its very conceptualisation of this 'past' is itself a simulation.[20]

A brief history of technology

> The museum took the place of concrete reality; the guidebook took the place of the museum; the criticism took the place of the picture; the written description took the place of the building, the scene in nature, the adventure, the living act. This exaggerates and caricatures the paleotechnic state of mind; but it does not essentially falsify it.
>
> (Mumford 1934: 181)

Perceptions of any environment or technology change over time. Consider reading, sometimes held up as 'real' and 'authentic' in contrast to new electronic media (Birkerts 1994). Yet reading has been criticised as an unreality:

> With the spread of literacy, literature of all grades and levels formed a semi-public world into which the unsatisfied individual might withdraw, to live a life of adventure following the travellers and explorers in their

> memoirs, to live a life of dangerous action and keen observation by participating in the crimes and investigations of a Dupin or a Sherlock Homes, or to live a life of romantic fulfillment in the love stories and erotic romances that became everyone's property from the eighteenth century onward. Most of these varieties of day-dream and private fantasy have of course existed in the past: now they become part of a gigantic collective apparatus of escape.
>
> (Mumford 1934: 314)

From stories of Renaissance mathematical perception, to Lorraine glass, to department stores described as panoramic culture, technological artifacts contribute to new ways of seeing (Friedberg 1993). For example, as Harold Innes writes in *The Bias of Communication*, the invention of printing was a threat to public life: 'Innes' insight was that newspapers actually created monopolies of information. The emergence of the "audience" spelled danger for public life, as it transformed people into essentially private readers and listeners' (Lyon 1994: 44).

Similarly, Oldenburg explains:

> What the tavern offered long before television or newspapers was a source of news along with the opportunity to question, protest, sound out, supplement, and form opinion locally and collectively. And these active and individual forms of participation are essential to a government of the people. An efficient home-delivery media system, in contrast, tends to make shut-ins of otherwise healthy individuals; the more people receive news in isolation, the more they become susceptible to manipulation by those who control the media.
>
> (Oldenburg 1989: 69–70)

Walter Benjamin describes how through historical changes in modes of perception, definitions of reality can change too. He explains how photography changes the way people see, from Muybridge's motion studies capturing what the eye cannot see to his lament of the loss of art's aura in an age of mechanical reproduction. Susan Sontag builds on Benjamin's ideas about perception. While acknowledging the power of photographs to closely represent the real, she calls for a more rigorous examination of the extent to which, by incorporating mechanised and mediated images and sounds into our daily experience, increasingly it is those images that constitute reality and authenticity (Sontag 1977).

So what does this all add up to? I do not doubt that new information technologies, together with many parallel social trends, both reflect and contribute to changes in people's perceptions of reality. I agree with Sorkin, Boyer, and Slouka that it is important to question what kind of biases a given technology might have for our perception. Yet to call cyberspace 'unreal' or inauthentic in comparison to so much else is deceptive. I am not suggesting that virtual and physical environments are equivalent. The spectrum of difference between them is important (Borgmann 1995; Light 1997). Yet recognising their interaction helps us understand how virtual environments become part of everyday reality. Our lives have always been mediated: by our environment, religion, gender, class and the existing technologies of our time (Berger and Luckmann 1966). While statements such as 'something unusual is happening today in the relation between the real and the imaginary, reality and its representations' (Soja 1996: 242) may be applicable today, they also would have characterised observations of 50 and 100 years ago.[21] The decline of cities, then, cannot be explained simply as a physical phenomenon attributed to the growth of electronic media. Rather, one might wonder whether amidst a shifting paradigm, we are witnessing an understandable, nostalgic reaction.

Reinvigorating city space

A central debate in the history of technology has focused on the question: Does technology drive history or does society shape technology? (Marx and Roe Smith 1994; MacKenzie and Wajcman 1985). The unidirectionality of this question also has come under scholarly fire (Hughes 1983). Architectural theorists face a related question: To what extent does society shape architecture or does architecture shape society? Certainly reformers have wanted to believe that architecture can inspire social change. One illustration is the New Urbanism. For example, Seaside's 'imitation of premodern, pre-automobile urban forms pivots on the notion that social relations and culture follow' (Audirac and Shermyen 1994: 165). Yet Jane Jacobs contested similar proposals by her contemporaries, writing that public spaces are only public spaces in the ideal sense when people make something of them. She describes many failed attempts to create public parks which just ended up as lonely squares of grass. Recent work has followed Jacobs in suggesting that 'Buildings do not dictate their own meanings, but change according to the ideas and actions of city dwellers' (Matthew Gallery 1996; Warren 1996).[22] This mutual shaping argument is central to understanding the 'architecture' of cyberspaces, where how people use information environments is as important as how these spaces are designed. Without this perspective, we risk falling into the technological determinist trap Leo Marx calls postmodern pessimism.

What is postmodern pessimism? Relating technology to a decline in quality of life is an old argument. It is presented in works such as Thomas Carlyle's 'Signs of the Times' (1835 (1829)), and novels such as Auguste Villiers de l'Isle Adam's *L'Eve Future* (1879). It is elaborated in Lewis Mumford's *Technics and Civilisation* (1934) and Sigfried Giedion's *Mechanization Takes Command* (1948). Each of these works laments the dehumanising aspects of new technologies. Yet while compelling, these are largely determinist arguments. Seeing life as dominated by large technological systems is implicitly a technologically determinist position (Marx 1994). As Leo Marx explains in his essay 'The Idea of "Technology" and Postmodern Pessimism' (1994), contemporary fears about the domination of life by large technological systems – he points to new electronic communications technologies – are themselves determinist:

> This outlook ratifies the idea of the domination of life by large technological systems, by default if not by design. The accompanying mood varies from a sense of pleasurably self-abnegating acquiescence in the inevitable to melancholy resignation or fatalism . . . In their hostility to ideologies and collective belief systems, moreover, many post-modernist thinkers relinquish all old-fashioned notions of putting the new systems into the service of a larger political vision of human possibilities. In their view, such visions are inherently dangerous, proto-totalitarian, and to be avoided at all costs. The pessimistic tenor of postmodernism follows from this inevitably diminished sense of human agency.
>
> (Marx 1994: 257)

It is this fatalist viewpoint for the future of electronic cities that is pessimistic, perhaps excessively so. Sorkin concludes his article 'Scenes from the Electronic City' by stating 'The question is whether or not we'll have any choice' (Sorkin 1992b: 77). He thereby suggests a determinist outcome. Perhaps time will prove Sorkin right. Yet to suggest we might have no choice is to accept the technological determinist position. The Internet is in its early stages and has an enormous number of potential future directions. It would be a self-fulfilling prophecy if critics who assume the worst avoid participation in network development just at the time they could have the biggest impact.

Cyberspaces

The story of online service providers resembles the story of the American landscape, with both municipal and commercial ('public and private') services sharing responsibilities for infrastructure. Civic networks and Freenets are different from private online services. While some commercially-owned networks impose speech restrictions (Light 1996) there are also private fora on which anything goes. So many scholars and social critics have weighed in on the negative aspects of technology for cities, it is easy to overlook concrete, promising trends in the other direction.

Critics such as Mike Davis rightly attack the exclusions of electronic spaces. Cyberspace is heavily dominated by élite groups, yet often presented as if open to all; its privatised spaces presented as if they are public (Light 1996). Concerns about the commodification of space are justified, as cyber-space is increasingly home to commercial networks, commercial sites and commercial transactions. Simultaneously, cybercritics point to the decline of place-based relations as a factor in urban decay, linking these to the rise of electronic communications. Yet to call this critique the whole story would be like looking at cities only in terms of their shopping malls, or looking only at the malls' trade literature and never examining how people actually behave there. These visions of cyberspace have not accounted for the increasing number of non-profit and grass roots organisations meeting online and in person to reinvigorate physical space. Examples of these projects abound; they even constituted the curriculum for a course and community service program at MIT (http://alberti.mit.edu/dusp/11.401).

Several examples illustrate the categories of cyberspace devoted to enhancing a sense of place. Combining commercial and municipal networks, there is now a proliferation of community-network information online (http://alberti.mit.edu/arch/4.207/anneb/thesis/toc.html). Looking at a single city offers a case study. For Chicago alone, my search was encouraging. For example, the Chicago Area Northside Neighborhood Online Network has been online since 1995 (http://www.tezcat.com/~neccn/cpbproposal.html#partners). It offers training and Internet access to community-based organisations throughout the city. 'Building on existing human networks in our communities, we have successfully trained residents and staff of over 60 community organizations, and built up a unique multi-racial, mixed-economic, mixed-gender pool of community users' (http://www.tezcat.com/~neccn/cpbproposal.html#partners). In the spirit of a community network that lets the locale direct its civic network, there will be a partnership with several other organisations. The community electronic network was initiated to overlap with already existing social networks in Chicago, organising around jobs and economic development, children, youth and family, issues of affordable housing and to offer computer access to residents.

Erie Neighborhood House, a non-profit, multi-service agency, operates the Erie Technology Center. This is:

> a comprehensive computer laboratory dedicated to the computer and information literacy of West Town residents with limited English proficiency and low educational achievement. Because the Technology Center works intimately with the other educational programs at Erie Neighborhood House, it is able to serve a wide variety of ages (age 5 to 85) in the effort to integrate traditional teaching/learning methods with current technology applications. Currently the new facility consists of a computer center, Internet access and space for layout and design work.
>
> (http://www.tezcat.com/~neccn/cpbproposal.html#partners)

In a similar spirit is NeighborTech, which works with Chicago's inner city neighbourhoods to help citizens become technologically literate. 'Through training classes, informational meetings and seminars, NeighborTech's Internet provider service, and our newsletter, we want to see low-income residents, not-for-profit agencies, and small businesses located in low-income communities getting connected and getting on-line.' These include Community Computing Centers and Hot Neighborhood Computing Experts (http://www.iit.edu/~nnet/). How can one possibly describe this use of cyberspace as anti-urban? It is precisely the opposite.

Chicago Community Networking (http://www.cnt.org/ccic/chico_net.html) is even marketing the 'Other Chicago' online:

> Geographic Chicago is intrinsically a fine place with many advantages such as abundant fresh water, a skilled work force, sociable people, diverse cultures, many transportation connections, fine colleges, beautiful churches, and great restaurants. We are proud of it. But lacking are humane, sound, democratic and fair social and economic policies. In spite of hardships, the poorer neighborhoods are filled with heroes, valiant efforts and people struggling for survival. There are interesting, entertaining, and educational grassroots places to visit in the 'other' Chicago. We invite you to see it and meet its people in cyberspace; then visit the 'Other Chicago' in the real world.
>
> (http://www.cs.uchicago.edu/cpsr/other-chicago/index.html)

These are a few short examples of proactive local uses of the Internet. Each educates diverse groups of users to do more than simply log on, and attempts to redress the élitism of cyberspace. These virtual spaces, created with concerns about the physical environment, are not dissociated from place. Nor are they cyberspaces of retreat. Online relationships can be strong, and may continue offline (Rheingold 1993; Turkle 1995). These spaces are not unreal for people who use them in their work, just as Orange County, Los Angeles, and Las Vegas – capitals of unreality to urban social critics – are real for the people who live there.

I am the first to admit the disparity between much of the rhetoric of virtual communities, virtual public life and the role of these new technologies in people's everyday lives (Light 1996). Yet any blanket condemnation of the inauthenticity and dislocation of new electronic spaces seems overdone. Like reducing complex differences among cities to an oversimplified single truth about 'the city', there is a risk in similar portrayals of cyberspace.[23] As in other eras, electronic communications and other technologies have multiple, sometimes contradictory effects (Fischer 1992). Writings about a generic 'cyberspace', like writings about 'the city', are troublesome since there are differentiated cities and differentiated cyberspaces.[24] In addition, since the practical uses of information technologies are constantly changing, perspectives on cyberspaces are likely to evolve just as perspectives on electricity, telephones, radio, photography and computers have evolved over many decades (Sontag 1977; Marvin 1988; Douglas 1986; Fischer 1988; Ceruzzi 1991).

It is critical to broaden our understanding of what cities and cyberspaces can be. A city stands for the disparate experiences people have in a physical space; no two people have the same experience of a single city. Similarly, people's experiences online differ, and no two people have identical experiences in the varied geography of the Internet. Just as a city does not mean one thing to all people, the 'Internet', and 'cyberspace', while singular words, are not monolithic things. Any theory that attempts to overlay a simplistic, unidirectional interpretation on interactions online is likely to find a counterexample.

The new geography of the Internet, as the various examples of community-building in Chicago illustrate, is a dynamic environment of many different places and spaces used for many different purposes. Writers of urban decline and their speculation that new technologies will accelerate this decline, become less compelling in the light of counterexamples occurring even now. Increasingly, users are harnessing technology to strengthen spatial ties within their cities. The pessimists' contributions are spunky and fun to read. Yet when they are situated in a broader debate within the history of technology and daily life, their pessimism must be balanced against an increasingly vigorous set of examples of how individuals are using cyberspaces to improve city spaces.

Acknowledgements

Thanks to Peter Buck and Nicholas King of Harvard's Department of the History of Science and to the editors for their comments. Thanks also to staff and students at the Department of Geography, University of Edinburgh, for many interesting conversations shared during the writing of this essay.

Notes

1 I use the term 'digital' here since virtual spaces are largely digital. Yet similar claims were made regularly about analogue simulations, such as Adorno's commentary on radio and phonograph records (Adorno 1990, (1934); Adorno 1945).
2 Translated and cited in Rubin (1979: 351).
3 In fact, the theme goes back hundreds of years as a rhetorical play in contesting a lost traditional way of life with an encroaching modernity. See Raymond Williams, *The Country and the City* (1973).
4 In this essay, simulation is defined in Baudrillard's sense – as a copy of something that has no original; an ideal.
5 Similar concerns about the privatisation of public space proliferate in architectural and geographical theory, especially in articles analysing shopping malls, theme parks and festival marketplaces (Crawford 1992; Davis 1992; Deutsche 1996; Sorkin 1992b; Goss 1996; Warren 1996).
6 Ann-Louise Shapiro has argued that Georges Eugène Haussmann's changes to Paris erased low-income workers from the centre of the city (Shapiro 1985). T.J. Clark has written that zoning the lower classes out was a critical part of enhancing

the spectacle for the bourgeoisie (Clark 1985). The impact of changes to Paris under Haussmann made certain activities of the working classes invisible by redistricting their areas and pushing them into suburbs, as well as subjecting street performers and caricaturists to censorship. Caricaturists like Daumier called this to public attention in their cartoons; for example he represented Napoleon III as a demolition worker. Here we find a similar concern about who is the public and for whom is the city.

7 For a recent critique of Seaside, see Al-Hindi and Staddon (1997).

8 These are, of course, not the only contested meanings of public space and the public sphere.

9 For further discussion see Section 3, 'Image Building and Main Street' in Francaviglia (1996).

10 This position is presented elsewhere, for example see Huxtable (1997: 103).

11 In a later article, Crawford moves in a different direction to criticise 'narratives of loss'. She points to how 'urban residents are constantly remaking public space and redefining the public sphere through their lived experience' (Crawford 1995: 4), even criticising her fellow contributors to *Variations on a Theme Park*.

12 In his article, Chitty discuses a range of positions on preservation and restoration.

13 As Dolores Hayden points out, women have not enjoyed this *droit de ville* (Hayden 1984: 210–211).

14 Chermayeff and Alexander write of television, radio, telephone and phonograph 'Transformed by electronics, the dwelling is no longer a refuge but an arena. It now serves as the marketplace, the forum, the stadium and school, the theater and movie house rolled into one' (Chermayeff and Alexander 1963: 95). This is not to say that the pair were against technology; at several points in their book they write of technology as a tool and the possibilities of computers in design.

15 That sometimes these stand in for all American cities is a problematic assumption in itself. McKenzie's book does not discuss Celebration, but it is a well-known example of the kinds of development he examines.

16 I suspect that like metaphors for cities (Vidler 1978), the city metaphor has influenced the kinds of project pursued in cyberspace; many are electronic analogies of physical spaces.

17 In Baudrillard's writing, a 'real' referent no longer exists.

18 Yet in another article he offers a more hopeful view of the relationship of reality and virtuality: 'The space of virtuality threatens only if it supplants the space of the physical. As supplement, as sheer augmentation, electronic means can put us in touch with fresh cohorts and global possibilities. As replacement, though, the risk is clear' (Sorkin 1993: 107).

19 The concept of 'the city' is also problematic since there are many different kinds of city and many different experiences of each.

20 Recent scholarship has called into question a previously accepted distinction between reality and virtuality (Hayles 1995; Cronon 1995; Davis 1992; Light 1997; Westwood and Williams 1997; Starr 1994). Simulations can inspire reflection on how what was previously conceptualised as real is itself a simulation. The point is that physical life and digital life are much more similar than commonly perceived. For example, older theorists such as Benjamin and Adorno looked at how external media shaped people's experience. Now theorists are pointing to a variety of internalised mediations, from race, class and gender to individual bodies (Hayles 1995). This perspective suggests no two people's view

of a single city or urban form are likely to be the same. For example, what is simulation in an American context may be perceived as authentic in other contexts. A recent *New York Times* article on the Polus Center, the first American-style shopping mall in Central Europe, describes how the developers sought to 'transplant the authentic ambience of an American mall' to Hungary (Perlez 1996). In new writings on cities the theme is that reality and virtuality are converging; cities become more virtual with concepts like 'imagined' cities and Soja's 'thirdspace' (Westwood and Williams 1997; Soja 1996). This is a slightly different formulation from the one presented by Sorkin and his contributors. They frame reality as collapsing into virtuality and disappearing altogether, such as in Margaret Crawford's chapter 'The World a Shopping Mall' which reflects on how the entire world is becoming a giant shopping mall.

21 A similar point can be made about preoccupations with a perceived loss of connection among city sites. Commentators write that 'What's missing in this city is not a matter of any particular building or place; it's the spaces in between, the connection that makes sense of forms' (Sorkin 1992b: xii); and 'Most important, we have lost our knowledge of how to physically connect things in our everyday world, except by car and telephone' (Kunstler 1993: 246). Similarly, when railroads and railway stations were first introduced, they seemed also to bring about a 'dislocation of the city' (De Boer 1993: 31; Schivelbusch 1986). The 'industrialisation of space', to use Schivelbusch's phrase – changed people's experience in the late nineteenth century. Now we are shifting to a new paradigm.

22 Umberto Eco wrote that architecture is not determinist; it oscillates between guiding you to particular behaviour, and not controlling you at all (Eco 1986a (1973)). Yet this does not seem to reconcile with his comments in *Travels in Hyperreality*, written two years earlier. For example, his discussion of Disneyland talked about the passivity of people in the Disney environment: 'An allegory of the consumer society, a place of absolute iconism, Disneyland is also a place of total passivity. Its visitors must agree to behave like its robots. Access to each attraction is regulated by a maze of metal railings which discourages any individual initiative' (Eco 1986a (1973): 48).

23 Malls and simulated spaces are not inherently dangerous in themselves, but in how they are tied into other aspects of life. It would be problematic if the privatised public space of a shopping mall were the sole place for people in a city or region to gather. The implications would be similar if the only online spaces were highly regulated public or private networks.

24 Certainly there is a proliferation of online worlds which are intended to simulate or replace the real. Yet these are not the only cyberspaces.

8 Geographies of surveillant simulation

Stephen Graham

Introduction: telematics and surveillant simulation

The history of technology demonstrates that, in the long run, interlinked changes across a range of technological systems tend to be more important than single technological innovations in facilitating social and spatial change (see, for example, Hall and Preston 1988; Beniger 1986). Thus, we might criticise the rather narrow preoccupation of much current social science work on virtual spaces, which tends to centre almost exclusively on the social construction of subjectivity in 'cyberspace' (for which read the Internet). Such a perspective, I would argue, often neglects the broader societal implications of a whole raft of current inter-linked innovations in computing and telecommunications.

In such a context, this chapter tries to connect a broad perspective on 'virtual geographies' with political-economic debates about surveillance, computerised simulation, and the socioeconomic restructuring of geographic space. My starting point is William Bogard's (1996) recent book, *The Simulation of Surveillance*. Computerised surveillance and simulation have both been subject to much debate within recent social theory and commentary (see Lyon 1994; Droege 1997). Approaches in social and cultural geography, and postmodern commentary more generally, however, have tended to separate treatment of surveillance from that of simulation. The former have usually drawn on the work of Foucault (1977) on the disciplinary and self disciplining underpinnings of modern societies (see, for example Squires 1994; Philo 1992), and on Jeremy Bentham's famous eighteenth-century writings on his Panopticon prison design (see Hannah 1997). The latter have drawn on Baudrillard's notions of the 'postmodern' shift towards hyperreality and orders of simulacra (see Baudrillard 1983; Kellner 1994; Soja 1989). Largely separate positivist debates on simulation, meanwhile, have addressed the technical issues surrounding the computerised simulation of everything from geographical systems (see Batty 1996, 1997a; Sui 1997), to planetary landscapes, biological mechanisms, human genetic processes, and cosmological space-time (see Hall 1993).

In contrast, Bogard's work is useful because it takes an holistic perspective of the *complex interactions* between computerised surveillance and simulation.

'It is simulation', writes Bogard (p. 9), 'that is the key to explaining the direction that surveillance societies are taking today, a movement that is more about the perfection and totalization of existing surveillance technologies than some kind of radical break in their historical development' (ibid.: 9). To him, computerised simulation and computerised surveillance are increasingly merging to be integrated systems of rapid, and invisible, social control. Thus, the closely allied disciplinary gaze of surveiller and the self-discipining practices of the surveyed analysed by Foucault and Bentham becomes decoupled and distanciated over space and time via telecommunications and computers. The surveiller 'no longer, properly speaking, *has* a gaze' (p. 57) as surveillance becomes predicated on both systems of computerised societal simulation and, increasingly, simulations of surveillance apparatus itself (with, for example, extending application of both real and mock CCTV cameras). Bogard analyses these interactions within a critical framework of the political-economic shifts that are underway in contemporary society. He thus balances his treatment of subjectivity and identity with a rich treatment of the implications of simulation and surveillance for social control, spatial structures and power relations.

To Bogard, the importance of broad, interacting technological systems is illustrated by considering the much-vaunted technological 'convergence' between computers, telecommunications, media, and bio-technologies based on the progressive digitalisation of information. Such technological blurring is potentially important for four reasons. First, it increasingly supports the interlinkage of wide ranges of terminal equipment across geographical distance into digital, multi-media 'telematics' networks able to deal with flows of digital data, sound, voice and (increasingly) still and moving images. Thus 'telematics societies' become technologically feasible, defined by Bogard as: 'societies that aim to solve the problem of perceptual control at a distance through technologies for cutting the time transmission of information to zero' (Bogard 1996: 9).

Second, advances in computing technology mean that the powers of digital technological systems for processing, manipulating, transmitting and storing data are increasing extremely rapidly. This means that systems supporting new orders of magnitude of automated data capture, monitoring and surveillance can be directly constructed to try and match vastly complex systems of social and economic behaviour extended across material spaces.

Third, computers are, in turn, moving from being essentially 'data crunching' devices to sophisticated visualisation and simulation devices, as is the case with Geographical Information Systems (GIS), digital mapping and remote sensing and Virtual Reality (VR) techniques within which whole, immersive, electronic environments become constructed (see Lister 1995; Druckrey 1994a). 'The transition from solid models to digitally-generated images has gone to completion in an astonishingly short time' (Stone 1994: 7). Computerised simulation and modelling systems now allow the huge quantities of data captured by automated surveillance systems to be fed directly into dynamic facsimiles of the time-space 'reality' of geographic territories (neighbourhoods, cities, regions, nations, etc.), which can in turn be fed into support new types of organisational change, spatial targeting, and urban and regional restructuring.

The final element of the technological jigsaw provides the geographical foundations for the fine-grained monitoring of the time-space dynamics of geographic spaces. This has been provided by rapid advances in geo-referencing technologies such as satellite remote sensing, the global web of Global Positioning Systems (GPS) satellites and digital telecommunications. GPS satellites can triangulate geographical locations, anywhere on the planet, down to one metre resolution levels. Together, these technologies allow locations and patterns of flow to be precisely defined, surveilled and virtually simulated against a global geometry of precise, digital, time-space coordinates (Abler 1993).

More and more powerful data surveillance thus becomes spatially visualised and operationalised through sophisticated GIS and, increasingly, Virtual Reality (VR) and computer monitoring technologies. Their development is fuelled by heavy research and development investment as geographers, surveyors and cartographers attempt to perfect the apparatus for 'cybercartography', 'cybergeography' and ever-more 'realistic' geographical simulations (Openshaw 1994; Pile 1994: 1818). New techniques which blend remotely-sensed data with digital maps and 3-dimensional virtual simulations further strengthen the connections between surveillance and simulation.

Ultimately, technological enthusiasts predict immersive, real-time, virtual simulations that are so intimately connected to surveillance systems that they can be taken to be 'mirror worlds', 'software worlds in a box' (Gelerntner 1991), 'intelligent environments', or 'virtual urban spaces' (Droege 1997). Gelerntner (1991) has predicted that, linked to a range of real-time surveillance inputs, software constructions will become such life-like metaphors for the 'real' world that they will be taken for 'software models of some chunk of reality, some piece of the real world going on outside your window'. In such 'Mirror Worlds', he writes, 'oceans of information pour endlessly into the model (through a vast maze of software pipes and hoses); so much information that "the model" can mimic the "reality"'s every move, moment-by-moment' (ibid.: 3). Indeed, it is widely argued that with current advances in GIS and virtual reality systems, simulated facsimiles will become more and more like the 'real' world. For Jacobsen (1994: 37), for example, 'the addition of virtual worlds to GIS will result in a hybrid technology, the living map, that enables users to naturally experience geospatial information and the world this information represents'.

Interlinked technological systems of data capture and surveillance, computerised processing and simulation provide multiple webs of highly capable, and speedy, systems of 'surveillant simulation'. These, argues Bogard, are actually less visible than the bureaucratic paper-based systems they're replacing (Bogard 1996: 3). Bogard's central argument is thus that surveillance systems can now provide the data inputs necessary to develop electronic simulations of 'reality' used by a number of powerful organisations such as the military, the state and large firms.

A good illustration of surveillant simulation comes from the military sphere, where, as Kevin Robins (1996: 55) suggests, 'surveillance and simulation feed off each other. And surveillance and simulation technologies together feed in to the control of a new generation of "smart", vision-guided strike weapons.' Thus, the first generation of Tomahawk cruise missiles carried internal programs with digital simulations of the terrain they were to follow to allow target identification generated by intense surveillance of sophisticated military satellites. Current versions of the missile have been upgraded to use the even more accurate Global Positioning System (GPS) satellites which allow global tracking and target acquisition to an accuracy of one metre.

But increasingly, I would argue, processes of computerised surveillant simulation characterise the operations of many large organisations in civil society and the private sector too, as military-standard control and communications webs are translated to civil markets. Of course, the surveillance of dominant states and organisations has long been based on simulations and socially-constructed categorisations, as with the use of cartographic representations to help create dominated colonial spaces (King 1996). But the computerised linkage between surveillance and simulation helps to reconfigure and intensify surveillance practices because simulations become continually updated representations cybernetically connected 'backwards' to extending webs of data capture and 'forwards' to (attempted) disciplinary and consumer practices. At the same time systems of surveillant simulation become less and less visible because of their complex, disembedded time-space geographies, based on instantaneous flows of images and data.

There are a widening range of examples where automatically-captured data and images are processed to produce electronic simulations of the 'real' world (visualised data bases, Geographical Information Systems, CCTV image banks, digital DNA scans, digital transaction and travel records, etc.). To the user organisations, these are then taken to *be* the 'real' world and are, in the next iteration, used to support the restructuring or targeting strategies of organisations, based on the fine-grained allocation of goods and services, or ever-more intimate patterns of social control and surveillance, in (near) real-time through the space-time fabric of nations, cities and regions (see Virilio 1993).

The growing nexus between systems of surveillance and those of simulation has major, but poorly explored, implications for geographical change, for social control, for patterns of inclusion and exclusion, for the development of visual culture, subjectivities, and for the spatial dynamics of the 'information economy'. This chapter attempts to start exploring these. It looks in detail at three areas where areas of surveillant simulation seem especially well-developed: in crime control and the electronic tracking of subjects; in retailing, banking and home telematics; and in road transportation.

Tracking, tagging, and CCTV: surveillant simulation as social control

My first example centres on the emerging links between surveillant-simulation technologies and crime and social control initiatives, particularly in cities. In these surveillant simulations the behaviour of human subjects can effectively be reduced to their time-space electronic trails or images, as their movements and behaviour are logged, tracked and, increasingly, mapped, using systems linking CCTV, computerised tracking systems, GISs and mobile and fixed phone networks. The rapid extension of such technologies across geographic space means that 'a person going about his or her daily routine may be under watch for virtually the entire time spent outside the house' (Squires 1994: 396). Tim Druckrey (1994b: 15) too notes 'the increasingly invisible dispersal of electronic tracking' technologies. Through wide-area systems, covering whole cities, regions, nations and international transport routes, the behaviour of human subjects may increasingly become aggregated into detailed time-space surveillant simulations offering radically new possibilities for tracking and social control.

A good example of the emergence of surveillant simulation as social control can be found in the wide-area public Closed Circuit Television (CCTV) camera systems now operating in the UK. More than 200 CCTV schemes are now in operation in the public spaces of the UK, most of which use analogue video technology backed up by radio, telephone and photographs of target subjects (Graham *et al.* 1996). Virtually every sizable urban settlement in Britain now has public CCTV; systems are also increasingly spreading to cover residential areas. CCTV is being seen as a new and cost-effective part of the local policy 'tool kit' for dealing with a range of urban problems – including the cutting of crime, improving consumer and business confidence in town centres and underpinning the economic competitiveness of urban areas. Wide area CCTV systems integrate state-of-the-art surveillance cameras – often with remarkable resolution and infra-red night time capability – via microwave or cable telecommunications links into systems for continuously surveying towns and cities. The extension of CCTV grids across urban Britain has been backed by the rhetoric of politicians and press, by heavy investment from government, police, local authority and public–private Town Centre Management (TCM) organisations and by considerable public support – though, in practice, such support remains far less unanimous than often presented in the media (see Norris and Armstrong 1997). Evidence is building that through CCTV people and behaviours seen not to 'belong' in the increasingly commercialised, and privately-managed, consumption spaces of British cities tend to experience especially close scrutiny. Research by Norris and Armstrong (1997), for example, shows that much of the scrutiny that results from CCTV tends to focus on young men who 'look' a certain way, and on certain minority groups, including ethnic minorities.

Currently, however, surveillance within CCTV is not linked to simulation;

rather, the human eye and brain of the operator, linked into police records and photographs, with all its subjectivity and discretion, become the route through which CCTV imaging is translated into disciplinary social action and attempted control. But technological developments towards the digitalisation of CCTV, seem likely to lead to much higher degrees of automation and a much greater reliance of linked surveillant-simulation techniques.

Analogue CCTV systems are crude compared to the digital systems now emerging which constitute much more sophisticated systems of surveillant simulation, with much greater control capabilities. Micro-cameras and digital facial recognition technology are developing fast, both for in-store security systems and wider city-centre networks, allowing much more extensive, automated, digital CCTV systems to be built. New, digital systems are algorithmically programmed to scan for certain 'unusual' events or targeted individuals or vehicles, thus withdrawing opportunities for human discretion in the tracking and monitoring of individuals. Digital CCTV will allow real-time, time-space searching for specific events to occur as well as retrospective, digital searching aimed at correlating behaviour patterns with patterns of crime (Norris *et al.* 1996).

Early examples of digital, algorithmic CCTV applications are already emerging. Certain UK rail stations now have 'smart' CCTV which automatically warns when specific crowd densities are met on platforms. The City of London now has an 'intelligent screen monitoring' algorithmic system for automated surveillance of its 'Ring of Steel' anti-terrorist cordon. Here, a stationary vehicle triggers an alarm in the control room as does a car heading down the street in the 'wrong' direction (Norris *et al.* 1996). In another example, Sydney airport will soon introduce a system which scans automatically and covertly for known illegal immigrants entering immigration (Norris *et al.* 1996). In a new experimental project, BT is also working with the Massachusetts Institute of Technology (MIT) and the major British retailer Marks and Spencer on a digital image and television-based computer system known as 'Photobook' which, using real-time cameras linked via advanced facial-recognition software to image databases of the faces of convicted shop-lifters, will alert security staff of the arrival of the presence of convicted shoplifters in Marks and Spencer stores. Accuracy is said to be 'greater than 90 per cent' (McKie 1994).

In the long run, BT anticipate major new telecommunications markets. For example, 'all commercial outlets in a town could be linked and an alarm be set off the moment a person who has been seen shoplifting in one store enters another' (McKie 1994). When backed by digitised face prints of the type now being developed by the UK's Driver and Vehicle Licensing Agency (DVLA), the potential for national face-recognition and monitoring systems in the UK operating through expanding CCTV seems a lot more than some paranoid dystopia (Davies 1995). The state of Massachusetts is already in the final stages of digitising the faces of its 4.2 million drivers, as a means of overcoming fraud (Davies 1995: 197). More prosaically, the designer of

'Photobook' dreams of a 'front door camera that announces the identity of the person outside' (Griffith 1996).

In the United States, the control and surveillance capabilities of telematics are being widely explored as tools for new methods of social control in cities, methods that go beyond the highly expensive option of simple incarceration in prisons. By 1991, over 4.3 million Americans had been under 'correctional supervision' at home (Gowdy 1994). The burgeoning costs of the American prison programme are leading to the widespread use of 'electronic tagging' for low-level offenders who are free to maintain some semblance of daily life through 'walking prisons' (Winckler 1991). 'Less dangerous offenders now are confined to the home, except to go to work and run errands, freeing jail space for more dangerous criminals' (Gowdy 1994).

Anklet transponders, linked to telephone modems, provide continuous monitoring of the location of offenders. Newer 'smart' systems promise a much more fine-grained and tailored control over their behaviour. For example, in a retailer, the 'arrival of an ankleted shoplifter would set off a silent alarm, and the system would identify the offender to the store management' (Winckler 1991: 35). When linked to wider urban surveillance systems, through city-wide radio networks – available by the year 2000 – the movements of all ankleted offenders could be correlated with the incidence of crime, in time and space, to help in conviction. 'Every place the offender went – and the time he or she was there – would be recorded and compiled and could then be cross-indexed against known crime scenes and times' (Winckler 1991: 35). Thus, within this GIS-based surveillance simulation system, the 24-hour electronic tracks of individuals could be correlated with time-space patterns of crime incidence to underpin unprecedentedly fine-grained mechanisms for social control.

Home teleservices, surveillant simulation and cybernetic consumption

My second case centres on the emerging linkages between surveillant-simulation systems and telematics applications in the sphere of consumption and, more specifically, the rise in retailing of home teleservices – interactive cable TV and phone, video on demand, etc. Here we need to consider the broader possible role of surveillant-simulation systems in mediating access to increasingly cybernetic, tele-based, consumption services, as technological trends seem likely to shift inexorably toward a consumption driven, 'information superhighway' dominated by very large media and consumption corporations (Mowshowvitz 1996). As trends towards home-based consumption based on telephone, the Internet, cable and broadband home networks, combine with the growing use of electronic cash (credit cards, smart cards and 'cyber cash' on the Internet), home-based shopping, banking and consumption systems are emerging which precisely monitor, in real-time, the consumption patterns of households. The much-vaunted experiments in interactive, broadband home telematics, such as the Time Warner interactive TV system at Maitland, Florida, are experimental precursors to the much wider roll-out of highly capable home media and consumption systems which are intrinsically based on building up surveillant simulations of consumers' behaviour (Burnstein and Kline 1995).

Rather than relying indirectly on aggregated or individual consumption data from the census and credit and information bureaus, as has been the practice in the postwar period (Pickles 1995), these systems actually build up their own surveillant simulations of *actual individual behaviours*, in real-time. Robins and Hepworth note that, 'it is in the nature of interactive telematics as process and control technologies that electronic transactions (television viewing, teleshopping, remote working) *must necessarily be recorded. The system is inherently one of surveillance and monitoring*' (Robins and Hepworth 1988: 169, emphasis added). Wilson suggests that the extension of such systems means that we are entering a new 'era of cybernetic' consumerism by integrating domestic, home-based and electronic/cash-free retailing and credit systems with logistics systems such as Just In Time (JIT) and with the information gained from junk-mail and on-line response. This leads inexorably to an extremely efficient 'cycle of production and consumption, since every consumptive activity will generate information pertinent to the modification of future production' (Wilson 1986: 26).

Whilst allowing for the freedom of choice that such systems offer consumers, individuals linked into these telematics systems are themselves engaged in generating 'Transactionally Generated Information (TGI)', so building up their own 'digital personas' – surveillant simulations for corporate use (Crawford 1996). This raises questions about how self-generated surveillant simulations, built up covertly and geared to the needs of large corporations, are also involved in the construction and control of subjectivities and identities. Who, in other words, owns one's digital persona; the subject, the data bureau or the Transnational Corporation? And what are the virtual geographies surrounding the data flows through which these surveillant simulations are continually constructed, updated, and refined? For Allucquère Rosanne Stone:

> out of the snail track of our passage through a world of myriad simultaneous opportunities for consumption, [providing corporations] build their own images of who we are, freed from the constraints of linearity of sense. Our doppelgangers are already free of the tyranny of localized subjectivity; they follow the geodesics of capital and of ideal citizenship. It's ourselves that haven't yet caught up.
>
> (1994: 7)

It is clear that the virtual geographies surrounding Transactionally Generated Information can have very real impacts on the material geographies of opportunity, constraint and restructuring. TGI is usually used for various forms of exogenous social control by credit bureaus and consumer service organisations undertaking restructuring based on so-called 'Data Warehousing'. TGI allows firms to track real-time consumption habits, preferences and practices; to identify poor credit risk individuals, households and areas; to individually target and deliver direct marketing campaigns; and to build up commodified information packages for reselling within the lucrative 'information marketplace' (Crawford 1996; Mowshowvitz 1996).

Three examples help to demonstrate the virtual geographies and surveillant simulations surrounding on-line consumer monitoring. The first is the apparently humble case of the supermarket customer loyalty card – currently a key route to personalised surveillance and cybernetic customer targeting in the UK and US food retailing industry. Where previously firms had to rely on crude estimates, such cards provide the technological infrastructure for mass, continuous surveillant simulation of customers by the managing corporation. Each time a customer with a loyalty card buys goods their card is 'swiped' through the Electronic Point of Sale' (EPoS) terminal at the checkout. This allows an individual profile of consumption habits to be built up over time, which can then be aggregated to provide a real-time simulation of throughput through all stores. In turn, this can feed into ordering, logistics, storage and supply chain management. It also provides the raw material for 'mass customisation' and direct marketing. Massey (1996: 26) suggests that in the UK 'retailers like Safeway and Tesco can now build detailed pictures of spending patterns based on data gleaned from loyalty card/swipes. Eventually retailers will be able to target customers with offers specific to them – potentially setting special price details accessed by individuals using self-scanners.' Such cards illustrate the essentially ambivalent nature of consumer surveillant simulation. Whilst they give more affluent (targeted) users access to discounts and services directly customised to their consumption patterns, such practices also raise concerns. Where does customised service become a social intrusion? What are the impacts of the re-selling of individual dossiers within the 'information marketplace', to support wider direct marketing for financial services and utilities? And what are the implications of direct surveillant simulation of consumer landscapes for retail geographies in the context of the spatial restructuring of grocery networks, the oligopolisation and internationalisation of markets and the increasingly careful exclusion of those groups and areas without the disposable incomes and bank accounts to make them attractive targets of customised services?

The second example, which hints further at the complex geographies and subtle processes of inclusion and exclusion that surround on-line consumer systems, comes from the integration of computer and telephone systems (known as CTI) in customer telesales centres. Such centres are now used by major retailers, banks, insurance companies, transport firms, airlines and utilities. Telesale centres service regional, national and even international markets from a single, technologically advanced, node – through the use of free or local call phone tariffs linked into corporate telematics networks. By automatically surveilling the source of incoming telephone calls, through a system known as 'Call Line Identification' (CLI) and linking this number into customer databases, such systems now allow callers to be sifted according to how 'good' a customer they are. In effect, the surveillance of the caller is automatically linked to a simulation of all known customers, to allow customers to be treated differentially. Thus, UK utilities are already able to answer the calls of 'good customers' (i.e. those that have paid their bills promptly) before 'bad customers' (those with a history of default who are queued), without either the operator or customer being aware that their prompt or slow service is directly shaped by automated surveillance systems linked to computerised databases. Such work processes also allow, of course, for intimate real-time work place surveillance. Managers can assess each individual worker's response rates and productivity levels and can secretly switch between telesales staff, listening in on calls.

The final example, that of video-on-demand technologies (VOD), is a much-vaunted system that allows consumers to 'order' selected videos and media products for personal transmission down phone or cable lines to their homes. Many VOD trials are currently in progress, with the hope that it will herald truly tailored and individual media consumption. But VOD systems also produce a continuous stream of information for a cable or telecommunications company about the detailed media and consumption preferences of individual households. For example, the telecommunications company, Bell Atlantic, are developing a computer system linked to VOD which will 'monitor the movies that a person orders and then suggest others with the same actors or theme'. The system would also 'enable advertisers to send commercials directly to customers known to have bought particular kinds of merchandise. Thus, people who bought camping equipment from a video catalogue might start seeing commercials for outdoor clothing' (Andrews 1994). In a similar way, 'real-time residential power line surveillance' (RRPLS) will use normal electricity wires and IT-based utility meters to build up unprecedentedly detailed profiles of the electricity use of households (Crawford 1996). Such is the sophistication of the technology that it can 'infer that two people shared a shower by noting the unusually heavy load on the electric water heater and that two uses of the hair dryer followed' (Crawford 1996: 57).

Road telematics: surveillant simulation as (differential) power over space

My third case is that of Road Transport Informatics (RTI). The control capabilities that new surveillant-simulation technologies bring are of central importance here in supporting a shift from 'dead', public, electromechanical highways, to 'smart', digitally controlled and, increasingly, privatised highways (Graham and Marvin 1996). Virtual electronic networks of automated sensors, CCTV, tracking and charging devices, computers and GISs are being laid over established road transport networks helping to undermine their 'natural monopoly' characteristics and so allowing private firms to operate them profitably (Robins and Hepworth 1988). Road networks, with all their complexity of flow and pattern, increasingly become surveillant simulations supporting new practices of commodification, control and exclusion which provide the basis for strategies which differentiate groups according to the power over space they are seen to warrant. Whilst traditional toll systems already operate in many places, the emergence of 'intelligent highways' supports the translation of whole highway networks into computerised, commodified systems which can be managed flexibly and developed privately for profit. Essentially, Electronic Road Pricing (ERP) systems enable the commodification of road space, allowing it to be allocated at a price, within markets, for profit by private firms. Hepworth and Ducatel argue that ERP will 'create the physical infrastructure needed to privatise road space and will also create an institutional structure for administering a privatised road system' (Hepworth and Ducatel 1992: 92).

Within Road Transport Informatics (RTI) systems, people and their vehicles are effectively reduced to their moving image and signature. The central question raised by the development of the resulting 'intelligent highways' is whose 'intelligence' becomes embodied within the new road telematics systems? Currently, the development of transport informatics tends to be biased towards the need to minimise time-space constraints and increase transport and telecommunications access for the most powerful groups: the corporate élites, business traffic, land and property development interests and other 'road warriors' (Massey 1993). There are close linkages between the control of space that enhanced mobility brings, and the basis on which particular groups are allowed to have access to new technologies to overcome urban congestion. Swyngedouw argues that 'road pricing, or other linear methods of controlling or excluding particular social groups from getting control over space, equally limits the power of some while propelling others to the exclusive heights of controlling space, and thereby everything contained in it' (1993: 323).

An excellent example of how surveillant simulation becomes implicated in the construction of new, dualised, urban highway networks and systems of increased *differential* power over space can be found in the construction of a new, private, commodified, highway network (number 407) around Toronto (Campion Smith 1996). Built to ease congestion on the world's busiest highway, to which it runs parallel, in-car transponders will automatically charge all users of the highway around $1 per 11 km trip, without the necessity of stopping them. Tariffs will vary automatically, to peak around rush hour commuting periods, so ensuring that use of the highway never exceeds pre-defined limits. Thus the free movement of traffic on the highway can be guaranteed, overcoming the time and financial costs of congestion. Traffic patterns and flows will be continuously monitored and data will be aggregated into a simulation model of traffic on the highway. The simulation will be used to assess the appropriate tariffs through linkage with demand forecasts; ultimately, such surveillant simulation should allow a cybernetic linkage between tariffs and demand, so reducing the likelihood that congestion will occur even with rising demand and car ownership. Cars without transponders will be automatically photographed and their owners tracked and fined through linkage with data bases in Drivers Licensing authorities. By the year 2000 over $100 million per year is expected in tolls; and speed limits may even be higher on the highway than for other state highways. The consortium which built the road is now selling off all the key development sites along it to the highest bidder, for malls, affluent neighbourhoods, business parks, and logistics creating, in effect, a second-tier land-use transportation system for the élite interests in Toronto. Those without the ability to pay the tariffs, meanwhile, will remain trapped in the congestion and lower speed limits of Toronto's public highway system.

Bias and contingency in surveillant simulation

It has already been widely argued that, with the rapid emergence of superimposed grids of surveillance in retailing, consumer services, the media, the state and transportation 'the modern citizen is objectified as a life-path comprised of information, as a 'spatialised dossier' (Hannah 1997: 352). But this 'dossier' is far from reaching some omnipotent Panopticon, some all-seeing 'Big Brother'; it always remains incomplete, fragmented, patchy, and unevenly developed across and between the 'life-paths' of citizens. Thus, 'in "real life" we face a variety of normalising machines, imperfectly co-ordinated, and each with imperfect powers' (Hannah 1997: 353).

In contrast, the importance of trends towards the widespread application of surveillant-simulation techniques is that they support increasingly co-ordinated, extensive and *comprehensive* systems of surveillance and social control. Technological developments linking surveillance with societal simulation, and the increasing horizontal co-ordination between 'dossiers' and sites of surveillance (credit bureaus, banks, retailers, utilities, media firms, transport operators, state and correctional agencies), seem likely to prefigure a rapid intensification of co-ordinated, comprehensive surveillance. Above all, it is becoming more and more difficult to escape, to lift a phrase from Bruno Latour (1993: 121), from the 'skein' of technological networks that undergrid the apparatus of surveillant simulation. With their widening horizons of automated data capture and their instantaneous geographical reach, it would seem that 'we are in a generalized crisis in relation to all environments of enclosure. [. . .] Societies of control are in the process of replacing disciplinary societies' (Deleuze 1988a: 4). Three key questions emerge here, each of which has important implications for broader debates about virtual geographies.

Surveillant-simulation and dystopian urban futures

First, do trends toward surveillant simulation necessarily prefigure some wholesale shift toward societies of dystopian social control and segmentation (as is so often implied in cyberpunk science fiction and critical social theory – see Burrows 1997)? In general terms, it would certainly seem that electronic surveillant simulations are being constructed to support decision making, business restructuring decisions, social control and the development of further iterations of surveillance by service organisations within and across geographic space. Within the context of a political economy dominated by a profit-driven, liberalised/privatised and internationalising corporate environment, surveillant-simulation systems are emerging as crucial techniques for bolstering profitability, flexibility and responsiveness. For retailers, banks and utilities, for example, GIS surveillance systems are increasingly being woven into processes of business process re-engineering and service restructuring. This makes it possible to drive service plans and the 'roll out' of investment across cities according to tight geo-demographic targeting criteria.

As cybernetic loops monitoring citizen behaviour become more sophisticated (through retailers' customer information collection, mail order, consumer credit, profiling agencies, home telematics systems, road transport informatics, wide-area Closed Circuit TV, etc.), it is increasingly becoming possible to replace *aggregate* geo-demographic spatial data sets (say, at post-code or census tract level) with individual sets based on *actual citizen behaviour or consumption*. Thus, computerised simulations of the geographic space of cities and regions become possible, simulations which ever more closely resemble totally panoptic, real-time simulations of the city (the best example here being CCTV). Such panoptic and cybernetic networks thus start to resemble the command, control and communications webs already developed in the military. In the consumption field, the process of targeting only reaches its limit as service enterprises attempt to compete for market share within increasingly liberalised markets (whilst, of course, gradually easing out of less-profitable commitments or obligations covering poorer groups and areas).

Surveillant-simulation technologies are also being developed and applied within the context of a strong supply push from an increasingly globalised complex of media, telematics and 'correctional' industries. What Bob Lilley and Paul Knapper (1993) call the 'corrections commercial' complex – i.e. the fast-growing complex of security, military and prison corporations – who are, post-Cold War, attempting to colonise civil markets, are also key players in this supply-side push. They are being further supported by the broader debates about the supposedly world-improving momentum of the 'information superhighway', the imperative to apply telematics uncritically to every aspect of civil life, and the pervasive crisis of public confidence in home, street and transport security.

The result, in advanced industrial cities, as Mike Davis (1992) has suggested in Los Angeles, will be the emergence of urban landscapes made up of many superimposed layers of surveillant simulation. Each layer might have its own finer and finer mosaic of socio-spatial grids; its own embedded assumptions and criteria for allocating and withdrawing services or access; its own systems for specifying and normalising boundary enforcement, through electronically defining the 'acceptable' presence of individuals in different urban 'cellular' space-times; and its own cybernetic loops of system feedback, within which systems of surveillance become ever more integrated into systems of simulation. As people leave a stream of digital tracks through their daily lives, their electronic personas become embedded into a web of surveillant-simulation systems; 'each of us will become increasingly isolated in our own separate technological enclosure or cell' (Crawford 1996).

Disciplinary control within cities, then, comes to rely not just on the Foucauldian array of physical structures, disciplinary controls and urban planning practices (see Driver 1984) but on pervasive webs of electronic systems, which assert disciplinary control by 'distributing bodies/uses in space, allocating each individual/function to a cellular partition, creating an efficient machine out of its analytical spatial arrangement' (Boyer 1996: 17). The self-disciplining effects of surveillant-simulation practices, whereby subjects actively work to position *themselves* in relation to such practices, become based on a whole apparatus of consumer information systems, real and mock CCTV, infrastructure control systems as well as the traditional architectonic and urbanistic practices. As Virilio (1987: 16) argues, cities are shifting from a state where physical barriers and walls controlled access and 'belonging' to a state where 'the rites of passage are no longer intermittent – they have become immanent' and are woven as automatic, cybernetic systems, into the urban fabric. Such electronic systems, with increasing degrees of automation, also threaten to provide silent, invisible, and pervasive networks of cybernetic social control, with unprecedented potential for exclusion. Norris *et al.* (1996: 13) warn that:

> those who cannot pay will be excluded from motorways; known trouble makers from football grounds; the unsightly casualties of 'care in the community' removed from the decorous order of city streets and shopping malls; known shoplifters and fare dodgers excluded from shops and transport systems. . . . If the growing divide between those who have and have not and those who are included and excluded is intensified through the use of new technology, there is a real danger that our cities will come to resemble the dystopian vision so beloved by futuristic film makers.

Such fears, that surveillant simulation will prefigure and support socio-spatial systems that are more socially polarised and exclusionary through invisible, automated social judgements are thus very real. There seems little doubt that systems of surveillant simulation are helping to underpin broad shifts towards more and more intensely polarised, even dualised, material geographical spaces, especially in cities (Burrows 1997; Graham and Marvin 1996; Boyer 1996).

The dangers of over-generalisation: contingency and appropriation

But, and this is my second question, is this the end of the story? Are the processes underway really so stark and simple? In fact, whilst acknowledging broad scale trends and biases, we also need to be wary of the dangers of over-simplified and generalised scenarios; of accounts that assume totalised, dystopian, geographic 'impacts' of surveillant-simulation techniques drawn from paradigmatic examples like Los Angeles (Amin and Graham 1997).

It is easy to read the accounts of Bogard or Mike Davis and assume the easy emergence of completely integrated, all-seeing surveillance. Such accounts, however, tend to dramatically oversimplify the reality of technological innovation, which is a great deal more 'messy', difficult, contingent and open to contested interpretations and applications (Thrift 1996a; Bingham 1996). We therefore need to be wary of easy generalisation and deterministic readings of technological 'impacts', whether they be utopian or dystopian in character (Graham and Marvin 1996).

As recent debates in Actor Network Theory (ANT) have demonstrated, the construction of new technological networks (including surveillant-simulation systems) will always be an essentially performative, socio-technical process involving the enrolling of complex hybrids of social and technical 'actors' across distance. This applies from the design of embedded algorithms, through the deployment and operation of telematics networks, to the ways in which such networks become involved in detailed changes of social practices. ANT provides a fully relational perspective which underlines the dangers of easy, deterministic generalisation. It is 'concerned with how all sorts of bits and pieces; bodies, machines, and buildings, as well as texts, are associated together in attempts to build order' (Bingham 1996: 32). Absolute spaces and times are meaningless here. Agency is a purely relational process.

Because of the ways they become linked into specific social contexts by human agency, technologies have contingent and diverse effects (see Collins 1995). Thus, what Pile and Thrift (1996: 37) call a 'vivid, moving, contingent and open-ended cosmology' emerges. The boundaries between humans and machines become ever-more blurred, permeable and cyborgian. And 'nothing *means* outside of its relations: it makes no more sense to talk of a 'machine' in general than it does to talk of a 'human' in general' (Bingham 1996: 17).

The importance of ANT is its implication that 'no technology is ever found working in splendid isolation as though it is the central node in the social universe. It is linked – by the social purposes to which it is put – to humans and other technologies of different kinds. It is linked to a chain of different activities involving other technologies. And it is heavily contextualised' (Thrift 1996a: 1468). The lesson of ANT is therefore that if we are to understand the virtual geographies of surveillant simulation we need to balance our macro, political-economic treatments with much finer-grained, micro-level treatments of how such technologies are socially constructed and their 'effects' contingent on social practice (Graham and Marvin 1996; Graham 1998b). A good example of this comes from CCTV. Far from being a technologically-determined process, Norris and Armstrong (1997: 4) have shown how the uses of urban CCTV are currently 'contingent on a whole range of social processes: whether the screens are being monitored, and if they are whether an incident is seen and then recognised as deviant; if it is seen, whether it produces a response and the nature of that response'.

The complex social appropriation of telematics also means that the same technologies can be constructed and appropriated differently by different interests, in different contexts and with varying results. Surveillant-simulation techniques might support resistance and transgression, as well as social control and regressive urban restructuring. Thus, community groups and activists might utilise GIS technologies to bolster their lobbying for improved services in their spaces. Ramasubramanian (1996), for example, outlines how GIS techniques were used in Milwaukee to prove that an insurance firm was effectively red-lining African-American census tracts in the city (itself using GIS techniques). And mass access to CCTV and video systems may actually help to make the exercise of public power more accountable on city streets (complementing, of course, the increasingly automated and algorithmic systems likely to be used by crime control agencies). Kevin Robins (1996: 139) suggests that, with the mass diffusion of consumer video, 'the city now constitutes a mosaic of micro-visions and micro-visibilities. With the camcording of the city we have the fragmentation and devolution of vision-as-control to the individual level.' Thus, attention needs to centre on the complex, ambivalent relations surrounding surveillant-simulation techniques and the many subjectivities they may represent – whilst, *at the same time*, being sensitive to the definite macro-level biases which still tend, overall, to shape their design, deployment and operation.

Virtual geographies/economic geographies

Finally, it is important that we consider how surveillant-simulation techniques become implicated in the elaboration and construction of new material geographies of employment, urbanisation, flow and development. Three points arise here. First, the proliferation of surveillance systems is about much more than flows of representations; the construction of virtualities and simulacra; of mechanisms for fine-grained control; of cybernetic processes of automation; and of community activism. It also fuels some of the fastest growing economic sectors of the 'information economy', with very different trajectories emerging for different places within informational divisions of labour (see Graham and Marvin 1996: ch. 5; Hepworth 1989).

Second, the economic flows and labour processes surrounding the growth of surveillant-simulation systems seem to accelerate the processes noted by Castells (1989, 1996), through which economic processes become more 'disembedded' from the physical and social landscapes which are their focus, operating instead through some telemediated 'space of flows' within 'Network Societies'. Thus, data warehousing and consumer marketing industries generate huge demands for sophisticated 'switched in' office space, located in places with good labour supplies, public subsidies and adequate transport, telecommunications and property infrastructures. Such back office, telesales and data processing zones tend to locate far from the main urban cores in lower-cost suburban, rural or even 'Third World'/Newly Industrialising spaces (Graham and Marvin 1996). The customer support infrastructures for utilities, telecommunications and transport firms now routinely operate on-line from cheap, distant, automated call centres far from the territorial 'patches' covered by their physical infrastructures. Customers ringing London Electricity, for example, are dealt with 250 miles north of London, in Sunderland. Lower level and routinised data processing functions may even be out-sourced to even more dispersed locations, employing (largely female) staff on pay per keystroke wages, sometimes in their own homes but more often within back office districts in peripheral cities like Milwaukee and Newcastle (Richardson 1994). Theoretically, flows of images from CCTV systems can now also be easily switched over broadband networks to cheap labour locations. The World Bank has seriously suggested that the CCTV systems covering US malls should be monitored in Africa to take advantage of low wage costs and offer 'developmental' benefits to the continent (Bannister 1994). Ironically, all these work processes employ their own surveillant-simulation techniques to support worker discipline and performance

Finally, though, higher value-added software industries which shape surveillant-simulation products and techniques require ongoing face-to-face innovation and the high level, multifaceted infrastructure and services of large, core, metropolitan regions. Such industries tend to cluster in creative 'information districts', either in campus-like sprawls around major metropolitan areas (as in the case of Silicon Valley) or, as with multimedia design, in gentrified inner districts in older urban cores (as with Trebica in New York and Soho in London – Castells and Hall 1994). Each, of course, also links into its own global geometries of flow by tying in high-level support personnel in newly industrialising, high skill nations like India (Castells 1996).

Conclusion

From this discussion three challenges for 'virtual geographers' become fairly clear. They must develop perspectives which can analyse how broad, interacting technological systems help reconfigure virtual and material geographies. They must balance notions of wide-scale, macro-level biases in technological development with analytical approaches which accommodate the contingency of social action. And they need to maintain holistic perspectives which don't over-privilege the 'social', the 'economic' or the 'cultural', but rather allow the multidimensional nature of virtual geographies to be unpacked and explored (see Lee and Wills 1997).

Note

A more extensive and detailed discussion of surveillant simulation, including more examples, and a broader theoretical discussion, can be found in Graham, S. (1998b) 'The spaces of surveillant-simulation: New technologies, digital representations, and material geographies', in *Environment and Planning D: Society and Space* (forthcoming).

9 Rural telematics

The Information Society and rural development

Christopher Ray and Hilary Talbot

Introduction

Is there a specifically *rural* dimension to the Information Society? Is there a category of the broader socio-technological phenomenon that might be labelled 'rural telematics'? How might academic study contribute to an understanding of what is happening? These are the sorts of questions that we set out to answer in this chapter.

Given this focus on rural areas, we are interested in what the literature on the Information Society/telematics has to say about the structuring of space and place. Noteworthy commentaries are to be found in Castells (1985), Pascal (1987) and Robins and Hepworth (1988). Although such accounts reject the simplistic or idealistic collapse of space (and time) imagined in McLuhan's (1964) 'global village', they nonetheless emphasise the importance of a globalisation perspective to any understanding of the trajectories involved. In particular, they stress the logic of liberal capitalism as it attempts to transcend politico-administrative boundaries in the hunt for profits and the creation of markets. The space-reducing capacity of information technology takes this into the new order of activity and effect that we have come to call post-Fordism, central to which is the potential for management/control to distance itself geographically from other functions of production. Having removed any remaining rationale for physical proximity between production functions and between production and markets, information technology enables companies to centralise control and localise (i.e. decentralise) production and services.

This leads Pascal, and Robins and Hepworth to reflect on 'the end of the *city*' as the city loses its historical role as a 'business incubator'. Production functions can be farmed-out to low-cost localities (low cost because of the supply of non-unionised labour, the availability of subsidies or whatever) but to a degree that is increasingly global. Control from the centre is enabled through the communication capacity of information technology.

The result, according to this line of reasoning, is that economies – and their associated socio-political structures – are being transformed. Without the imperative of physical proximity, the predictions are for demographic and industrial location dispersal into lower cost – i.e. rural – locations. This trajectory of dispersal/distancing is being intensified by the availability of low-cost, practically universal networks of information flow. And, what is more, the distancing effect together with the imperative of liberal capitalism is leading to a new scale of international and interregional division of labour that is in a continuous state of mobility and which increases the vulnerability of regions and rural areas.

Thus, the literature on the Information Society tends to predict a post-Fordist future in a dynamic interplay of globalisation on the one hand, and the resurrection of locality on the other. This 'complex interplay between centrifugal and centripetal forces' (Robins and Hepworth 1988: 165) offers the prospect of the socioeconomic restructuring of rural and urban space. The implication is that rural areas – as long as they remain, in relative terms, low cost, flexible economies – will witness an increase in income streams, albeit at the price of vulnerability and dependency on external forces.

Accounts of telematics also talk of the 'rebirth of the local' – through the trajectories of decentralisation and individualisation of labour – as having implications for social relations. Post-Fordism suggests the emergence of a new work culture (see Casey 1995) in which the household and the individual find themselves as key units of production. The various forms of teleworking and, indeed of re-emerging home-work in general, may either lead to life-enhancing, flexible work or to the 'sweated labour' of low pay, contract or piece-rate employment (e.g. Castells 1985). Equally, however, telematics may enhance the socioeconomic vibrancy of local communities as increased opportunities and less commuting allow work to return to its historical domain physically and functionally overlapping that of the domestic.

Telematics is also said to promote 'community', through the enhancement of networks of communication within localities ('community networks'), between households and various levels of policy and administration and between spatially-separated individuals with common interests ('teledemocracy') (Fernback and Thompson 1995; Beamish 1996). In much of this literature, there is a theme of telematics and the Information Society as tools that local communities can employ in order to protect their chosen way of life, including the perpetuation of a sense of 'local community'. But, more generally, accounts of the neo-Fordist/Information Society are foretelling a re-merging of economic and social space at the local level.

Thus, the literature has so far emphasised the 'centrifugal–centripetal' dichotomy: globalisation together with decentralisation (individualisation, the household, the local community). The 'end of cities' is predicted as units of production are dispersed, increasingly, to non-urban locations. Wherever mobile capital lands and lays the high technology egg, then economic growth will flourish: high technology functioning as 'the engine of new economic growth and play[ing] a major role in the rise and decline of regions and metropolitan areas' (Castells 1985: 12). While rejecting the technological determinism of the accounts of futurologists such as Toffler (1980), there therefore remains within the literature a sense of the inevitability of the process, driven by the playing-out of the logic of post-Fordism.

But this would be to ignore major barriers that confront this process happening 'naturally'. It fails fully to account for the policy design/ implementation process itself. Neither does it include the effect on the trajectory of the Information Society – reactionary or mediating – of the existing social, demographic, economic, political and cultural structures. Above all, it has failed to address the problem of the geographically unequal provision of infrastructure. As a result, there seems little analysis of the way in which local and regional territories are seeking to mount strategic action in order to engage with the information momentum even though Castells (1985: 15) at least has accepted that the actual outcomes of telematics and the Information Society 'will be mediated and fundamentally modified by economic, social and cultural processes' (Castells 1985: 15).

It is our objective in this chapter to illustrate this through a concrete example. We offer a theorisation of the recent experience of one particular rural area as it began the process of negotiating its stance in relation to information technology. Our use of a case study means, inevitably, that the analysis is time and place specific. It is also an interim analysis in that, as we will show, both the contexts and meanings constructed are fluid and evolving. Indeed, in our conclusion, we note the possibility that our analysis reflects an interim, coping strategy by the territory. But we justify our use of a local area case study in that it has enabled us to ground the intoxicating imagination of much of cyber-rhetoric in terms of the perceptions and actions of people and organisations who have to manage the interface between such rhetoric and the sociocultural-economic implications for the people, communities and enterprises within localities. Moreover, our case study area is, in many ways, representative of rural areas throughout the European Union that are seeking to improve their socioeconomic well-being.

The conceptual framework for this chapter is rural socioeconomic development: because that is a key interest for us as academics, but also because our case study analysis reveals that the (European/global) telematics agenda came to be subsumed within a territorial/rural development agenda. We argue here that, over a comparatively short period of time, a rural development perspective emerged to counter those elements of the telematics rhetoric that were seen as potentially threatening to the interests of the territory.

Consequently, we have devoted the first section of this chapter to rural development theory and its component endogenous, exogenous and local/ extra-local approaches. We then turn to the case study before offering our concluding remarks. The analysis has been particularly informed by the use of the theory of actor networks as our main analytical device.

Rural development theory and telematics

Our purpose in this chapter is to reflect upon the idea that telematics can be an agent for socioeconomic development in rural areas. The core of our analysis is a case study of the North of England or, more precisely, the *rural North* – a sparsely populated, remote, rural area which in 1996 embarked on the process of developing a telematics strategy. But we begin by establishing the general context of the study: rural development theory and policy within the European Union.

What does 'rural' mean and why might it be a significant category? Although most people would recognise a rural area when they saw one, academics have found the business of arriving at a definition of rurality an elusive and ultimately pointless task (see, e.g. Philo 1992; Murdoch and Pratt 1993). Rather than rehearse that debate here, we will be using the term rural to refer to areas of low (and often declining) population density, faced with problems of geographic and economic peripherality, including those areas whose below-average socioeconomic performance and over-dependence on the primary sector have led to their being targeted for development programme assistance by the European Union.

Two broad types of policy model have been used to improve the performance and well-being of rural areas. The exogenous model is based on the premise that the forces for development exist external to the target territory (the 'top-down' approach). According to this model, national governments and the European Union would design policy intervention in accordance with their broader agendas and then 'impose' standardised solutions on their component areas. Policy as it affected rural areas, and particularly with the advent of the European Union, was dominated by a sectoral approach, i.e. the *agricultural* sector. Agriculture policy itself was driven by a modernisation (exogenous) ethic which was productivist and which imposed standard technologies that largely overrode geographical diversity (Van der Ploeg and van Dijk 1995).

The critique that emerged in the 1960s of aspects of modernity (centralisation/peripheralisation, environmental degradation, cultural homogenisation, etc.) in general and the experience of 'Third World' aid led to doubts being expressed about the net effects of the exogenous approach to development policy. Within the specific context of the European Union, socioeconomic convergence between regions/rural areas remained elusive, the capacity of agriculture to provide employment declined inexorably, and capital became increasingly mobile so that – rather than converging on the European average – many rural economies were becoming increasingly fragile.

In response to this, the endogenous approach began to emerge. This privileges the local area/region with the ability to generate and control its own development. Fundamental to endogenous development is the enablement of the local *territory* to define its problems and to solve those problems as far as possible through the mobilisation of local resources (indigenous enterprise, local human and physical resources, the local institutional environment, communities, etc.). Put simply, the approach seeks to fix the means for social and economic development in the locality so that the accruing benefits are retained locally. Accordingly, this requires policies and action that emphasise the distinctiveness and integrity of the territories in question.

In 1988, the European Union instigated a reform of its Structural Policy so as to move away from the sectoral approach embodied in the Common

Agricultural Policy and towards the targeting of specific rural areas (European Commission 1988). The European Union has become a major player in rural development policy and funding. The reformed approach to rural development was to adopt the principles of the endogenous model but in a somewhat ambiguous way in that the endogenous-territorial model was rationalised within the broader agendas of the Union, both economic (convergence so as to enable the Union to operate as a Single Market without internal barriers to the movement of capital, people and enterprise) and political (the building of pan-European institutions and a European identity). Furthermore, the European Union and national governments retained the power to set the rules for the use of Structural Funds, to approve which local organisations would be allowed to operate rural development programmes, and to approve the contents of territorial development plans (see also Ray 1996a).

Thus, 'pure' endogenous development seems here to be a myth in that both exogenous and endogenous approaches appeared to be operating simultaneously. This led observers to attempt a theory of 'beyond endogenous and exogenous models' (Lowe *et al.* 1995) by the application of actor network theory (Callon 1991) to rural development. This theory focuses on the dynamic, and not necessarily equal, power relations that form between the local level (the territory and its component individuals and organisations) and the extra-local level(s). It allows for the possibility that endogenous *and* exogenous forces will be operating in any given situation interactively. The emphasis is thus to look at: 'how local circuits of production, consumption and meaning articulate with extra-local circuits' (Lowe *et al.* 1995: 93). The theory conceptualises sets of relationships (networks) brought into being through various types of intermediaries ('actors') that could include: 'texts, technical artefacts, human beings, money' (ibid.: 100–101). The theory is that these local-to-extra-local relationships become a mechanism to construct meaning; or more specifically, they translate the inputs from the actors into a meaning of development particular to the network. Thus:

> An identity network has the potential for enrolment into the development process and for the 'translation' of agendas and can thus be seen to be constitutive of the development process.
>
> (Ray and Woodward 1998: 30)

Thus, network theory highlights the embeddedness of networks (socio-economic relationships, policy networks, etc.) within regions and rural territories. It provides an analytical tool for tracing the locations of power and which groups are able to participate in defining and benefiting from development policy and action. Moreover, it allows for the mapping of overlapping networks as they coalesce on to certain territories and development programmes and the recognition of how some actors/agendas will be indigenous to the local area whilst others will not.

Having established the present state of rural development theory, we can now turn to look at telematics and to reflect upon the notion of telematics as an agent of rural development. It is a notion that increasingly is being included, although sometimes rather uncritically, in the vocabulary of rural development policy statements and programmes. For example, the set of British Rural White Papers published in 1995 by the Department of the Environment (England), the Welsh Office and the Scottish Office included sections which identified telematics as a tool for rural development. Development programmes funded by the European Union and national governments also invariably acknowledge the need for telematics activity (e.g. Ray 1996b). Rural development agencies in the United Kingdom all have policies to promote telematics activity and, at the sharp end, the 'industry' has its own campaigning organisation in the guise of the Telework, Telecottages and Telecentres Association (TCA).

All these feed upon a more general rhetoric. An important player in the cultivation and promotion of this rhetoric is the European Union itself: specifically, the advocacy of Commissioner Bangemann as formalised in the report of the High-Level Group on the Information Society (Bangemann 1994) and an élite 'think-tank' established by the European Union, the Information Society Forum (ISF 1995). According to this rhetoric, we are set on a course – a revolution – towards a post-industrial 'Information Society' in which telematics will come to pervade all aspects of economic, sociocultural and political life; the advent of this new reality is imminent and inevitable, the technological momentum unstoppable, the nettle of the cultural turning point demanding to be grasped.

The role of this rhetoric as an actor in territorial rural development will be explored in the case study that follows. For now, we will focus on the ways in which the rhetoric apparently resonates with many aspects of the rural development perspective.

Central to the rhetoric is the capacity that telematics is said to possess to overcome the disadvantages of geographical location. Enterprises can improve their access to markets through the use of 'telemarketing'. Physical remoteness and small scale no longer have to mean that enterprises are limited to local markets. Telematics also introduces new possibilities for enterprise creation in which physical location ceases to be a 'factor of production', as in companies that provide graphic design and Internet services. Not only do these telematics businesses appear to overcome the disadvantage of physical peripherality, they help to 'modernise' the identity of rural enterprise and rural areas whilst still conforming to the very small scale, self-employed ethos of rural areas. Telematics, then, presents itself as a means to transcend space and to build vibrant, diversified, local economies.

But, if the rhetoric talks about transcending space, it also suggests that this can be accompanied by a valorisation of *place*. Place, in rural development, often translates as 'community' (another semantic minefield whose dimensions have been frequently explored). 'Community' is said to benefit through the possibilities of teleworking from home and this resonates with the rural development agenda in its promise of removing some of the push/pull factors that result in migration out of the locality and improving the social and economic vibrancy of localities. 'Community' is also enhanced through the provision of services into locations that can no longer, or never had been able to, maintain a level that is now regarded as necessary for a reasonable standard of living; i.e. telematics can respond 'to social needs and raising the efficiency and cost-effectiveness of public services' (Bangemann 1994: 24). Thus, it enables people to remain in place, to build their communities and for work to come to the community.

'Community' may also be enhanced through the valorisation of local identity and the European-level rhetoric is particularly strong on this point. The regions of Europe, it is said, will acquire 'new opportunities to express their cultural traditions and identities' (ibid.: 5) and, 'once products can be easily accessible to consumers, there will be more opportunities for expression of the multiplicity of culture and languages in which Europe abounds (ibid.: 11). Individuals, groups and territories will be enabled to re-create and communicate their cultural identity to the 'outside' through the medium of telematics. As a result, localities become valorised, the sense of rootedness is enhanced, and the raised quality of life helps maintain a vibrant local population.

The telematics rhetoric does not stop at the possibilities for linking the local with the 'outside'; it also talks of internal linkages, within the territory. Rural areas can, by definition, find that endogenous social and economic activity is hindered by low population density. Telematics offers itself as a way of creating local communication webs of individuals, enterprises, voluntary associations and official bodies. Physical communities can thus be overlain, and enhanced, by virtual communities: 'geographical communities will enjoy internal means of communication more efficient than any since the town meetings of Ancient Greece. At the same time, new virtual communities are already being created via the Internet, bonded together by multitudes of shared interests' (ISF 1995: 17).

The possibilities of enabling the territory and its components to generate their social and economic development rests on the understanding of telematics as a form of information flow but one which is controllable by local areas. This flow can be either a source of information that allows for the promotion of the territory itself, or it can draw in information from the outside to feed training and education and provide intelligence about markets and policy environments. Telematics appears to offer the possibility of transforming rural areas whilst, at the same time, helping to preserve their essential sociocultural characteristics. Even more intriguingly, it claims to be able to reduce the problems caused by geographical, economic and political peripheralisation whilst, at the same time, enhancing the vibrancy and identity of particular rural places.

'Telematics for the Rural North' case study

We now turn to the case study – the development of a telematics strategy in the rural North of England. In actor network theory terms, telematics rhetoric, as described above, would be portrayed as a factor exogenous to the area and its development strategy. However, our analysis of the case study reveals that the telematics rhetoric was mediated by the network, the network translating the rhetoric into an agenda for the rural North.

In 1994, development practitioners in the rural North were largely ignorant of Bangemann's vision and policies, although there was some evidence of local telematics activity. The thrust of these exogenous policy messages only began to filter through to local actors during 1995, mainly via an intermediary organisation called Northern Informatics. In parallel, other actors, such as the Rural White Papers and the development plans of the major telecommunications providers, were also at work in promoting the opportunities apparently afforded by telematics.

Northern Informatics (formerly called NiAA) was formed in early 1995 as a partnership of influential organisations from the public and private sectors in the North of England. Its stated aims were to 'establish the North of England as a key site in the global information network; improve communications and information access for everyone in the region; attract and create new employment opportunities through information services, support economic development initiatives and revitalise the existing economy' (early Northern Informatics publicity brochure). It stressed the need for co-operation between organisations 'where the whole of the combined effort will be more valuable than the sum of its individual parts'.

Initially, Northern Informatics operated through 'sector' working groups that represented voluntary partnerships of member organisations but growing awareness of the need to develop a specifically rural perspective led to the formation of a 'rural sector group' animated by rural development experts from the Centre for Rural Economy at the University of Newcastle. These efforts meant that by late 1995, what could be described as Bangemann's vision of telematics was beginning to be discussed in the rural North. Following a meeting of local rural interests (public sector organisations, telecottages and businesses offering telematics services) under the auspices of the Northern Informatics rural sector group, a view emerged of the importance of telematics to rural development, but with much scepticism about the likelihood of many benefits reaching the rural North. The view was that if the rural North was to seize the potential of telematics, a strategic approach was needed. The Rural North Group was set up to take this forward and, after much deliberation and consultation within the region, they produced a policy statement early in 1997 (Talbot 1997a).

At this stage three obvious 'actors' were active in the embryonic network out of which was to emerge a territorial, rural telematics response (Figure 9.1): 'Bangemann rhetoric' and European Union policy; Northern Informatics

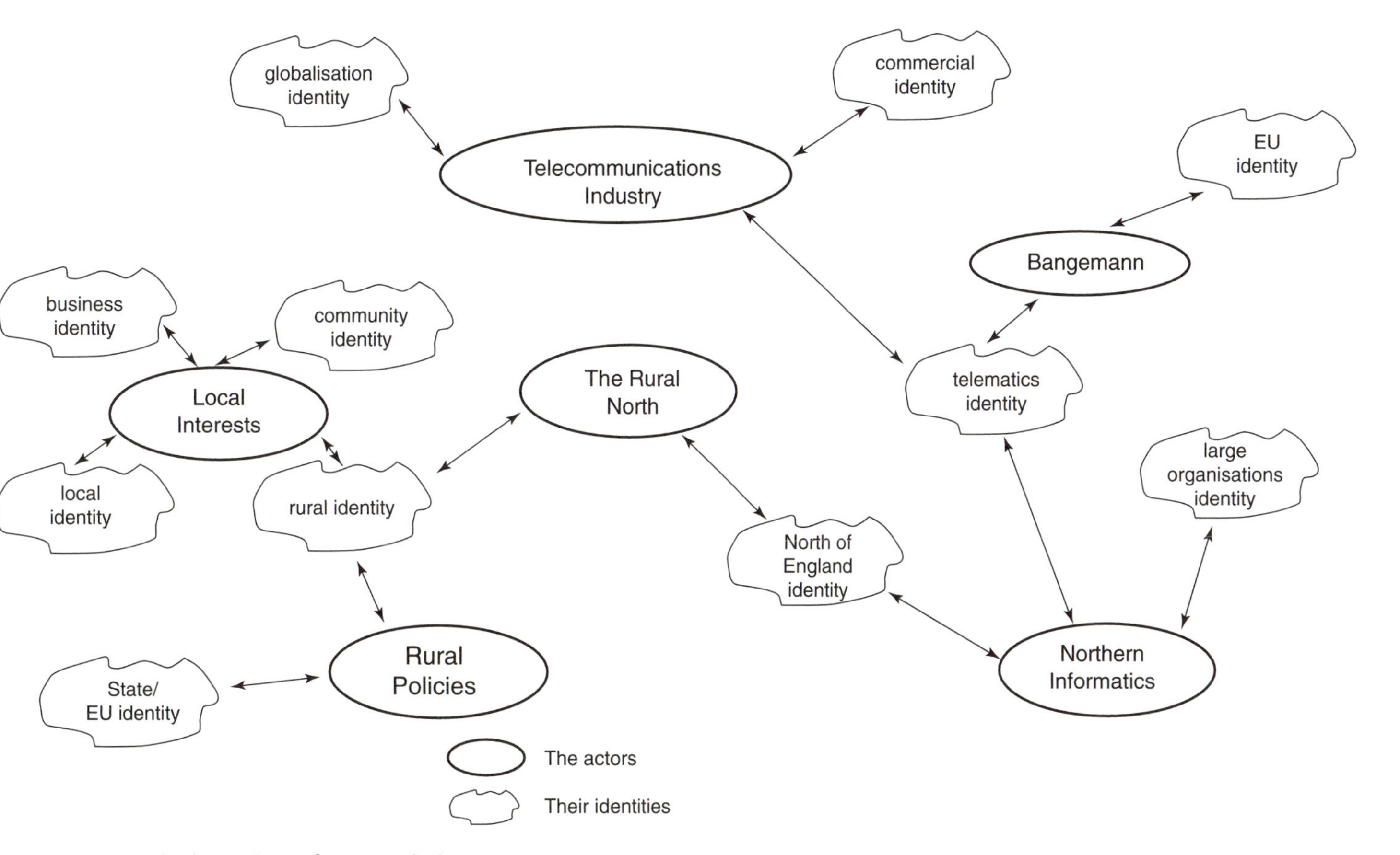

Figure 9.1 Initial relationships of actors and identities

(representing a regional focus); and the Rural North Group (an emerging *rural North* focus).

However, other actors were also at work, including one which could be called the 'rural policies' grouping. The English Rural White Paper, *Rural England*, published in late 1995 was a seminal policy document for those involved in rural development, being the first rural policy statement in England in half a century. When referring to telematics it echoed much of the Bangemann enthusiasm, suggesting that rural areas would be where: 'The computer, the fax, the e-mail and the explosion in telecommunications may have their most radical effect' (DoE/MAFF 1995: 59). This publication represented one of a number of rural policies that gave prominence to the use of telematics in rural development. Other influences included the European Union Structural Funds and county level programmes.

Another group of actors were the telecommunications providers and their regulators, the 'telecommunications industry'. These providers were supportive of, and implicated in, the Bangemann rhetoric: not only were they keen to promote the use of telecommunications, but, as a recently denationalised industry, they also reflected the broader European agenda of a liberalised single market. The mid-1990s also saw the fixed service regulator, OFTEL, developing its agendas, and the roll out of centrally controlled franchises for mobile phones and cable television. The final important actor group discernible in influencing the development of telematics in rural areas was found 'on the ground'. These were local economic development advisors, small businesses and community workers who were already active in the use of telematics, and thus conversant with the local issues (the 'local interests group').

Thus we can identify a number of actors who became involved in the process of developing telematics in the rural North. Of these, Bangemann and the telecommunications industries were clearly exogenous forces acting upon the region whereas the 'local interests group' had endogenous connotations. The roles of the other actors were somewhat more ambiguous. Following network theory, we can conceptualise each of these actors as representing their own agendas. Bangemann was the advocate of telematics and the Information Society but there was also an obvious link to a European Union agenda, such that his policies are premised upon trade liberalisation and the creation of a European identity. The 'telecommunications industry' is driven by a commercial and a globalisation ethos. Northern Informatics took, in common with the Rural North Group, the North of England as its geographic focus. Part of its agenda was also the cultivation of a powerful partnership of major organisations in the region. The Rural Policies group and the Local Interests group shared a rural and localist agenda, albeit at different levels, but the Rural Policies group also included wider national and European Union agendas.

At the conceptual hub of this emerging policy network was the Rural North Group, set up with the specific intention of developing a strategy statement for telematics in the rural North. It provided a link between the telematics identity of some actors, the rural agendas of others, and the northern focus of Northern Informatics (see Figure 9.1). The Rural North Group came into being early in 1996, a group of self-elected representatives of local interests with the task of developing a strategy for telematics development in the rural North. At this point, it clearly had links to three existing identities: telematics/Bangemann rhetoric, the North of England and rural agendas.

The group's relationship to the telematics rhetoric, built predominantly around Bangemann, was complex from the start. It recognised the important potential of telematics to rural development, but was sceptical about the benefits reaching the rural North. The group's deliberations did not fundamentally change this initial position, but served to crystallise their perspective – a position that sought to incorporate telematics into a rural development perspective – and to reinforce the view that a strategic approach was required.

The group's strategy document certainly did not reject all Bangemann's message, recognising that telematics 'has the potential to improve the competitiveness and enhance the services of the rural North' (Talbot 1997a: 1). But the local view was that it was the issue of access that had to be prioritised. The constraints to access were threefold: rural people and businesses were not psychologically/culturally ready to exploit the potential of telematics; the local provision of equipment and applications was very limited; and the telecommunications infrastructure was less developed than in urban areas, the economics of low population rural areas with awkward terrain making them unattractive to telecommunications providers. The strategy also counterbalanced many of Bangemann's celebrations of the benefits of telematics. For example, it explained how links to global markets (a benefit to peripheral areas in Bangemann's report) could work in two directions: rural firms could access global markets, but businesses outside the rural area would also be able to access these local markets, which hitherto had been served largely by local firms. In so doing, it highlighted the possibility that not only could telematics bring benefits, it could equally be an exogenous force that threatened the stability of rural areas.

The Rural North Group decided that the rural territory needed to devise a strategy to manage telematics. The group argued that telematics was a tool that had to be put to use to meet rural development objectives: 'Telematics' potential will only be realised if the local issues are understood and addressed, but it must not be seen as a solution for all problems' (Talbot 1997a: 1). Hence there were clear indications that, by the time the policy statement was produced, *rural* identity and the development agenda of the territory were taking precedence over the *telematics* agenda for the Rural North Group. The group had been set up under the auspices of Northern Informatics, giving it an obvious strong link to a northern regional identity but, over time, differences appeared in the relationship between the two. The large organisation orientation of Northern Informatics led to tensions with the less powerful actors in the rural region, and with the small scale activity typical of rural development. The Northern Informatics partnership was also dominated by an urban perspective on telematics and the development of its larger members. At the same time, the Rural North Group was recognising the specific rural dimension to some of the constraints identified. In the words of the Rural North Group's policy statement, there was the distinct possibility that the rural North could become the 'have nots' of the Information Society. The shared identity of the North of England was important in initiating the process but, by the time of the policy statement, this link had become more tenuous as the rural agenda came to dominate to the point where the group's agenda became more one of a generic *rural* response to telematics rather than a northern, regional one.

At this stage, our analysis is that the rural voice emerged essentially as a form of resistance to exogenous forces, including the perceived urban bias of the main telematics actors. Through an interplay of local, extra-local and intermediate actors, a territorial strategy emerged. Gradually the Rural North Group crystallised interests that represented the 'rural' in the rural 'North' as a counter force to the Bangemann (exogenous) and urban biases within the network. However, a further strand to this 'rural' identity emerged in the final section of the policy statement, 'Taking the Agenda Forward'. The Rural North Group recognised that it lacked the power to implement its strategy autonomously. It therefore decided to adopt a strategic and lobbying approach to the development of telematics in the rural North. It made an appeal to rural organisations in the North suggesting that this was a 'shared responsibility'. These organisations were called upon to work together; to review their policies; and to support actual projects. They were invited to form a rural voice to promote rural issues, to raise awareness of the potential of telematics and the constraints in realising that potential, and to lobby national and European government in relation to telecommunications provision.

The network identity that was emerging was marked by a duality: a rural North focus that could be used in the recruitment of the support of major local actors; and an identity that emphasised the *rurality* of the response, as a lobbying stance when negotiating with the extra-local. As a result, new relationships began to emerge – fuelled by the telematics rhetoric – which reinforced the rural North identity and concerns, in one sense, creating a 'distance' between the rural North and the nationally/globally powerful exogenous actors, and the urban dominated northern identity. This realignment is depicted in Figure 9.2.

However, the policy statement of the Rural North Group also drew attention to the powerless nature of this rural voice: 'Many of the rural organisations in the North are small, local and poorly resourced' (Talbot 1997a: 1), emphasising the need to recruit more powerful organisations in the region which had some responsibility towards rural areas (such as County

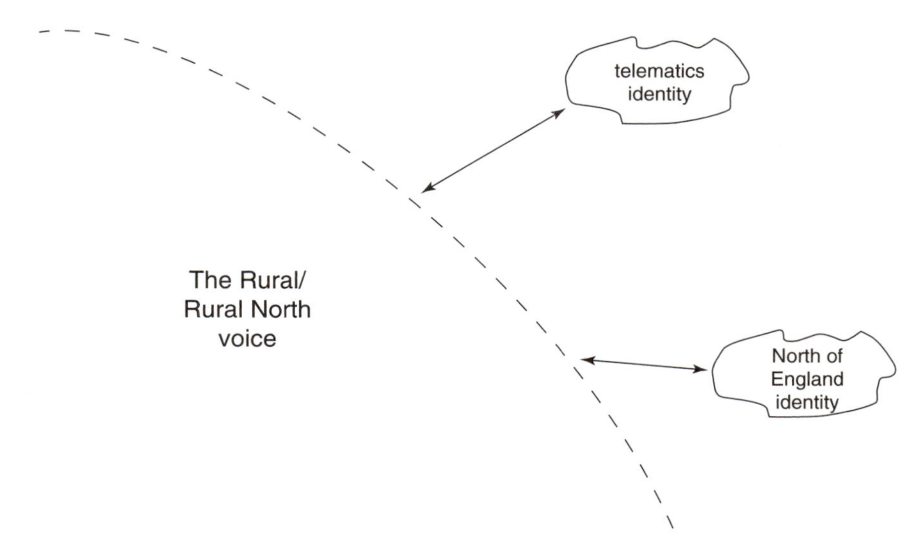

Figure 9.2 Realignments of actors

Councils and Training and Enterprise Councils) to help promote rural issues, raise awareness, work together and lobby government as part of a rural voice. During 1997, the Rural North Group was active in promoting its agenda through presentations and discussions with rural fora, through the media and through the publication of the policy statement. The Rural North Group was in this sense recruiting new actors to the network and adding 'weight' to the rural North lobby and, interestingly, the group did not allow its regional identity to limit its attempts at recruitment but rather sought to influence such national rural organisations as the Confederation of Rural Training and Enterprise Councils. With the agreement of the Group, the published report (Talbot 1997b) treated the rural North as a case study area, presenting the findings as generic issues for rural England.

Conclusion

Our characterisation of telematics rhetoric has shown how it appears to resonate with a great deal of the current perspective on rural development theory and practice. Telematics presents itself as being able to transcend the problems of geographical peripherality with businesses using the technology to market their goods, rural economies restructuring towards products and services for which geographical distance from markets is not a disadvantage, and localities benefiting socially and economically from teleworking (either from home or through telecottages). The potential is for the simultaneous transcendence of space and valorisation of place.

However, the territorial response described in our case study history was to juxtapose these potential benefits with a portrayal of telematics as an exogenous threat to the socioeconomic well-being of the area. In the eyes of the territorial actors, the liberalisation agenda and the urban bias in the telematics rhetoric represented potential forces for the further marginalisation of the rural area. For all the possibilities of place valorisation, left to its own dynamic, telematics was also seen as threatening demographic and economic outflows as telematics businesses became footloose and services such as banking and retailing lost the rationale for maintaining a local physical presence.

However, by using network theory we can see that telematics rhetoric, starting as an exogenous actor, produced, in an unintended way, a deliberate, territorial response. The space-reducing rhetoric triggered a raising of territorial awareness and the need for a strategic response. The rhetoric was thus a key input into the local response. The nature of the response (also influenced by other rural policy dynamics) was as a coalition of territorial interests that cultivated a *rural North* voice.

More precisely, these territorial interests formulated their stance in relation to telematics in terms of their *rurality*, interpreting Bangemann's telematics rhetoric of positive benefit into a number of interrelated, and sometimes conflicting, strands. The rural North's telematics rhetoric celebrated the potential benefits that telematics could bring to rural areas; in line with its more 'bottom-up' approach, it presented telematics as a *tool* for rural development that had to be controlled, used appropriately; it raised concerns about the denial of access to rural people; and it gave stark warnings of the negative effects on rural areas of some exogenous forces – for example, banks withdrawing from market towns and distant firms accessing rural markets. For much of the rural North group's rhetoric, telematics was being interpreted as a set of rural/urban and local/extra-local dichotomies.

But the notion, central to network theory, of local/extra-local circuits of meaning helps us to take the analysis forward from one of local resistance to outside forces. Awareness of the dichotomies resulted in a re-focusing of the conceptual environment and mode of action away from a telecentric view to a territorial/rurality perspective. But then, confident in its new identity and rationale, the Rural North Group could proceed to re-negotiate its relationship with the extra-local and far more powerful actors: particularly, the state, the European Union and the telecommunications providers. The territory could now begin to negotiate access to the resources it needed and the type of telematics development that it saw as appropriate to local needs.

We remarked in the introduction that our analysis may well be interim, that we may have taken a snapshot of a process that is still working itself out. In their attempt to translate telematics into an issue of rurality and the potential to intensify exclusion, the territorial interests that made up the Rural North Group may have been in the process of removing the local territorial focus altogether. Was the group, then, an ad hoc policy network?

The case of the Rural North Group is, as a story of the forming and reforming of alliances and relationships, hardly unique and is, in fact, typical of the emerging style of rural policy trajectories. It is possible, therefore, that the 'rural North telematics' perspective will come to be subsumed within further rounds of re-territorialisation of rural development programmes.

In conclusion, the case study has shown how the telematics and Information Society rhetoric came to be portrayed ambivalently, both as an external threat and as an opportunity that could be seized. Policy and political factors are combining with this rhetoric to bring into being new territorial spaces within which a range of agendas are seeking strategies to cope with, or control, the socioeconomic impacts.

The demise of cities and the consequent geographical dispersal of economic activity, foretold in the post-Fordist literature, contrasts with the way in which the actors in our case study saw the issues. Indeed, our rural actors emphasised the urban bias in the telematics rhetoric to the point where telematics was seen as inherently favouring urban areas. This was the moment of resistance: for the rural actors, telematics rhetoric meant urban advantage and this called for an interventionist stance. As a consequence, the issue of social restructuring was transformed into the issue of the unequal provision of infrastructure, training and education: without intervention to generate investment in favour of rural areas, local actors argued that the net result of the Information Society might be an intensification in social decline and difficult to predict restructuring of settlement patterns.

10 Internauts and guerrilleros

The Zapatista rebellion in Chiapas, Mexico and its extension into cyberspace

Oliver Froehling

> The novelty of the EZLN is not that it has inserted itself into satellite communications, so that they say today that the Zapatistas are more Internauts than Guerrilleros. [The novelty] is a redimensionalization of the political word that, paradoxically, returns to look at the past.
>
> (Subcomandante Marcos in Le Bot 1997: 349)[1]

Introduction

Chiapas, Mexico's southernmost state, is also one of its poorest. Its condition has been aptly described as a 'rich land and a poor people' (Benjamin 1989). Rich in resources like oil and tropical hardwoods, and a major producer of hydroelectricity and coffee, Chiapas reigns at the bottom of most social indicators. The majority of the rural population is characterised by absence: lack of electricity, of health care, of schools, and sometimes of food, whereas the very small upper class which has had a hold on state politics and the state economy for the last 150 years is characterised by opulence. The state thus exhibits a severe polarisation between a small, rich and urban minority who benefit from the resources and a severely marginalised rural population (Schmidt 1996: 30–35). The marginalised population in the state is composed of a majority of indigenous people, speaking as their first language various Mayan languages (Collier and Lowery Quaratiello 1994). This state, especially the area of the Lacandon jungle on the border with Guatemala, was to become the centre of an uprising against the Mexican government that has had a profound impact on Mexico since its inception on 1 January 1994 (Esteva 1994).

While it could have been just another peasant rebellion in Latin America, with a Marxist/Maoist ideology, suppressed by a tough military response applauded by the US and protested by some internationalists, this uprising had a decidedly different character. It enjoyed a much larger following than other guerrilla conflicts in Latin America. The popularity of the EZLN (*Ejército Zapatista de la Liberación Nacional*, or Zapatistas) extended way beyond the small confines of the state of Chiapas, drawing a large national and international basis of support. No doubt this was owing to a number of factors, not the least of which were the new political juncture after the cold war and the very different, non-dogmatic style of this new guerrilla, prompting the label 'Postmodern guerrilla', or 'the first guerrilla of the twenty-first century'. A major reason for this classification has been the role of the Internet in this uprising, specifically the use of this burgeoning high-tech medium by the indigenous rebels, which lead some observers like the RAND Corporation researcher Ronfeldt to declare this as a prime example of 'netwars', the new information warfare that supposedly will be common fare in the twenty-first century (Cleaver 1996; Wehling 1995).

Indeed, the Internet, which has expanded rapidly since 1994, became an important tool in the hands of Zapatista supporters and opened up new possibilities for organising support. But the fact that a large national and international basis of support was sought is not very new in itself (Adams 1996). What was new was the way in which the Internet, through its speed and connectivity, allowed the constitution of an international basis of support. Zapatista supporters on the Internet could produce news about the Zapatistas at a low cost and rapidly circulate it. Whereas the national media in Mexico have a long history of state control, the Internet could circumvent this control. That the uprising occurred during a time of rapid Internet expansion proved to be extremely fortuitous, but this emphasis on the Internet and its use by Zapatista supporters could easily lead to an overestimation of the impact of technology. Since most net postings originated outside Mexico, and net access within Mexico was and is very limited, the sphere of the Internet had a strong international rather than a national dimension. It did not start as a conscious effort at information warfare by the Zapatista army, but as a net of supporters, many of them located outside Mexico, who took up existing information structures and formed new connections. This international support worked in favour of the local struggle of the Zapatistas in Chiapas, because it helped to reshape the context of the uprising and spawned new connections outside the Internet. In this sense, the Internet certainly played an important role, not because it was a 'high-tech weapon' (Robberson 1995) that was first appropriated by the guerrilla, but because it allowed for the constitution of a wide web of supporters who were concerned with acting on behalf of the Zapatistas in a variety of ways.

Background

The causes of the uprising are easy to grasp: the abject poverty of the population, exacerbated by the gradual withdrawal of price supports and government services during the years of structural adjustment dating back to 1981; a general institutionalised economic crisis in which the tumbling of coffee prices in the early 1990s proved to be a tough blow to the small peasant economies, in which coffee is a major cash crop (Harvey 1996); the social transformation of the communities which came about with increasing migration, increasing influx of the national media and the extension of the world economy; and the racism of government and the national society where 50 years of programmes of indigenism have tried to eradicate indigenous cultures and languages, and where the presence of indigenous people, making up at least 10 per cent of the national population and a majority in the southern states of Oaxaca and Chiapas, is largely invisible, except for the casual cultural programme or tourism advertisements. Added to these factors was an increasingly renovated Catholic church injected with liberation theology and the arrival of a small group of guerrilleros from northern Mexico in the early 1980s (Le Bot 1997). These guerrilleros transformed their ideology through the long years of organising and mixing with the indigenous communities, resulting in a new type of guerrilla that bases itself more in indigenous traditions than in Marxist dogma. This mixture exploded on January 1 1994, when an army of between 5,000 and 15,000 masked indigenous rebels, one third of them women, in a well-planned and well-exercised operation took seven towns in the state. For a brief period they occupied one-third of the state's territory, with it the town of San Cristóbal de las Casas, a major tourist attraction in Chiapas. This event immediately circulated around the world, first in the major news-networks and then on the Internet. The main Internet media were e-mails and postings to discussion lists, managed by *listserv* or *majordomo* software in a variety of places, and postings to newsnet groups. Later, throughout 1994 and 1995, when the world-wide-web itself developed through its increasing commercialisation, Zapatista supporters also started to create web-pages with background information, links, Zapatista communiqués, pictures and even more sophisticated products such as animations and sound arrangements.[2] The indigenous uprising, precipitated by the backward conditions in Chiapas and carried out by an army of indigenous peasants, found allies working on the cutting edge of technology, diffusing the uprising around the world on the Internet (Froehling 1997).

The shooting war in Chiapas lasted only 12 days, after which a cease fire was declared by the government. There has since been an uneasy truce, interrupted by a government offensive in February of 1995 which displaced the Zapatistas from the majority of their territory but failed to capture their leaders. The main activity since the start of the uprising has been constant negotiation between the rebels and the government. Surprisingly, the government never denied that the basic Zapatista demands of 'housing, land, health care, work, bread, education, information, culture, independence, democracy, justice, liberty and peace' (Rosen 1996) had a basis in the obvious injustices of the state. The negotiations have led to the signing of one partial accord in 1996 which, however, has not been complied with by the federal government to date (August 1997), leading to the departure of the Zapatistas from the negotiating table in September 1996. With the military conflict at least formally suspended, the conflict was transformed into a 'war of ink and Internet' as the Mexican secretary of foreign relations had called it.

A controversial statement, since the brutality of a military occupation, with constant harassment at check-points, destruction of food supplies, rape and other practices of intimidation belie the picture of a purely intellectual dispute (Pineda 1996). It does signal however the importance of the battle for public opinion both on a national and international level, a battle that so far has been won by the Zapatistas.

The Mediascape

> There are people that have put us on the Internet, and the zapatismo has occupied a space of which nobody had thought. The Mexican political system has gained its international prestige in the media thanks to its informational control, its control over the production of news, of the newsanchors, and also thanks to its control over journalists through corruption, threats, and assassinations. This is a country where journalists are also assassinated with a certain frequency. The fact that this type of news has sneaked out through a channel that is uncontrollable, efficient and fast is a very tough blow. The problem that anguishes Gurría[3] is that he has to fight an image he cannot control from Mexico, because the information is simultaneously everywhere.
>
> (Subcommandante Marcos, in Le Bot 1997: 349)

From the beginning, Zapatista demands based in the conditions of rural Chiapas have included claims to necessary changes at both a national and international level (Harvey 1996). Their first attempt to carry their demands to a national level was military: after their army had occupied the cities in Chiapas, it was to split in two, one half retreating into the jungle to protect their villages, the other marching toward Mexico City, an action during which 'the world was going to collapse on them' (Marcos, in Le Bot 1997: 219). This suicidal action was to create a media event of such magnitude that it could not be ignored, taking the war away from the base communities and toward the national level. It was however made unnecessary by the immediate media attention, and the declaration of the cease fire. During the ensuing negotiations, the scale of demands has constantly been a point of contention, with the government trying to 'localise' the conflict. For the Zapatistas, national support from social movements and indigenous groups was therefore necessary to provide evidence for the national relevance of their demands. The building of coalitions at this national level worked well through contacts with existing groups and the rallying of support through more traditional means, such as fax, telephone, and sympathetic newspapers. International support on the other hand served to provide constant visibility to protect them from military annihilation. International visibility thus became an inherent part of their media strategy (O'Tuathail 1994).

From the beginning, the conflict had attracted national as well as international attention. The Mexican independent press, pushing its limits under the regime of President Carlos Salinas de Gortari, jumped on the coverage of this uprising, despite government attempts to contain it. While Mexican television networks are dominated by state interests and therefore provided rather limited coverage of the events, the print media provided a much broader analysis of the uprising, often sympathetic to the EZLN. Especially important was the Mexico City daily *La Jornada* which continuously published reports and opinions on Chiapas, as well as EZLN communiqués, and thereby doubled its circulation within three weeks (Schmidt 1996: 28). In Mexico, oppositions to the end of land reform in 1992, and the neo-liberal policies of the Salinas regime had found a point of coalescence.[4] Internationally, the capture of San Cristóbal de las Casas, a popular tourist destination, and the coincidence with the inauguration of NAFTA immediately rallied progressives in many countries around the cause of the Zapatistas. The opposition against NAFTA had already led to the formation of electronic communities and cross-border relations. These structures, though defeated by the passage of the NAFTA in October 1993, still existed and immediately mobilised in favour of the Zapatistas (Carr 1996; Cleaver 1996). The NAFTA connection provided one of the crucial symbolic links, and little did it matter that the date of the uprising was chosen primarily because of the expected lack of resistance by local police and army due to a New Year's day hangover rather than because of its symbolic significance (Marcos 1994). The words 'NAFTA is a death sentence for the Indians' were circulated widely and repeated over and over again.

This amount of international attention is widely credited with forcing the Mexican government to stop the shooting war, and to protect the Zapatistas from annihilation (Cleaver 1996). It would, of course, overstate the case to say that the international protests by themselves caused the Mexican army to stop, but they provided an environment in which actions were monitored from multiple sides and this 'reverse panopticon' made it easier for established organisations in, and outside, Mexico to put pressure on the Mexican government (O'Tuathail 1994). An illustration of the government's sensitivity to international attention is provided by an incident in 1993. When the army had discovered Zapatista camps in 1993, the information was suppressed because NAFTA, the hallmark of the Salinas policy, was still under discussion in the US Congress, and any attention to existing guerrilla armies, or lack of control, was thought to jeopardise its passage, immediately conjuring up easily exploitable stereotypes of third world chaos (Ross 1995a: 27). The well-executed Zapatista uprising on New Year's Day 1994 created an event of a magnitude that could not be ignored, precisely because of the attention focused on Mexico with the inauguration of the NAFTA. While the traditional mass media soon moved on and found other topics of interest, the print media on a national scale and the Internet on an international scale provided a continuous illumination of the events in Chiapas.[5]

Aside from information about the uprising, there were increasingly calls for action combined with the addresses and fax numbers of Mexican consulates, as well as US officials in order to let the Mexican government know it was watched and to use US leverage to influence the events in Chiapas. These fax campaigns were supplemented by direct protest actions in front of Mexican consulates, most recently in February of 1997 in a concerted effort of protests in front of 36 consulates in the US in support of constitutional reforms in Mexico (Bellinghausen 1997).

In Europe, these media attacks on the manifestations of the Mexican state were extended to involve political parties and governments, such as in Italy where parliamentarians produced inquiries and signed letters, published in Mexican newspapers, in their support for a just and peaceful solution to the conflict. These efforts resulted in the need for a media campaign by the Mexican government. In the last months of 1996 it tried to publicise its efforts for a peaceful solution, an attempt that backfired when it rejected the proposed constitutional reforms at the end of 1996, causing consternation in European governing circles (Bellinghausen 1997).

These actions within the media space were reinforced by direct contacts with Chiapas. One type of flow that connected the Internet community to Chiapas were peace caravans and peace camps, organised by human rights and church organisations. Aid caravans collected money and materials in the US and then delivered them to communities in Chiapas located in the zone of conflict. These aid caravans, also provided eyewitness reports of the consequences of the low intensity war in the area. These reports were often reposted on the Internet, and provided powerful statements rallying support for the communities. Permanent peace camps were established in a number of communities by a variety of organisations, in order to put international observers in the conflict area to provide for direct support and visibility and to decrease the possibility of violent retribution by government troops. Throughout the conflict many prominent intellectuals and media stars made their way into Chiapas to meet with Zapatista leaders. Prominent among them were Danielle Mitterand, the widow of the late French President, an Italian parliamentary commission, and the US film director Oliver Stone. The effect they produced was to give additional credence to the cause of the Zapatistas and to keep the uprising a media event. In many of these activities, Internet organising and the posting of reports were crucial to provide up-to-date information to the international community that was connected.

The Zapatistas actively promoted the formation of this national and international support network. They first reached out to national civil society in the summer of 1994 with the *Convención Nacional Democratica*, and then, after the international support had become visible and taken into account, called the *Intergalactic Encounter for Humanity and Against Neoliberalism* in the summer of 1996, taking place in a newly constructed area in Zapatista territory called *La Realidad* (reality). This international gathering attracted about 3,000 participants from five continents and 42 countries, and the Internet played an important role in the rapid organisation of this encounter (Acción Zapatista de Austin 1997). The encounter, aside from the attempt to form an international network of organisations, also provided the Zapatistas with the benefit of international visibility. A result of this encounter has been the effort to create an Intercontinental Network of Alternative Communication (Spanish acronym: RICA), in which groups that are loosely connected in their concern about Chiapas and the world politics of neo-liberalism in general can exchange information and co-ordinate strategies. This iteration also demonstrates the effect of coalition building in which the initial event, indigenous uprising in Chiapas, retreats behind the broader issue of the struggle against neo-liberalism, an umbrella that was actively promoted by the Zapatistas themselves (Marcos 1997b).

The next iteration, growing out of the *Encounter* of 1996, was the *Second Intercontinental Encounter* in Spain in 1997, already largely organised through information on the Internet. Here the formation of an international network based largely on Internet communications became a central topic of discussion. The *Second Encounter* brought together about 5,000 participants from very different groups throughout the world. The Zapatista principles where certainly a cornerstone of the discussion, which moved on explicitly to include international organising strategies in order to constitute the *International of Hope* (Bloque 7 1997). What is obvious in the encounters is the necessity to reinforce 'mere' cyberspace connections with other connections, forming a web of electronic, face-to-face and other connections (through print media for example) to reinforce and build a number of overlapping communicative links in different media spaces. The international encounters served this goal effectively, developing links between different groups that only knew each other through their web-pages or e-mails. The central discussion about the constitution of a network was then almost redundant, since the discussion of how to form a network, taking place through e-mail, faxes, phone calls and personal contacts in itself already constituted the emergence of a net of new communicative relations.

The constitution of an international network through the Internet thus facilitated a variety of actions or return flows back to Chiapas, as well as constant protests and symbolic attacks on the manifestations of the Mexican government around the world. Many of these actions were organised through other media as well, but the Internet through its reach and through its speed greatly facilitated timely action in this area and supplemented other organising strategies. Its role lay not only in producing information, but in producing effects outside its narrow techno space. It was not that the war was displaced into the Internet, rather it was extended by Zapatista sympathisers into the new medium (Froehling 1997).

Beyond the hype: Zapatista strategy and the Internet

The information on the Internet did not emanate from a central site under Zapatista command in Mexico, but rather from multiple sites in contact with each other throughout the world. The majority of web-sites in support of the Zapatistas are not in Mexico, but in the US, followed by Italy and Mexico

(Acción Zapatista de Austin 1997).[6] This reflects the distribution of Internet access among the population, which is highest in the US, whereas in Mexico Internet access is very limited and mostly centred on Mexico City. The Zapatista cause has also surprisingly drawn an incredibly high level of support from Italy, where prominent politicians, trade unions and political parties have collected money and participated in support actions. The media support of the Zapatistas is created by its supporters in and outside Mexico guided by the events and actions in Chiapas, and most notably by the writings of Subcommandante Marcos, the Zapatista's military leader and spokesperson. It is not a centralised effort at information warfare, but a more or less co-ordinated effort of supporters in different places, with different agendas (churches, human rights groups, left political groups) that converge around the issue of the Zapatista uprising. Cyberzapatistas are everywhere but they are not controlled by the Zapatistas in Chiapas.

This is important because it contradicts the hype that has surrounded the role of the Internet in the uprising, leading to the image of the Zapatistas directly communicating with the world (Robberson 1995), and using the Internet and a 'portable laptop computer to issue orders to other EZLN units via a modem' (US Army, in Swett 1995). Given the extreme state of material deprivation in the core area of the Zapatista uprising, which includes an absence of roads, electricity, telephone and communications in general, it is extremely unlikely that EZLN units in the jungle will find a telephone to plug in their modem, aside from the necessity of Internet access providers and the danger of interference. These statements are part of a hype that has as its issue not the uprising but the celebration of technology. There is no evidence for a direct presence of the EZLN on the web, rather the web has been used by supporters of the EZLN to co-ordinate actions, disperse information and relay EZLN communiqués (Froehling 1997).

This hype was partially caused by the timing of the uprising. It was in the first year of the uprising, in 1994, that the Internet hit the private market and this new medium became a focus of traditional media outlets, which focused on this technology of which nobody quite knew what to expect and what it was good for. It was the year when many magazines first started to produce web editions, and in their printed editions provided information on this 'medium of the future'. *Time* and *Newsweek* for example ran a variety of articles on cyberspace and began to run their special *Cyberwatch* or *Cyberscope* columns, simultaneously reporting on and advertising this new medium. The appealing combination of men and women with guns, backward indigenous people and high tech media resulted in a number of articles in popular magazines on the use of the Internet (imagined or real) by the Zapatista guerrilla (e.g. Robberson 1995; Watson 1995). Thus the uprising, with a number of underlying causes and continuing pain and suffering caused in the region, was transformed in the mainstream media into a technology fetishisation, where the focus of articles on use of the Internet was not an analysis of the Zapatistas, but rather a promotion of the Internet through the underlying message 'Even indigenous guerrilleros are using the Internet, shouldn't you?'.

But what exactly is the Internet? At its basis, it is a set of horizontal connections between different machines, or servers, with a continuous production of new links while old ones fall into disuse. This type of arrangement finds its analogy in nature in the form of a rhizome, a subterranean stem without a definite beginning or end that continues to grow in all directions, constantly building new connections while old ones die. It is different from the arborescent structure of the tree, that grows from the roots and develops its stem and branches through binary divisions (Deleuze and Guattari 1987: 3–25, 506–507). Because of the obvious similarity, the Internet has often been described as a rhizome (Escobar 1994; Wark 1994a; Cleaver 1996; Grimes and Warf 1997). Space in this rhizome is not an absolute, geometrical space but a relational space composed of flows which produce de- and reterritorialising effects.[7]

The Internet, because of this rhizomatic, non-hierarchical character requires a different strategy from that directed directly at the traditional media. As there is no central command, a struggle to occupy a centre is futile, since there is no centre. The multifarious relations through the lists and web-pages can only be countered through the constitution of a different net, or the attempt to interrupt communication links or insert false information. Disruption however proves to be very difficult, since the Internet has also become a vital tool for international commerce. The other strategy tried by the Mexican government was to insert different information, supportive of the government point of view, into newsgroups. But neither these attempts, nor the development of web-pages by various government agencies could counter the information provided by Zapatista supporters (Cleaver 1996). It resulted simply in the constitution of a different net with little connection to the net of the Zapatista supporters, who mostly ignored these new sites, or provided links with (non-favourable) comments.

This structure of the Internet might also explain the curious coalition between the Zapatistas and Internet aficionados. As Cleaver (1996) points out, there is a similarity in the form of the Internet organisation as a rhizome and the organisation of the Zapatistas and their supporters. While the core of the Zapatistas is an army with a hierarchical organisation, the affiliation of different villages in the core area, and their connections to supporters throughout Mexico and the world are organised like a social movement, that is, like a rhizome. Many procedures in the Zapatista villages follow a model of base democracy in which decisions are made by consensus, and connections between villages are horizontal without a superior authority (Collier and Lowery Quaratiello 1994: 254–84). The problems faced by the Zapatistas, and the problems of part of the Internet community that sees itself threatened through government regulation and commercialisation, are in many ways similar, so that the Zapatistas could strike a chord and got support from Internet aficionados who transformed Marcos into a cyberpunk (Cleaver 1996).[8]

Constructing cyber-coalitions

> When the EZLN appears it has to dispute certain historical symbols with the Mexican state. The terrain of symbols is an occupied terrain, especially when it comes to the history of Mexico. Once one enters into the terrain of symbols, of language, one has to enter fighting in order to occupy a place. . . . There is at first a dispute over the historical image of Zapata which permits a first encounter in which the EZLN restates the political language in other terms. Not to invent a new language, but to resemantisize, to give it a new meaning, and a new meaning to the word in politics, and overall to history in politics. And therefore it takes recourse in the old, the tradition of the indigenous people, in their cultural tradition to find old personalities, old ideas, and in confrontation with the new ones continues this new language of the Zapatistas. I am talking about a postmodern language, which, paradoxically, nurtures itself in the historically premodern in order to constitute itself the way it is. This language begins to search for its own terrains of struggle, the terrain of the press, of the symbols, and occupies the spaces that appear. A new space, a novel space, which was so new that nobody thought that a guerrilla could turn to it, is the Information Superhighway, the Internet. It was an unoccupied terrain. One supposes that its objective is commercial, the flow of capital across the computers and satellites in this globalizing world. And yet the human side is about to open.
>
> (Subcommandante Marcos in Le Bot 1997: 348–349)

Clearly, the simple mixture of technology (the Internet) and of an uprising is not enough to explain the amount of international attention and involvement. Mexico's other guerrilla movement, the EPR (*Ejército Popular Revolucionario*) who violently appeared in 1996 also has a web site,[9] but has not managed to gather any significant amount of international support. Marcos' analysis above of the extension of the Zapatista ideology describes the importance of common symbols in building coalitions in- and outside cyberspace. It began when supporters took up the messages emanating from Chiapas and redistributed them. These messages found themselves far from Chiapas and in computers around the world, giving them a new context but also moving people to act, to once again go beyond cyberspace. There are a number of factors that made this international coalition possible and help us understand its popularity.

First, there is the geopolitical juncture with the end of the cold war and communism. The left had found a movement it could support, after all the confusion and disarray resulting from the collapse of 'actual existing socialism'. On the other hand, the end of the cold war had also made it impossible to brand the Zapatistas as 'Russian supported communists', since communism had been officially defeated. Public opinion could be convinced of the sincerity and the 'just cause' of the Zapatistas, especially given the fact of their very different presentation (O'Tuathail 1994). With the appearance of Zapatista communiqués and writings, with more information about them and the beginning of the cease fire, the Zapatistas entered into a dispute about symbols with the government. These were the symbols of the fatherland, indicated by the invocation of Zapata, a national hero from the Mexican Revolution, in their name, the ample use of Mexican flags and a discourse that takes back popular culture and heroes from Mexican history, carefully avoiding any dogmatic political statements.[10]

On an international level, it was a number of key signifiers that managed to sustain this coalition. There were the issues of poverty and indigenous people, sometimes evoking romanticised images of indigenous life and a defence of traditional ways. Another issue that gave it appeal to a wider audience was the role of women. Prominently among the declarations published on the first day was 'A women's revolutionary law', outlining basic demands for women's rights, among them: equality in participation, right to education and reproductive freedom, reflecting the basic demands of an army of which one-third were women.[11] These were not of course 'just symbols', but very real issues, that could be taken up and related to the situation at home by people everywhere. The emphasis on women's rights, in conjunction with statements about gay rights and other prominent issues managed to convey the image of a guerrilla that was fighting for every just cause possible. The Zapatistas' opposition to NAFTA and to an obvious dictatorship that had stolen the last presidential election of 1988 added to the appeal, creating an obvious and appealing David vs Goliath situation.

A central role in the representation of the Zapatistas is played by their spokesperson, Subcommandante Marcos (Gómez Peña 1995). He, as a Mestizo spokesperson for an indigenous movement, provides many of the signifiers that knit together a wide coalition of supporters outside Chiapas. He uses different styles, humour, self-criticism, references to literature and indigenous culture as well as making connections to other social movements that have very little to do with the direct causes for the uprising in Chiapas. These signifiers were picked up and rapidly circulated through e-mail and www-sites across national and ideological boundaries. This posturing and refusal to be defined according to traditional ideological boundaries appeals to a much wider and diverse audience than any ideological manifesto ever could. The Zapatistas are conscious of this fact and try to refuse a definition as long as possible, calling themselves humorously a *desmadre* (a slightly vulgar term indicating total disorganisation) when pressed for a definition of their politics (Le Bot 1997: 302).

All of these signifiers at this particular juncture in world politics managed to draw together an international coalition of supporters, who in their turn contributed to influencing the events in Chiapas. This working back of international media on the events of a country is not something new or something very original. In the politics of protest, events are often staged to conform to media expectations (Luhmann 1997; Hardt and Negri 1994) and the effect of enlarging the scale of protest through media is a central objective of the politics of protest (Adams 1996). It is, however, not only protest politics but also government policies that result in this type of politics, as exemplified by Carlos Salinas de Gortari, President of Mexico from 1988–1994. Through his close relations with the international press he managed to create the impression of a resurgent and structurally sound economy, which was hailed by some as the second Mexican miracle (Esteva 1994). This impression depended on the positive evaluation of the foreign press, which was then presented as news by the Mexican press and contributed to this image internally and externally, even though wages were falling and there were severe difficulties in this type of economic recuperation, built essentially on foreign financial investment. The bubble burst when the Zapatistas filtered through the military cordon surrounding them in December 1994, peacefully occupying a large part of Chiapas. This was the final straw to send the Mexican stockmarket down, and with it stockmarkets all over Latin America, causing the 'tequila effect' as it was baptised by the American press. One could certainly take this as an illustration of the butterfly effect from chaos theory, in which a small event like the movement of a small guerrilla army in a backward province of Mexico 'causes' the crash of stockmarkets in Latin America, from Mexico to Argentina. The point is however that the international level, which is construed through the media, of which the Internet is certainly a part, is not a separate form of politics 'out there' somewhere, but is an integral part of how national and regional politics work.

Supporters of the Zapatistas were able to construct a sense that their information was 'the truth', due to their basis in a very appealing signifying chain and a climate of general distrust of government information as emanating from a dictatorship sustained by press manipulation (the PRI, or *Partido Revolucionario Insitucional*, has ruled Mexico since the 1920s). This was corroborated when the economic crash of 1995 was publicly blamed by international investors on the provision of false economic information by the Mexican government. This distrust was combined with the popular stereotypes of Latin American governments as inherently corrupt, an image carefully nurtured over the last 10 years during the war on drugs, in the US media to explain the overall lack of success of this policy. As such, not only was there a construction of alternative information in the Internet, but this information could also claim to be true, i.e. free of government censorship. Through the new medium, and the fact that reports came directly from eyewitnesses and scholars, 'truer' information could be provided. Years of press abuse and manipulation, and the effect of the Salinas simulacra had to be paid for by ceding the media territory to the Zapatistas, who were guided by a clever self-representation that circulated rapidly around the world.

A last word

I have so far stayed away from a discussion of the nature of the Internet, trying to demonstrate that the net is not inherently good or evil. More than anything, it is a site of struggle and can be used by social movements for their purposes as much as it can be used to transfer money, read newspapers, sell products or look at pornography. In and by itself, the Internet is not useful. Its utility lies in its interconnection with activities in other spaces, in its ability to facilitate different and further reaching social relations.

An important effect of the net has been to enable the Zapatistas to sustain a struggle against the odds by challenging the containment efforts of the Mexican government. The new political territory configured by the Internet provided an environment in which multiple national and international actors sympathetic to the Zapatistas could be drawn into the conflict in order to pressure the Mexican government and influence its actions. This possibility had been opened through the integration of information flows necessary for the movement of international capital. For the Zapatistas, this has meant the possibility of broadening their scope, at the price that the direct issues that led to the uprising have been displaced first by the figure of Marcos, and then by a widening of political concerns, to an intercontinental struggle against neo-liberalism (Froehling 1997).

The extension of the Chiapas struggle into the Internet thus resulted in new alliances, an increased scale of the conflict drawing in new actors, and thereby a new way to influence the events in Chiapas as well. Without this objective of forging connections to influence certain events, without the explicit goal to use the space of the Internet to act outside the Internet, all progressive talk on the Internet simply remains a virtual revolution. The reification of this new space as in itself liberatory is only a new exercise in fetishisation that neglects the fact that the Internet, like all technology, is a social relation that is constituted through other social relations in different spaces. In Chiapas, the success of the Internet should therefore not be measured in the number of web-sites or discussion lists, but in the multiple effects produced in other spaces outside cyberspace.

Acknowledgements

I would like to acknowledge gratefully the Institute for the Study of World Politics, Washington DC for their Research Fellowship, as well as the University of Kentucky for the Dissertation Year Fellowship, both of which made part of the research and writing possible.

Notes

1 All translations of Spanish sources by the author, unless otherwise indicated.
2 For an excellent example of this type of site, see ACTLAB at http://www.actlab.utexas.edu/~Zapatistas/index.html
3 Gurría is the Mexican Secretary of Foreign Relations, who had made the statement about Chiapas being a mere 'war of ink and Internet'.
4 The reform of Article 27 of the Mexican constitution in 1992 essentially ended land reform and removed protection from communal land holdings, thereby

opening up the possibility of the destruction of the land base of many peasant communities

5 Some examples of URLs of Zapatista sites are:
List of Zapatista sites:
http://www.eco.utexas.edu/faculty/Cleaver/zapsincyber.html
Zapatista Homepage: http://www.peak.org.~justin/ezln/
La Jornada: http://serpiente. dgsca.unam.mx/jornada/index.html
Index of Chiapas95: http: //www.eco.utexas.edu: 80/Homepages/Faculty/ Cleaver/ Chiapas95.html
and Homepage of the Intercontinental Encounter:
http://www.physics.mcgill.ca/~oscarh/EncuentroLaRealidad.9604/

6 An Index of Zapatista websites and materials on the net can be found at: http://www.eco.utexas.edu/faculty/Cleaver/zapsincyber.html

7 See Froehling 1997 for further elaboration of this argument.

8 For an image of Marcos as a cyberpunk see http://www.eco.utexas.edu/faculty/ Cleaver/zapsincyber.html

9 The EPR's appearance first confused observers, some of whom took them for a government organised scheme to discredit the Zapatistas. This theory was dropped however when the EPR, in a concerted action, in one night attacked military and police installations in five states, killing a number of military personnel and soldiers. The relationship between the EPR and the EZLN is rather strained. Its web-site adheres very much to traditional guerrilla politics and therefore is rather boring, at http://www.xs4all.nl/~insurg/

10 An excellent collection of Zapatista communiqués interviews can be found in *¡Zapatistas!*

11 A special web site on Zapatista women can be found at http://www.eco. utexas.edu:80/Homepages/Faculty/Cleaver/begin.html

11 Gender and the landscapes of computing in an Internet café

Nina Wakeford

Introduction

Increasingly cultural criticism is questioning essentialist perspectives on technology which link it inescapably with masculinity, men's activities, and the absence of female participation (Stabile 1994; Ormrod 1995; Grint and Woolgar 1995; Terry and Calvert 1997). Indeed the categories of 'gender' and 'technology' may not be as discrete as is suggested by a formula in which gender is superimposed on technology, or vice versa. Feminist theorists of technology are moving away from approaches characterised by 'explorations of how pre-existing social relations of patriarchy express and shape technology' (Ormrod 1995: 31). Calvert and Terry begin their recent volume *Processed Lives: Gender and Technology in Everyday Life* (1997) with quotations from two feminist theorists who complicate the separation of the two terms gender and technology. First, for Haraway, a machine is not an 'it' to be animated, but '[it] is us, our processes, an aspect of our embodiment' (1991b: 180). Second, for de Lauretis, gender itself, amongst other factors, 'is the product and process of various social technologies' (1987: 2). In this new positioning, technologies and genders are mutually constituted, and cannot but be touched by other factors in our embodiment and social practices such as sexuality and race, which until recently have been absent in the social studies of technology literature. 'To do otherwise is to reify gender as binarism and technology as "thing"' warns Ormrod (1995). She proposes that the task for feminist sociology is to articulate the imbricated relationship between gender and technology. She comments:

> feminist sociology on technology must be able to show how relations of power are exercised and the *processes* by which gendered subjectivities are achieved. It must therefore attend to the range of discursive practices and the associations of (durable) materials, meanings and subjectivities with which gender and technology are defined and differentiated.
>
> (1995: 44)

It follows that the definitions and differentiation of gender and technology cannot be fully resolved before any subsequent programme of research takes place, and neither will such research enable concrete demarcations which can be projected, unchanged, into the future. Shifting the focus from technology in general to the specifics of the Internet, Ormrod's suggestions seem particularly helpful, not only for sociology but also for allied disciplines engaged in the critical study of 'technocultures'. The Internet has been presented as the exemplary technology of the future and of postmodern subjectivities (e.g. Kroker and Weinstein 1994; Dery 1996). Yet despite claims that it is a gender-neutral space, it has enabled alliances of materials and meanings which have contradictory gendered subjectivities. I shall discuss these contradictions through the representation of the Internet as a set of intersecting landscapes of computing in which gender is produced, represented and consumed.[1]

Rationale for studying a 'real' place

The early research on gender and Internet cultures was stimulated by the claim that gender and other aspects of social identity might become irrelevant in the new worlds created by information and communication technologies. This belief was built on the premise that computer networks allowed users to be physically invisible to other users. The most transformative visions were offered by futurists and utopian thinkers (Stone 1991; Rheingold 1993; Dery 1996; Turkle 1995). Particularly amongst communities of practice inspired by Science Fiction, the deeply entrenched inequalities associated with certain social identities were treated as if they might disappear as the significance of the 'real' body diminished. Amongst virtual world programmers this was frequently translated as a more widespread fantasy of transcending the body (Stone 1991: 113).[2] In the new electronic networks which were conceptualised as *spaces* (rather than mere conduits), identities, freed from the restrictions of embodiment, were presented as malleable. Surveying the accounts of cyberspace, Stallabrass concludes:

> The greatest freedom cyberspace promises is that of recasting the self: from static beings, bound by the body and betrayed by appearances, Net surfers may reconstruct themselves in a multiplicity of dazzling roles, changing from moment to moment according to whim.
>
> (Stallabrass 1995: 15)

Even in the early literature the ideas of multiple gender roles and 'gender swapping' (Bruckman 1996) were used to exemplify this promised escape from body and appearances. Moreover the declining significance of the body was associated with social change beyond electronic environments, a theme of many manifestos of the 'digital age' (Winner 1996). Amy Bruckman, a computer scientist and creator of on-line MUD environments reports:

> Gender swapping is one example of how the Internet has the potential to change not just work practice but also culture and values.
>
> (Bruckman 1996: 318)

Despite these aspirations, the subsequent research suggests that a utopian vision of gender-free or even gender-equal electronic space is far from being realised. Early media reports highlighted deception leading to 'rape online' (Dibbell 1996) or 'computer cross-dressing' (Stone 1991). Researchers from a variety of disciplines concluded that even where a multiplicity of roles is possible, traditional images and experiences of gender persist in most Internet fora (Herring 1996; Kendall 1996, forthcoming; Reid 1996). Some women have created pockets of resistance on the Web or in discussion lists (Wakeford 1995, 1997a, forthcoming), but overall the territories of the on-line world reflect the unreconstructed ideologies of the population of 'white male cyberboors' (Winner 1996).

Studying an Internet café builds upon the existing research on gender on-line by exploring how gender operates in a 'real' place where the Internet is both produced and consumed. Observations and interviews during my fieldwork at 'NetCafé' suggested an approach which borrows metaphors of spatiality from cultural geography to explain gender in terms of its production as part of *landscapes of computing*. Landscapes of computing are defined as the overlapping set of material and imaginary geographies which include, but are not restricted to, on-line experiences. The choice of an Internet café was influenced by its role in relation to the Internet and computing more widely. The café is a *translation landscape of computing* where the Internet is produced and interpreted for 'ordinary people' who consume time on the machines, and/or food and drink. My initial question was 'How do the staff achieve this production and interpretation?' Additionally there was the possibility that NetCafé might perform other translations alongside those which were explicitly intended. The business goal of management is to sell the Internet experience – using a computer to access the Internet at a café – as a product. However NetCafé was also chosen because it explicitly foregrounded gender.

Integral to NetCafé's translation landscape of computing was the production, mediation and consumption of gender as a component of the product. NetCafé was the product of the collaboration of two women and an Internet Service Provider (ISP) run by their male partners. The official history was that the two female directors thought up the idea when they first met. That Sunday morning they sketched the idea of the 'world's first cybercafé' on a napkin at the kitchen table. The café opened its doors six weeks later with the aim of encouraging those who did not traditionally use computers, particularly women, to access the Internet. Customers were encouraged, using imagery and discourse which suggested that NetCafé was not the home of the white male nerd, to see the Internet as a place where women could participate. Less clearly visible were the ways in which gender was embedded in the daily production of the café environment. It was these complex and shifting quotidian practices which were investigated.

The staff, customers and machines who inhabited NetCafé encountered material and imaginary landscapes of computing culture beyond that of translation. First most participate in *on-line landscapes*, both visual and textual,

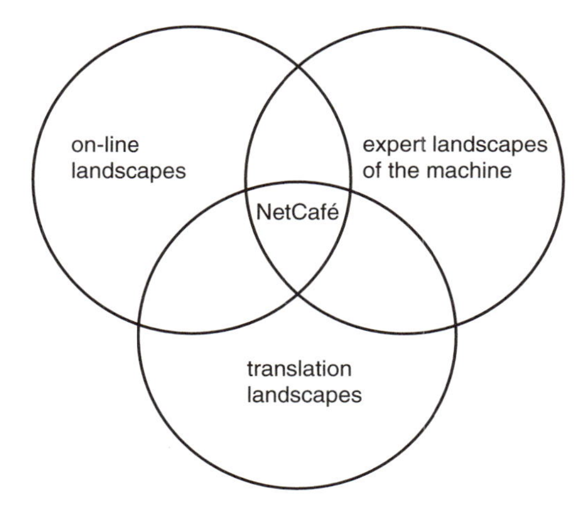

Figure 11.1 NetCafé's location within three landscapes of computing

which may or may not involve interaction with others. These are the spaces which are frequently described as 'cyberspace' or 'the virtual'. Second, they encounter *specialist landscapes of the machine* presented to them by those who set up the network, ensure compatibility of hardware and software configurations, and are called on when the other landscapes of computing are in jeopardy (i.e. the computers crash). NetCafé is a site where these three landscapes – translation, on-line, specialist – converge, as shown in Figure 11.1. Customers can log on to a machine, access a textual or graphical environment, participate in computer mediated communication and ask for help (and expect explanation as part of the 'product') in a place where expert knowledge is emphasised. NetCafé is also a place of employment, where many of the characteristic features of the workplace such as surveillance are evident. NetCafé employees, customers and machines also engage in 'workplace geographies of display' (Crang 1994) as part of the ongoing creation of the landscapes of computing, and the construction of gender. Although I suggest here that distinguishing landscapes of computing is useful analytically, the typology is not exhaustive and none of these landscapes is completely coherent, stable or entirely discrete from the others. Indeed the points at which they intersect and clash may provide the most fruitful sites to study.[3]

Gender and on-line landscapes

The themes highlighted in current research about gender and on-line interactions situate the practices observed in NetCafé.[4] Interactions both exist within, and help to define, on-line landscapes: the spaces of e-mail fora, discussion groups, real-time textual exchange in MUDs and IRC, the Web, and 2D–3D environments. Little of the existing work addresses the tangible situated spaces of computing and these on-line landscapes. So far most research has focused on the discursive and symbolic peculiarities of Internet culture.[5]

The metaphors used to characterise on-line landscapes at their most abstract are drawn from discourses which are implicated in particular constructions of gender. In the USA, terms such as 'frontier' have been used to describe on-line space as a new (electronic) Wild West. Critics such as Miller speculate that this imagery appeals to ideals of 'individualistic masculinity' (Miller 1995). Miller also worries that the characterisation of 'frontier' may encourage a particular set of fearful responses by women to these networks. Also attentive to the discourses of gender, Stallabrass suggests that 'the urban boy's idea of the street' is an integral part of the literary cyberpunk landscape, the influence of which is significant for many builders of on-line technologies (Stallabrass 1995: 6; see also Dery 1996).

Gender swapping

Shifting the focus from these grand narratives of cyberspace to the constituent on-line landscapes and their spaces of practice, it is clear that each has its own distinctive architecture and conventions of behaviour (Wakeford 1995). Furthermore these landscapes are stratified by cost, availability of access, system compatibility and assumptions about technical skill. In terms of the discussion of gender, those which stand out are the real-time interactive textual spaces of MUDs.[6] More than any other on-line landscape MUDs are portrayed as being transgressive through gender performances. Seemingly the use of MUDs changes the way we can 'experience' gender:

> For the participants, MUDing throws issues of the impact of gender on human relations into high relief. Fundamental to its impact is the fact that it allows people to experience rather than merely observe what it feels like to be the opposite gender or to have no gender at all.
>
> (Bruckman 1996: 322)

How does this experience of gender happen? In this textual environment, on-line users ('player' or 'character'), are frequently requested to provide a description of themselves as well as (nick)names. Part of this description is a 'gender flag' (Reid 1996). This flag may control the set of pronouns which are used by the MUD program to refer to the character in subsequent interactions. Kendall calculated in 1994 that 21 per cent of 8,541 characters in GammaMOO were designated as female (Kendall 1996). However, she also describes the wide range of vocabulary and pronouns which have been generated outside the traditional – the usual male/female, he/she. In Gamma-MOO these include: '"either" which uses the s/he and him/her convention; "splat", similarly, uses *e and h*; and "plural" uses they and them' (Kendall 1996: 217). Some writers insist that these new languages can displace traditional gender categories altogether. McRae, writing of sexuality in MOOs, comments:

> The spivak gender available on MOOs, for instance, has a unique set of pronouns: *e, em, eir, eirs, eirself*. It has encouraged some people to invent entirely new bodies and eroticize them in ways that render categories of female or male meaningless.
>
> (McRae 1996: 257)[7]

Even taking on board the difficulty of matching these gender designations with material/biosocial bodies who choose them, many of the accounts suggest that men are more likely than women to experiment with alternate gender roles (Kendall 1996; Turkle 1995). Often particular constructions of race, gender, ethnicity and body are combined when extended descriptions of several text lines or more are permitted. For example one user describes their character as 'a beautiful, female, felinoid . . . Her Asian features are accented by her bright blue eyes' (Puss-n-Boots character in Kendall 1996: 214).

It seems from the existing accounts that any displacement of gender is happening through the combination of traditional or stereotypical tropes, and through mechanisms which strenuously resist the mapping of MUD-gender to the experience of material/biosocial bodies. This is clearest in the ways in which other aspects of MUD personae are articulated:

> Choices of race are more likely to be between Dwarvish, Elvish and Klingon than between Asian, Black and Caucasian; choices of class are more likely to be between Warrior, Magician and Thief than between white or blue collar.
>
> (Reid 1996: 331)

These options for 'race' or 'class', located within subcultural knowledge of gaming and fandom, displace the discussion of traditional classifications and the vocabulary of social movements. Researchers have indicated the racial and class location of the majority of users is 'affluent and white' (Reid 1996; see also Kendall 1996; Turkle 1995). Gender, race and socioeconomic position are all muffled by a set of MUD-specific cultural practices which discourage exporting classifications 'back' to other lived experience in any simplistic way. With gender this becomes complicated as the linguistic displacement is continually undercut by an insistence that the 'authentic' gendered self, rather than other aspects of social identity, be revealed (Kendall forthcoming). Nevertheless the way in which a user is accountable to a revealed 'real' gender rarely allows for a discussion of the complex social experience of a gendered material/biosocial body.

MUD cultures of gender are localised, and often gender conventions are unique to a specific MUD. In environments where long-term relationships of social support have evolved there is very little hidden play with gender roles on an ongoing basis, although certain users may be known to participate as characters of the opposite sex (Kendall forthcoming). However even in these cases, when a user adopts a persona of the opposite gender, the user develops a reputation not as a woman, but as a man-with-a-female name. In cases where such gender swapping is revealed without prior warning (e.g. Stone 1991), the result is not to promote a cultural shift in values about gender, but to increase construction of social risk within the electronic space.

Gender-based experiences of communication

Although from the point of view of computer code and spatial imagery MUDs are architecturally more complicated than some other on-line spaces (Curtis 1996), one feature which is shared is the reporting of sexual harassment of female users. This is one of the gendered patterns of communication in on-line landscapes which draws the widest mass media attention. MUDs have been disproportionately highlighted due to early and widely publicised incidents (not uncontroversial in the MUD communities themselves) including 'A Rape in Cyberspace' (Dibbell 1996: 375). However as Brail indicates, sexual harassment exists on a continuum of daily 'wanna fuck' e-mails from strangers to those with obviously 'female' names, to the electronic 'stalking' which Brail herself endured (Brail 1996). It is difficult to assess the extent of this intrusion or the overall significance which users attach to it. It highlights another way in which the experience of on-line gender is influenced by specific and localised on-line landscapes, in this case that space provided by the ISP. Those who have university (.ac or.edu) accounts are much less likely to receive the quantity of untraceable 'wanna fuck' e-mails as those with commercial ISPs, particularly America Online (AOL) whose users in the USA report receiving up to thirty unsolicited mails per day.[8]

Gendered experiences of e-mail communications are not limited to unsolicited postings to females. The research in this area casts some of this variation as 'posting in a different voice', following the work of psychologist Carol Gilligan (Herring 1996). Focusing on the divergent styles of posting on discussion groups, mailing lists, usenet newsgroups, and also in the cultures of Frequently Asked Questions (FAQs), Herring has found that on-line utterances are male oriented and male dominated (Herring 1996).

Although there is often a day-to-day perception amongst users that men are abusive or 'flame' in on-line interactions, Herring suggests that this is an oversimplification. She identifies flaming as the most extreme form of an 'adversarial style' which can also include a superior stance, posting long and/or frequent messages and participating disproportionately in a discussion. This style is predominantly employed by male users. Herring found another style, which she calls 'attenuated/supportive'. She concludes 'This style is exhibited almost exclusively by women and is the discursive norm in many women-only and women-centred lists' (Herring 1996: 119). Herring maps this discovery on to the contrasting ethics and moral norms of men and women on most lists and newsgroups. She cites the response to her survey which was distributed on-line. Men were more likely than women to flame Herring about the questionnaire itself. In their responses to the questionnaire men exhibited three values which were almost entirely absent from women's responses: freedom from censorship, candour, and debate. An orientation of 'freedom from censorship' celebrates an anarchistic value system, where flaming is a kind of corrective justice. Candour values honest and frank expression of any opinions. To value debate means 'confrontational exchanges should be encouraged as a means of arriving at deeper understandings of issues and sharpening one's intellectual skills' (ibid.: 129) Together these three factors constitute the logic through which some users infringe the values of politeness which might otherwise prevail. Herring's study suggests that men seem to distinguish between good adversariality ('debate') and bad adversariality ('flaming'), whereas women do not make such a distinction treating all such adversarial encounters as hostile.[9]

Other researchers have found that the participation of men and women varies *between* discussion lists and newsgroups, even when both on-line landscapes feature the same subject. Clerc has studied forms of media fandom on the Internet, and found that amongst fans discussing the same television programme electronically, women were more likely to participate in discussion groups than usenet newsgroups (Clerc 1996). According to her survey, 42 per cent of female respondents posted at least once a week to lists compared to only 22 per cent of the male respondents. In newsgroups, the figures were 28 per cent for women and 37 per cent for men. Clerc concludes that the specific on-line landscape is the crucial factor in women's participation. The difference is clearly the format and not the [television] series: for example, although women are a very strong presence on STREK-L[the Star Trek discussion list], they are less than a third of the posters on the newsgroups rec.arts.startrek. current and rec.arts.startrek.misc. This is also true of the Dr Who newsgroup and list (Clerc 1996: 83).

Although her study is of a specific set of fandom fora, Clerc's is one of the few studies which also looks at parallel cultural practices off-line. Women are more likely to participate in traditional forms of fandom (print fanzines, conventions) than in any of the spaces available on-line, even where they have the opportunity to contribute electronically. Clerc's work demonstrates the risks in assuming that on-line cultures can be mapped (back) elsewhere in any straightforward way, and highlights the way gender operates within on-line landscapes to transform the nature of fandom which occurs there.

Cyberfeminism

Another strand in writing about gender and on-line landscapes has stressed the new types of feminist activism and possibilities of resistance to male power available through these new technologies. Cyberfeminism, as advocated by

Sadie Plant (1996, 1997), celebrates women's connections with machines. In her words cyberfeminism is:

> an insurrection on the part of the goods and material of the patriarchal world, a dispersed, distributed emergence composed of links between women, women and computers, computers and communication links, connections and connectionist nets.
>
> (1996: 182)

In this view the Internet is a new space within which gender does not disappear but can reinstate itself to the advantage of women. Situating her claim in the work of Irigaray, and making extravagant claims of technical capacity, Plant argues that by their very nature virtual worlds 'undermine both the worldview and the material reality of two thousand years of patriarchal control' (1996: 170).

Plant's prophecy is that the networks of the digital 'matrix' promote nothing less than a new sexual revolution (1997), yet she does not specify the on-line landscapes which are involved in this social movement. Although it is widely acknowledged that Plant has been the key figure bringing women and cybernetic futures to public attention, particularly in the United Kingdom, her lack of attention to the experiences of specific women users has been criticised by feminist computer scientists (Adam 1997). From a political point of view Squires warns that cyberfeminism may not serve materialist feminism if it perpetuates 'technophoric cyberdrool', and cautions:

> we would do well to insist that cyberfeminism be seen as a metaphor for addressing the inter-relation between technology and the body, not as a means of using the former to transcend the latter.
>
> (Squires 1996: 195)

Although cyberfeminism has been taken up by some women in on-line landscapes such as Web publishing, the most common overt resistance to the relentless practices of white masculinity is local and specific, such as the Australian groups VNX Matrix and geekgrrl (Plant 1996; Wakeford 1997b; VNX Matrix 1998). Their rebellious strategies subvert the conventions of the technical systems, such as confusing search engine users who are looking for 'babes on the Web' by using 'grrl' instead of girl (Wakeford 1997b). Elsewhere women have created 'safe space' in the form of private lists on the basis of interest or identity (Wakeford 1997a, 1997b, 1998; Wincapaw forthcoming). Hall has suggested that different forms of cyberfeminism coexist on one women-only discussion list although her version of cyberfeminism is locally specific and far more modest than that of Plant (Hall 1996). Clerc's study of fandom uncovered the 'Star Fleet Ladies Auxiliary and Embroidery/Baking Society', a private discussion list created by a woman user who expressed dissatisfaction with the women's roles on Star Trek and was harshly criticised for it on the public list (Clerc 1996). The use of irony is a common theme in the names assigned to private lists. Women who want to talk to 'real' women learn that they must seek chat rooms on commercial services with names such as 'sensible footwear' not 'girlchat'.[10]

It is clear that on-line landscapes cannot be characterised by a single set of conventions relating to gender. In some spaces, gender assignments are expected to be internally consistent and non-transferable. Elsewhere, particularly where access is restricted in women-only spaces, gender is part of a trusted representation which does not recognise the on-line/elsewhere distinction (Wakeford 1998). In yet other places displaying a male persona rather than a female persona will attract more attention to utterances (Herring 1996). Everywhere in on-line landscapes gender is part of a skilled practice, either in its denial or in its promotion. Confusion around gender occurs where the user is not aware of the degree of trust (or scepticism) to have towards the representations of identity which are presented there. As environments seeking to imitate 3D spaces emerge, ready-made graphical images of 'women' are produced as characters and represented by 'avatar' persona (Schroeder 1997; McDonough forthcoming). Although the range of ways in which gender can be represented appears to be increasing in these spaces, many on-line landscapes continue to be stratified by a very limited repertoire of gendered representations which circulate within them. In the following section I contrast such spaces to the 'real' spaces of NetCafé, and examine how gendered representations were created, maintained and subverted. Even within this more restricted physical space, the routines and repertoires relating to gender are presented as multiple and sometimes contradictory, although the means by which the representations are achieved at NetCafé are not primarily textual as in many on-line landscapes.

Gender and translation landscapes

The daily activities at NetCafé involve participation in on-line landscapes, but many of the practices do not happen on-line. Unlike many domestic or institutional settings of computers, Internet cafés are locations in which the explicit process of consumption of the machine includes attempts at its contextualisation and interpretation as part of the product which is purchased by a customer. In Internet cafés the product could be conceived as having several interrelated components available for consumption, including:

- the machine as an isolated computer;
- the machine as part of a local network within the café;
- the machine as part of a global network;
- the systems/technical infrastructure (e.g. speed of network connection);
- the staff and their embodied knowledge;
- the café atmosphere/ambience;
- the café decor;

- the café location; and
- the food and drink.

Each of these components is itself complex and may combine incongruous elements. The café decor, for example, indicates the ways in which the café places itself in relation to other local or national businesses and structures of finance (advertising in windows, free postcards and flyers, promotions, etc.) as well as being the partial outcome of staff claiming space to represent their own version of 'the cybervibe'.[11] While not suggesting that every actor in an Internet café will perceive or consume the same product, in general the key feature of these spaces is the combination of a series of familiar experiences (buying coffee, sitting in a café, observing norms of sociability, etc.) with the often more unfamiliar encounters with the computer.

Spatial organisation of NetCafé encounters

During four months of 1996 I worked at NetCafé, an Internet café in Central London.[12] In this section I describe the basic layout of the café and how the interactions between machines, customers and staff were staged. The encounters which occurred at NetCafé show, in Ormrod's terms, 'how relations of power are exercised and the *processes* by which gendered subjectivities are achieved'. They also illustrate how the Internet operates as a process by associating people, machines and spaces and the relationships between them. NetCafé was in many ways a highly organised and stratified space which sought to portray itself as a place where gender would not matter.

NetCafé was organised on four levels, only one of which was the café 'floor'. As a consequence most of the café was 'offstage', out of public view. The café floor was the central focus for public attention, and the place where most of the customer-staff-machine interactions were staged. In this space the tasks of staff or 'cyberhosts' were to sell Internet time on the networked computers arranged on high tables around the edge of the café, and to serve drinks and light snacks from the service counter which faced the front door of the café. The counter was the location of the monetary transaction, and it was the spot where time on the machine was 'booked'. If customers encountered a machine or a cyberhost before having been to the counter, they were directed to it. Many of the activities which took place around the counter were comprised of the cyberhosts organising the customers into manageable and ordered sets of interactions, including indicating where they should wait if a machine was not yet available and encouraging them to purchase 'something [to eat or drink] while you wait'. Customers were not encouraged to loiter around the counter, displacing the site of the majority of the extended cyberhost–customer interactions to the sides of the café where the computers were located.

One of the first things which I noticed about the café, and which drew both customer and staff comments, was the colours of the interior furnishings. The walls were a yellow-lime green, the benches and tables were orange or lilac. Most other horizontal surfaces (the counter, the divider between the café floor and the café manager's space) were matt silver metal. The counter front was a deep crimson red, which matched the shelving units behind it, housing bottled drinks on glass shelves. Above the counter were six clocks, each telling a time in a city/timezone: New York, London, Paris, Moscow, Tokyo and Sydney.

In a windowless basement below the café floor, and accessible only by a steep set of stairs were the 'Training room' and the 'Staff room'. The former reflected the layout of the café above: computers on counters around the edge of the room. It was only used when 'group trainings' were in progress. The latter consisted of a brick alcove which one had to half crawl and half climb into. Once inside it resembled a small wine cellar since there was no natural light. This was the place which cyberhosts were expected to use when they were 'on break' from the café floor.

In the storeys above the café floor there were two sets of offices. On the first floor, a set of small rooms was clustered around a tiny reception area. These offices housed staff who organised Internet training sessions, 'Events' (corporate product launches which rented the café space, with or without training sessions), 'Public Relations' and some technical support staff associated with a Webcasting experiment. By contrast most of the technical staff (networks and systems) worked on the top floor of the building, which for a time also provided space for an Internet (print) magazine, and an ISP. The ISP expanded in the middle of the fieldwork period, relocating the 'management' component of its staff to offices 200 yards down the street. For most of my time there, the top floor was devoted to the ISP technical support hotline and 'systems', those who provided the technical backup for the café and the ISP. Between all these levels of the building there was one stairway, off which there were toilets. Above the top floor office was a roof, which staff used as a place to get away from the other levels. During the summer months, it was used during breaks far more than the official staff room below. As a result, on their way up and down the stairs, cyberhosts often saw and interacted with other levels of the café workings which they would not have otherwise encountered if they had travelled only between the café floor and the basement.

The day-to-day routines of the café floor centred around the hours of opening to the public. Most days the café began business at 11am, although office staff on other levels and the cleaner were often in the building by 8.30 am. Occasionally corporate 'events' meant the café had to be up and running at 9 am. The cyberhosts and the technical support staff worked overlapping shifts, either 'early' or 'late', since the café closed at 10 pm. Many of the other staff worked from 9–10 am until at least 8 pm. Trying to see as many of the various activities as possible, I typically spent between 10 to 14 hours at the café, on six or seven days a week. It was not unusual for staff, particularly cyberhosts, to also spend their after-shift time at the café, mostly on-line at an unoccupied machine. The hours which I spent at NetCafé were not out of the range of a regular employee. I divided my time between the floor, where I 'helped out' with customers, and the other levels of the café, where I wrote up my notes on my portable computer (less intrusive than pen and paper in this setting) while chatting to other staff. During certain hours of the day, and particularly later on in my stay when the café manager's office was remodelled, I occupied a desk on the first floor of the building, in the same room as one of the directors and the training team. I spent blocks of time sitting upstairs on the top floor with the technical support staff, and listening to the operation of the helpline. I also conducted taped interviews with the director, cyberhosts and technical staff about their impressions of NetCafé and their other Internet and computing experiences.

The typical manner in which encounters between customers, cyberhosts and the machines took place in NetCafé, as noted from my early observations, was as follows:

> A customer enters the café by the only door, walks up to the counter where they are booked on to a machine. They pay for their machine time in 30 minute chunks for £2.50 (discounted for students). They are directed to a machine (numbered) or to a table to wait for a computer to become free. Once on the machine, they may request the assistance of a cyberhost, who, when not serving customers, hovers around the machines or the counter. All cyberhosts wear t-shirts bearing the café name and that is how they are identified. A dress code, which is not rigidly enforced, also requires dark trousers or black jeans. If they are called upon to do so, cyberhosts are encouraged to help customers with any problems while on the computers, but 'introductions to the Internet' (known as an 'intro') for novices are officially limited to five minutes. The machine is introduced to the customer either by indicating in its direction from the counter, or, if an 'intro' is going to be offered, the cyberhost accompanies the customer to their seat, and stands alongside the customer while explaining the basic procedures of e-mail, newsgroups, the Web and real-time chat. The end of the purchased time on the computer is signalled by the cyberhosts who call out first names in the general direction of the machine to which the customer has been assigned. Often because machines crash or because a customer requests a different location, the user is not at the original machine. Calling out to the café floor enables staff to track this movement, which is particularly important if the customer wants to buy more time. Most customers depart soon after their Internet access time is finished.

Customer profiles

In one of my last weeks at NetCafé I conducted a survey of customers over a 7-day period, collecting 694 questionnaires. From these questionnaires the following general profile of NetCafé customers was generated. Although in most spaces inhabited by numerous computers available for public use (such as computer labs, Turkle 1984) men overwhelmingly outnumber women, in NetCafé they made up just under half of the customer respondents. Only half of the respondents were in full- or part-time employment, a figure which was largely explained by the fact that one-third of the total customer base evident from the survey were students. The age profile of the respondents also reflected the student population: 41 per cent were in the age group 18–24 years, a further 37 per cent were in the group 25–34 years. The customers had a high level of education, over half at the time of the survey already held some kind of university qualification. Although it appeared to be a local population – two-thirds reported that they lived in London – in reality many of these were short term visitors to the capital, and the largest identifiable group in this situation were American students on summer programmes in the United Kingdom.

In terms of activities at the café, half of all customers had come into the café for the first time and one in ten were using the Internet for the first time. Over half of returning customers are 'regulars' which I defined as coming into the café at least once a week. At the extreme of this spectrum of use, 3 per cent of respondents came into the café more than once a day. For the most part, whether they had used it or not, three-quarters of the respondents had access to the Internet elsewhere, and roughly one in five had web-pages either for themselves or related to their employment or educational institution. The survey illustrated an interesting separation of NetCafé from its on-line presence. The great majority had heard about NetCafé through word of mouth or through traditional print media, rather than through its web-pages or Web broadcasting presence.

'It's Showtime' (again): gender and display at NetCafé

The Internet is frequently portrayed as a globally networked assemblage of digital units: text, images and sound. NetCafé produced the Internet as a local *as well as* a global phenomenon, a non-digital as well as a digital place. It exhibited a strong sense of physical presence through the articulation of the café floor as a space where the Internet happened, the careful maintenance of a distinctive 'cyber' interior and exterior design, and the management of staff roles as cyberhosts. NetCafé also purposefully drew upon and reworked components of London subcultures, particularly distinctive nightclub music and fashion.

In this section I will illustrate this practice of creating the Internet through the concepts developed by Crang in his work on the performative nature of restaurant employment. Although restaurants represent a somewhat different set of social norms than do cafés, Crang's emphasis on the dimension of display in restaurant work resonates powerfully with the way in which NetCafé operated to create its product. Gender was integrated into this product through a relentless (although not uniformly intentional or overt) presentation of the machines in association with gender and gendered spaces. Sometimes gender appeared to be tied closely to the traditional segregation of routine duties which might have taken place in any service-oriented workplace, rather than constituting a conscious articulation between gender and technology unique to NetCafé. Nevertheless this service work was integral to the translation landscape of the Internet café. Ultimately it cannot be isolated from the production of gender and networks of machines.

Crang's article, 'It's Showtime: on the workplace geographies of display in a restaurant in Southeast England', reports on an extended period of participant observation at 'Smoky Joes', one of a chain of restaurants which emphasised dining as an event, or even a theatrical experience. The phrase which captures this appeal to performance, 'It's Showtime', appeared in the staff handbook which detailed the employees' roles and responsibilities. Crang points out that the restaurant was 'marked by the co-presence of production and a particular form of consumption' (p. 696). At NetCafé the workplace geographies of display also demonstrated this co-presence. The café not only produced a particular kind of consumer/producer experience relating to computers, but also one about gender.

In his analysis Crang suggests that the workplace geographies of display aimed at restaurant diners operated through six mutually determining axes of 'sociospatial relations of consumption'. At NetCafé the processual logic of the translation landscape could also be characterised as operating through sociospacial relations of consumption and (re)production. Each of Crang's 'sociospacial relations' can be found in NetCafé's environs: imaginations of interactional settings; spatial structures of interactions settings; forms of communication; ethoses of the product; organisation and authority relations; identity politics. Each set of these relations can be used to illustrate the way in which gender was produced, represented and consumed in this site.

Imaginations of interactional settings

NetCafé existed not only as a physical place where encounters with the Internet came about, but also as a result of the imaginative geographies in the minds of managers, staff and customers. Although there was a strong vision from the directors that the café was a place where gender did not determine the kind of product which was consumed, an 'equal access' mission, during many of the activities on the café floor the practices of 'doing gender' alongside 'doing technology' were complex. Imaginations of interactional settings, as Crang points out, are important not just for the geographical metaphors which they draw upon and generate, but also for the part they play in regulating social practices.

The space which management and staff tried to construct as a place where women would be encouraged to envisage a connection to the machines tended to be subverted by the customers themselves, and to a lesser extent by those managing the cyberhosts. Customers tended to re-inscribe specific roles as they encountered cyberhosts. The latter often commented upon the way in which, regardless of the gender of the customer, the female cyberhosts would be expected to serve the coffee, and the male cyberhosts to help with the machines. In several of the interviews cyberhosts commented upon this as a practice which interrupted the process of running the café: all floor staff were theoretically interchangeable, and there was no division of labour based on specialist coffee making or computer skills. One expressed frustration that customers didn't understand that 'they don't get their own cyberhost'. Neither were they expected to select a cyberhost based on gender (or on any other basis). During my participation there, the female cyberhosts had on average a longer employment at NetCafé than their male counterparts, and had greater expertise in solving complex technical problems. The irony of customers trying to make alliances between the male cyberhosts and the machines was not lost on those who had to translate the machines on a day-to-day basis.

Amongst themselves the cyberhosts talked about the unfairness of the gendered treatment, but it was difficult to reverse given the norms of polite service deemed appropriate in the café. Occasionally, the male cyberhosts would deliberately appear to be unavailable when the customer signalled for help, therefore forcing an interaction with a female cyberhost. One day a male customer asked a female cyberhost, alone at the counter, a technical question about the machine's capabilities. As she was answering a male cyberhost arrived to check the machine bookings, and the customer repeated the question word for word at the new arrival. The male cyberhost firmly pointed out that his question had already been answered by his colleague. However this kind of action was seen to be risky and was not the norm amongst the staff.

Spatial structures of interactional settings

The spatial structures of NetCafé were intertwined with the ways in which it worked as a landscape of translation and a place for encounters. Although the café floor was the site of the public encounters, the business as a whole was made possible by interactions and collaborations by staff on all the other levels of the building. Looking at NetCafé in its entirety, the gendered division of labour in producing the landscape was apparent to a greater extent than if I had concentrated my observations on the café floor alone. In this way the spatial structure of the building displayed the different kinds of employment which occurred on each level. The stairway between the floors became a main site of interaction for staff and in particular a way in which the products of the translation landscape were negotiated and refined out of the public gaze.

These spatial arrangements also interrupted the attempts to promote gender neutrality around the machines. This was highlighted in the contrast between the café floor and first floor areas, and the top floor of the building where technical support and system administrators were located. During the middle of my fieldwork one of the female cyberhosts bumped into me as I descended the stairs from a period spent on the top floor. She wrinkled her nose in mock disgust and told me that she avoided going upstairs because 'it's like a men's urinal'. The top floor area was indeed mostly populated by men. In the technical support area, where all but one employee were male, each member of staff sat at a section of a long bench next to a computer and phone. Around the computers lay notepads, fast food wrappers, manuals, novels (usually Science Fiction) and small objects such as plastic animals or statuettes. Under the computer benches my feet would sometimes find the discarded parts of motherboards, disc drives and tangles of cables. Even though the technical support group operated as a highly efficient on-call team for problem-solving, in general the cyberhosts associated the way in which artefacts were arranged on the top floor as 'a mess', and distanced themselves from the 'scruffy' technical support by contrasting their 'cooler' NetCafé t-shirts. However they knew that the operations of the café floor were dependent on the willingness of the top floor technical support to react immediately if there was a system failure, and so generally kept their views to themselves. Nevertheless masculinity was represented on the top floor as both 'a mess' and technically superior.

Forms of communication

The forms of communication by which the NetCafé produced encounters also crucially affected the ways in which the products were consumed, as Crang suggests for Smoky Joes. The range of ways in which communication could take place in NetCafé included the wide variety of forms of face-to-face contact between staff, customers and machines, as well as the intersection with on-line landscapes in which communication could take the forms of textual exchange and, for a short time during my stay, a live video of the café broadcast over the Web.

The public relations (PR) team played the most significant role in the presentation of the café as a local and global phenomenon, and as such constituted a form of communication influencing the sociospatial relations of consumption. PR were extremely successful at constantly reworking the history of the café in media presentations, particularly through images of its two female founders. The two women were photographed in glossy women's magazines sporting 'cyber' clothes, and featured in several general audience computer and computing culture magazines. One of the women commented to me that through such publicity they were making the Internet photogenic. Many of the publicity shots in the volumes of PR records displayed how easily the media had manage to inscribe a feminised NetCafé as the 'social' side of computing by portraying the women holding coffee cups near computers. In contrast the male technical directors tended to have been photographed in physical contact with the machine, usually touching the mouse. PR also organised several events in which the cybervibe of NetCafé was embodied within an artefact: a sculpture of red lips holding between them a computer disc. This selection of part of a body (lips) made female (bright red gloss) and linked with a bodypart of a machine (the disc) was used as a trophy for awards given out by the café and as a logo in the interior decoration.

Ethoses of the product

Just as the imaginations of interactional settings were not always shared amongst all participants of NetCafé, the way in which the product was defined was itself an outcome of struggles between different representations of objects, people and practices. As Crang reminds us, in any encounter where a product is being sold or exchanged there are 'politically contested definitions of the product being provided' (1994: 697).

At least in the eyes of the directors the product was a mediated experience with a computer and/or nourishment where the *absence* of gender as a marker was a constituent part. Yet gender was also used as a resource for PR who produced images of women and machines as a hook for publicity. As suggested above (pp. 192–3) the directors' objective was also complicated by the day-to-day practices of 'doing gender' resulting from customers' imaginings. Yet part of the reason that most cyberhosts had applied for employment at NetCafé was that it did *not* present itself as a traditional café or a computer 'nerd' hangout. Rather it was a cultural venue with relatively greater status than a common café, and which produced a computing 'cybervibe' which included imagining oneself as associated with a key venue in which Internet and multimedia developments would take place regardless of gender. The cyberhosts saw themselves as producers and guardians of this 'cybervibe' through their bodily appearance, through the soundscapes which they facilitated through the music system and through their local knowledge of other London 'goings-on', in particular the 'techno' and 'drum 'n bass' music scenes. In this way definitions of the product embedded assumptions about, and experiences of, local London genders as much as they did about technology.

Organisational and authority relations

Many of the authority relations in NetCafé could be mapped directly on to the spatial structures of the building. Directors and middle ranking staff were physically 'above' those under their supervision. If a cyberhost had 'gone upstairs' it usually meant to see someone higher up the ladder of pay, status and power.

Although it was possible to enter the upper floor offices via a side door to the building, most staff whose work directly impinged on the day-to-day running of the café would reach the staircase by walking through the café itself. Even when the café was relatively empty of paying customers this gave the impression that the location was bustling, as staff paused for a chat with colleagues or collected a cappuccino to transport upstairs. On the other hand the technical staff and many of those working on the top floor preferred the side entrance to the building, so avoiding seeing (or being seen) by those interacting on the café floor. As the top floor staff were predominantly male, the flow of people overrepresented the number of female staff in the building and circulating around the café. In this way the organisational and authority relations became part of the geographies of display and the sociospatial relations of consumption as they rendered visible or hid the gendered practices (or the imaginings of those practices) which constituted NetCafé. One of the aspects of female work which was hidden was the early morning cleaning. The cleaner, the only employee not to have an e-mail address, played a crucial part in invisibly translating the machines into a workable form when food and drink had been spilled. She would also report on the state of relative disarray on the various levels of the building and would relay messages and sightings of staff who were needed on other floors.

Identity politics

As is clear in the preceding sections, the interactions at NetCafé are bound up with the identities of the participants. The machines developed individual reputations for working or being 'sick'/'difficult'. The cyberhosts managed to create an atmosphere of diverse participation since as a group they were located in many national cultures: Mexican, Italian, French, Spanish, Cypriot, British and American. The identities of the customers were signalled when the cyberhosts called out their names at the end of their allotted time. Some customers also become known as 'regulars', altering the way in which the encounter proceeded. Regulars were unlikely to need explicit translation of the product, and were more likely to consume food or drink while 'hanging out' with cyberhosts on the café floor. Frequently regulars had considerable computer expertise and would become a capable aid for a cyberhost dealing with a difficult question from a customer. None of the identities mobilised at NetCafé were immune to such repositioning in relation to one another.

A great number of the ways in which identities figured in NetCafé interactions was through the appeal to gendered practice and discourse. Much of this was invigorated by the café staff themselves, but the way in which the category 'women' could be used strategically for profit fluctuated over time. When NetCafé first opened much was made in the press coverage and at the café itself of the women-only trainings which were offered. In the beginning these trainings were full, but a year later they had been halted, the training manager told me, 'from lack of interest'. 'Women' became represented in the bodies of the cyberhosts and in 'the lips' artefacts (see below, p. 198). The female directors continued to be asked to share their opinions on women and technology at events around the world, and special trainings were held when a women's group was tied into a cause which was of particular interest to them. Since my fieldwork period a women's night has opened once a week in NetCafé, where a well known lesbian club DJ spins tracks while the computers are available for use. The flyer for the night advertises 'Sweets and Toys for Girls'. This event was conceived and organised by one of the female cyberhosts, who noticed the lack of customers on that night of the week. It has been a huge success in bringing paying customers into NetCafé at an off-peak time, and also in re-introducing the café as a place where public events happen with a predominantly female clientele. Furthermore it has created a space at NetCafé within which lesbian sexuality is able to be articulated in connection with the London club music scene and technology, manufacturing an instance of 'cyberqueer' (Wakeford 1997a).

Reflections on bodies, gender and the landscapes of computing

At NetCafé, representations of gender appear to be achieved at least partially through the 'doing' of technology. However technology cannot be equated with the computers alone. Rather we can recall Ormrod's view that technology is constituted by both discursive practices and alliances of materials and meanings. In Netcafé, the computers function via networks of social relationships which bring together disparate participants from several levels of the building. Put another way, the technology exceeds the boundaries of the machines. It leaks into the 'cybervibe', the interactions between cyberhosts and customers, and even the names given to other products in the café ('cybersalad' for example). One of the achievements of NetCafé as a translation landscape of computing is that it enables participants to scatter discursive representations of the Internet, the 'cyber', and global computer networks upon a range of encounters and artefacts which had not previously been recognised for their alliance with the technological. The processes of enacting the translation landscape of computing are a way of doing the Internet and a way of doing gender. In this section I reconnect the findings of my fieldwork at NetCafé with the previous work on on-line landscapes by returning to the way in which the gendered body is invoked. Framing my thinking is Adam's insistence that the connection of embodiment and technological systems must be taken seriously by feminist theorists and allied critics (Adam 1998). If the question of embodiment for feminist theory rests on the role of the body in producing knowledge, and Adam agrees with many theorists that this is the case, then the task here becomes one of articulating the kinds of bodies which inhabit NetCafé and specifying the knowledges which they produce (or are restrained from producing). The turn to bodies, as I will show, in fact leads the focus back to the metaphors of spatiality with which I began my discussion of NetCafé.

One cluster of gendered bodies which produce knowledges is composed by processes which consciously intermingle the gendered body with representations of technological artefacts and discourses. At NetCafé, three instances of this commingling can be cited. First, the bodies of the cyberhosts were used to display the competing definitions of the product. Cyberhosts were stratified by management on the basis of how 'cyber' they were, and this became manifest when the most 'cyber' of the staff was chosen to be pictured adorned with new merchandise such as t-shirts bearing the NetCafé logo. In this instance 'cyber' status was conferred by the fact that the woman in question had long green dreadlocks and several unconventional visible body piercings. Inspecting the manner of using staff bodies to promote NetCafé (a practice over which they had a questionable amount of negotiating power) revealed that in all but one instance it was women who were chosen to model merchandise. This tendency was promoted partially by the popularity of the 'skinny-t' tight t-shirt for which the expected market was exclusively female, rather than the unisex traditional t-shirt. Second, the bodies of the two female directors were used to create knowledges about gender and technology which were exported to the pages of glossy magazines as indicated in the previous section. These directors' bodies came to stand for NetCafé and in a more limited way, for the Internet. The portraits attempted to portray a mixture of idealised gendered bodies with hip 'techno' clothing. However, the opportunity for these women to create their own definition of their embodied relationship with computing remained outside the fashion shoots where the clothes, make-up, location and posture were determined by the conventions of the magazine rather than their own choices. Third, in the statuette of the lips holding the floppy disc a stylised representation of gender is purposefully brought together with an object metonymic to a computer. This statuette represents a fusion of body *parts*, and so I equate it to the first two examples with some hesitation. Yet these body parts appear to generate similar knowledges of the inescapable joining of women and technology (although I never managed to ascertain if the designer had reflected upon whether the lips were eating the disc or if they were spitting it out). The fact that the statuettes were presented as trophies for women in multimedia indicates that the imagery was also intended to be intelligible outside the internal semiotic system of the café.

Returning to the existing literature on gender and on-line landscapes, it is clear that bodies also figure in the deliberations of the extent to which physical presence matters in these spaces. However, the kinds of bodies which appear in the discussions of on-line landscapes tend to be restricted to linguistic performance or representations with limited circulation beyond the spaces in which they are created. The knowledges produced by these bodies come into view through a way of 'doing' gender (predominantly via textual input) which cannot be easily equated with that found amongst the bodies in NetCafé. Within NetCafé, at least in the three examples cited above, it is difficult if not impossible to avoid connecting women's bodies to material/biosocial everyday realities of being female and the knowledges produced by experiencing these realities. As explained in the previous section, whatever version of the café mission or the cybervibe was appropriated for the cyberhosts' own use was also subject to interruption by customers who reinscribed notions of gender and skill on to some bodies but not others in the course of their interactions. Of course, all users of on-line landscapes have material/biosocial realities, unless they are 'bot' (robotic) imitations within the software, and yet in most current accounts their bodies are figured in such a way as to obscure the role of the gendered body in producing knowledges outside that of the on-line landscapes. This is clearly the case in some of the futurist predictions about escape from physical presence, but is also evident in later research which has not questioned the process by which 'virtual' gender becomes 'real' and vice versa.

Another way of explicating this point is to describe an alternative means by which bodies were productive of knowledge at NetCafé. At this site bodies also manufactured knowledge in the course of their *movement through* the physical spaces of the café as well as the on-line landscapes which could be accessed there. Bodies-in-movement produce and incorporate accounts of their journeys as they encounter durable materials and discourses in the landscape of translation. Customers walked around the café floor interacting with both machines and cyberhosts, consuming machines and food, experiencing the decor and music and hearing the history of NetCafé. In this process they generated stories of how to do gender and the Internet through mobile bodies. Cyberhosts and other staff also moved through the building, creating descriptions of the levels of the café operations in terms of gender and technological expertise, including constructing the type of masculinity on the top floor among technical support staff. The female directors' bodily transit between NetCafé and other public arenas was integrated into the meanings of doing technology at the café. When the media attention on NetCafé was at its height, having transportable bodies which represented women and computing was crucial in the way the café was able to become a profitable translation landscape of computing without having a huge advertising budget. Last, the bodies of machines (or their body parts) were carried around the building particularly between the floors on which repairs were executed. As the machines moved around, so gendered meanings were made about who would fix a broken computer and who needed to be flattered or cajoled into doing so. Sustaining an approach of exaggerated gratitude was crucial when the need for those with detailed computer networking knowledge was at its most acute: at exhibitions and off-site trainings. Machines which travelled to participate in such activities were at most risk of suspending the image of NetCafé as a translation landscape of computing by not functioning at all, or, more commonly, by 'almost' working (for example having a very slow Internet connection, or by displaying broken Web links).

In conclusion I would like to suggest that the formulation of bodies as travelling within NetCafé directs us back to the utility of spatial metaphors and to landscapes of computing as ways to focus on specific material and imaginative geographies. For the study of gender in relation to technology it seems particularly apt to follow material/biosocial bodies in order to reach an understanding of how gender might be differentiated from technology in landscapes of computing such as are apparent at NetCafé. It might also be fruitful to follow material/biosocial bodies in landscapes of computing where it is less obvious to do so, such as on-line landscapes. This approach acknowledges Ormrod's recommendation that the feminist sociology of technology should move away from approaches that isolate technology from patriarchal social relations unless one is involved in the social shaping of the other. In NetCafé, by taking seriously the range of materials and meanings which were being used, a complex interplay of gendered representations and experiences was found which cannot easily be assimilated into the old rubric of technology as inherently masculine. Rather, as NetCafé's daily activities unfolded, the Internet was translated as a place where new alliances for gender were being forged at the same time as these alliances were being interrupted by old stereotypes through which gender and technology are still often understood. These alliances and their interruptions were as dependent on the local cultures of place and space as they were on the landscapes of computing.

Acknowledgements

I would particularly like to thank Pamela Giorgi, Tom Delph-Janiurek and Sasha Roseneil for reading and commenting on earlier drafts of this paper.

Notes

1 I use the term 'landscape' to refer to a collection of spaces and representations which are constituted by, and at the same time constantly reproduce, significant features in the cultural world of computing and the Internet.

2 As Stone (1991) notes, the desire for transcendence itself is socially marked: 'Forgetting about the body is an old Cartesian trick, one that has unpleasant consequences for those bodies whose speech is silenced by the act of our forgetting – usually women and minorities' (1991: 113).

3 For example, those who build the on-line landscapes are often those inhabiting the specialist landscapes of the machine. However the ways in which the social assumptions of this population are built into software and clients are often ignored or seen as irrelevant to these programmers themselves (McDonough forthcoming).

4 Space does not permit me to outline fully here the extensive research which falls under my term 'specialist landscapes of the machine': the spaces of computer programmers and other computer scientists, systems and computer network specialists, technical support workers and others 'close' to the machine (cf. Ullman 1996). This is the subject of other written work.

5 Although the new ESRC programme on 'Virtual Society?' (http://www.esrc.ac.uk) includes projects which take location seriously, e.g. 'Social Context of

Virtual Manchester' (Dr P. Harvey, Dr S. F. Green and Dr J. Agar, University of Manchester).

6 Pavel Curtis, an early MUD developer, defines a MUD in the following terms: 'A MUD is a software program that accepts "connections" from multiple users across some kind of network (e.g. telephone lines or the Internet) and provides to each user access to a shared database of "rooms", "exits", and other objects. Each user browses and manipulates this database from "inside" one of these rooms, seeing only those objects that are in the same room and moving from room to room mostly via the exits that connect them' (1996: 347).

7 When I first encountered the 'spivak' gender, I assumed it was a reference to Gayatri Chakravorty Spivak, the feminist postcolonialist critic who wrote of 'strategic essentialism'. However McRae (1996) reports that the spivak category was invented by Michael Spivak in *The Joy of TEX: A Gourmet Guide to Typesetting with the AMS-TEX Macro Package* (American Mathematical Society 1990).

8 Based on interviews with 15 students at University of California, Berkeley as part of evaluation of Cafe MOOlano (cf. Thorne, S. and Wakeford, N. 'Participat-ability in digital interaction: a cultural view of foreign language learning through real-time MOO conferencing', unpublished paper).

9 Herring is careful to point out that she does not mean to include 'all men' in one category and 'all women' in the other. Rather she states that the extremes of each behaviour were strongly gendered (1996). Yet her position, emerging out of the discipline of linguistics, could be read as remarkably essentialist.

10 This emerged from interviews with female users of America Online.

11 The term 'cybervibe' was used by two of the female staff of NetCafé one quiet afternoon when they had time to investigate new digital camera technology installed on one of the machines. With this, they took their own photos and using a photo-manipulation software, wrote 'the cybervibe personified' over their portraits and printed them out for other staff.

12 This research was enabled by a Postdoctoral Fellowship funded by the ESRC. The project 'Women's experiences of computer-mediated communication on electronic networks' is funded by ESRC Grant H53627502195.

Part III

Thinking and writing the virtual

12 The virtual realities of technology and fiction

Reading William Gibson's cyberspace

James Kneale

Cyberspace and science fiction

The word 'cyberspace' is rapidly becoming an academic and journalistic ubiquity. The information spaces of the Internet and the World Wide Web command increasing attention from the media as we enter an era of 'cyberculture' (Dery 1992). Users of personal computers find it hard to imagine where their documents 'are' in the seemingly non-existent space accessed through their workstations (Turkle 1984). In offices and Internet cafés, urban and electronic spaces come together (Graham 1996). And yet these dataspaces bear only the slightest resemblance to cyberspace, the science fictional geography created by William Gibson in the short story 'Burning Chrome' (1986/8)[1] and developed throughout the Sprawl Trilogy of novels: *Neuromancer* (1984/93); *Count Zero* (1986/7); and *Mona Lisa Overdrive* (1988/9). I think this background is important – not because I wish to reclaim the original site for the word 'cyberspace', but to stress that the production and consumption of ideas of cyberspace take place in many very different contexts. This paper aims to explore some of the meanings given to cyberspace in one particular context: the writing and reading of Gibson's 'cyberpunk' science fiction.

Cyberpunk, a subgenre of science fiction, acquired critical and popular notoriety between the mid-1980s and early 1990s. It usually depicts a dystopian near-future world dominated by corporate capital and drastically reconfigured by new technologies: body alterations, new forms of media and above all cyberspace. Cyberpunk has been hailed as postmodern science fiction (Bukatman 1993b; McCaffery 1991; Csicsery-Ronay, Jr 1991b); as a cultural form with a privileged insight into contemporary culture (Jameson 1991); but also as 'the vanguard white male art of the age' (Csicsery-Ronay, Jr 1991a: 183; see also Ross 1991, and Gregory 1993).[2] There are many other science fictional representations of informational spaces that could be studied beyond Gibson's cyberspace, and it is often argued that these offer more interesting and progressive constructions; however, Gibson is certainly the most important cyberpunk author in terms of influence and popularity.

In fact, Allucquere Rosanne Stone places Gibson's *Neuromancer* at a crucial point in her 'virtual systems origin myth'; it 'provided . . . the imaginal public sphere and refigured discursive community that established the grounding for the possibility of a new kind of social interaction' (1991: 95). While Gibson's cyberspace may seem to be a long way from the real-world virtual spaces mentioned earlier, Stone writes that '*Neuromancer* in the time of Reagan and DARPA is a massive intertextual presence not only in other literary productions of the 1980s, but in technical publications, conference topics, hardware design, and scientific and technological discourses in the large' (p. 95).

In this chapter I discuss the writing of cyberspace, its textual form and the interpretations of several science fiction readers interviewed in-depth in groups in 1992 and 1993.[3] I argue that Gibson writes cyberspace as a 'thin' space, in which speed and movement are the key metaphors for spatialised experience. The readers I interviewed felt that Gibson's depiction seemed vague, which I explain as a common response both to thin spaces and to science fiction. They were not disoriented for long, though; one of the ways in which they made sense of this ambiguous space was to rationalise away its more unusual aspects, describing them through technological metaphors. In particular, several of the discussants developed an understanding of cyberspace through their own experiences of information technology.

In a sense, as the SF writer Marc Laidlaw points out, representations of cyberspace are themselves technologies, tools used by authors and readers to make sense of this space:

> I have no particular interest in, or understanding of, technology as such . . . All you should really ask a writer about is *writing* and its technologies: narrative styles and strategies. Happily, it is here, in a discussion of literary technique, that the virtual realities of technology and fiction can intersect.
>
> (Laidlaw 1993: 648, emphasis in original)

Like Laidlaw, I am more interested in the intersection of these writing technologies and virtual technologies than in the nature of virtual reality itself. As far as Gibson's cyberspace is concerned, I would like to argue that Gibson's use of writing technologies allows the reader to make sense of his virtual ones, and that readers have their own uses for these technologies. Before I can expand on these ideas, I need to explain how Gibson produces cyberspace.

Cyberspace: conceiving the inconceivable

> Cyberspace. A consensual hallucination experienced daily by billions of legitimate operators, in every nation, by children being taught mathematical

> concepts . . . A graphic representation of data abstracted from the banks of every computer in the human system. Unthinkable complexity. Lines of light ranged in the nonspace of the mind, clusters and constellations of data. Like city lights, receding . . .
>
> (*Neuromancer*, §3: 67)

> He was thoroughly lost, now; spatial disorientation held a peculiar horror for cowboys [*cyberspace operators*].
>
> (*Neuromancer*, §17: 249)

Cyberspace, also known as the matrix, is Gibson's virtual dataspace, in which the combined knowledge of his information society is represented as virtual objects in an infinite space, organised as a regular grid.[4] Users interface with cyberspace through their computers to perform operations upon this data. These operations, like all activities in cyberspace, are spatialised, as users move through the matrix, shift from one location to another and enter and leave databases. These spatial metaphors represent ways for Gibson, his readers and others to make sense of the 'nonspace' of information, allowing them to create imagined geographies of the Internet and other dataspaces.

In fact, Scott Bukatman argues that in the following quote Gibson 'makes his own project explicit' (1993b: 152):

> all the data in the world stacked up like one big neon city, so you could cruise around and have a kind of grip on it, visually anyway, because if you didn't, it was too complicated, trying to find your way to a particular piece of data you needed.
>
> (*Mona Lisa Overdrive*, §2: 22)

As Bukatman points out, 'Cyberspace is a method of conceiving the inconceivable' (1993b: 152).

Reading worlds: reading, geography and science fiction

Examining Gibson's cyberspace necessitates an analysis of the relationship which joins authors, texts and readers, a relationship within which literary meanings are created and transformed (Radway 1984). Meaning is created between authors and readers, between the words written on to the page and the practices readers use to make sense of them; the creation of meaning can therefore be said to take place as part of a *dialogue* between them (Bakhtin 1984; Holquist 1990; Voloshinov 1973).

One of the clearest ways in which this dialogue is expressed in print is in the form of *conventions*, ranging from styles of address to formal and narratological structures. Conventions therefore represent agreed meanings between authors and readers and are visible *within* the text as part of the texture, structure or style of the narrative. Authors deploy them to suggest to the reader that the novel should be read in a particular way. Obviously, the reader must be familiar with these conventions and these strategies are open to contestation, but the fact that they are recognisable within so many texts suggests that they are often widely accepted.

Examining Gibson's cyberspace necessitates consideration of two sets of these writing technologies: those which concern the genre of science fiction, and those which concern space.

Every genre is based upon a different set of conventions, which further constrain the operations which can be brought to bear upon texts. I would like to argue that the science fiction genre is characterised by a tension between two opposed discourses, fantasy and scientific realism, and that conventions from both can be found within it. The crucial difference between the two is that the fantastic attempts to speak of the impossible, while scientific realism attempts to render experience unproblematically through appeals to rationality and scientific knowledge.

Studies of the fantastic note that it operates between the real and the un-real, using the latter to defamiliarise the taken-for-granted (Jackson 1981). Tzvetan Todorov (1973) argued that this in-betweenness could be seen within the text, in the form of conventions of hesitation. The reader hesitates to make sense of the text and cannot resolve the tension between the real and the un-real because the author presents both as equally plausible. Characters in the text are also unsure as to what is going on: they experience this hesitation themselves (p. 33). The conventions which enact this hesitancy in the text encourage the reader to become unsure of commonsense reality.

Science fiction depends upon the fantastic because it is set in an unknowable future (or past) and often in an as yet undiscovered place. Yet its descriptions are generally plausible rather than impossible, and consistent with scientific principles:

> Regardless of its setting in time and space, SF depends upon transgressions of what its readers think of as reality. To justify those transgressions, it establishes images of reality on grounds essentially theoretical.
>
> (Samuelson 1993: 198)

Scientific realism depends upon the fantastic, but goes on to resolve the hesitation between real and un-real. Where the fantastic asks questions, scientific realism gives answers. This operation is also visible in the text, as conventions offer readers the chance to make sense of the estranging worlds of the fantastic (Malmgren 1991; 1993).

But this process of translation is not guaranteed. While SF is generally successful in resolving the fantastic, authors and readers are able to use these conventions in unexpected ways. The conventions of scientific realism are subverted when the reader refuses to believe in the scientific explanation, preferring the strangeness of the fantastic. This subversive potential remains latent within SF texts so that moments of impossibility can be created in the practices of writing or reading it. One example that will be explored later is the coexistence of Haitian *vodou* and North American rationality within Gibson's cyberspace.

The second set of conventions which must be examined are those which produce representations of place. Although Daniels and Rycroft suggest that 'the novel is inherently geographical' (Daniels and Rycroft 1993: 460), it should be recognised that the spaces of the novel must be actively created by authors and readers, and are textualised as conventions. This allows us to see that there are many different ways of producing space in fiction, including the extended set-piece descriptions of many nineteenth century realist authors (see Tuan 1978), the modernist strategies of Dos Passos (Brosseau 1995), and the estranging non-spaces of fantastic fiction (Jackson 1981).

Gibson's representation of cyberspace uses two different strategies: realistic, or what Lennard Davis (1987) calls 'thick' spaces, which present extended descriptions; and attempts to convey space through the textual embodiment of the experience of a place. Marc Brosseau has usefully distinguished between these types as the geography *in* the text and the geography *of* the text respectively (Brosseau 1995: 95).

The former strategy is familiar to us from studies of Hardy and other realist novels, who produce literary landscapes through meticulous description. The geography *of* the text is a more complex concept and reflects the decline of these 'thick' spaces in modernist fiction. In Dos Passos' *Manhattan Transfer* the spatial experience of moving through New York is represented through the use of conventions which give the urban experience textual form. A walk through the city can be represented as collages which reproduce the 'spatial and temporal succession of the elements of the urban landscape' (p. 100). Brosseau's term for these strategies, which depend upon movement, is 'kinetic description' – description where 'the daily paths of an individual can be described in rhetorical figures' within the text (p. 101).

I would suggest that in the terms I introduced above, Gibson rarely 'thickens' cyberspace, concentrating instead upon the 'geography *of* the text'. In other words, his use of kinetic descriptive styles textually represents the spatial *experience* of cyberspace, rather than providing static set-piece descriptions. There are many possible reasons for this, including Gibson's well-known ignorance of computers (Bukatman 1993a), but I would suggest that the reason is generic. While it can be ordered, cyberspace is too fantastic a space to be comprehensively detailed and thickened in the style associated with realist fiction. So how can it be depicted?

Writing cyberspace: kinesis and fantasy

Cyberspace is experienced through movement, particularly in terms of speed:

> Headlong motion through walls of emerald green, milky jade, the sensation of speed beyond anything he'd ever known before in cyberspace . . .

> 'Christ', Case said, awestruck, as Kuang twisted and banked above the horizonless fields of the Tessier-Ashpool cores, an endless neon cityscape, complexity that cut the eye, jewel bright, sharp as razors.
>
> (Neuromancer, §23: 302)

> Bodiless, we swerve into Chrome's castle of ice. And we're fast, fast. It feels like we're surfing the crest of the invading program, hanging ten above the seething glitch systems as they mutate.
>
> ('Burning Chrome': 200)

Gibson creates an impression of speed and movement through the rhythm and pace of these descriptions. In *Neuromancer*, Case cuts back and forth between cyberspace, the real world and Molly's experiences as they are transmitted to him through simstim technology (1984/93 §4: 77–87). This represents a new and disorienting extension of Dos Passos' fragmented city, adding cyberspace to the collage of spaces presented in the text (Bukatman 1993b: 148).

Gibson also makes this experience ambiguous through synaesthesia (sensory confusion). Cyberspace is experienced as strange, impossible: smell, touch and taste are simulated and conflated. Two examples make this clear: 'Case's sensory input warped with their velocity. His mouth filled with an aching taste of blue' (*Neuromancer*, §23: 303); 'Cold steel odor and ice caressed his spine' (*Neuromancer*, §9: 140).

So it is possible to see that Gibson's depiction of cyberspace produces a fantastic textual space. However, this must be situated within the dialogue with realism which characterises science fiction. While I would argue that the kinetic style and devices like collage and synaesthesia can produce fantastic descriptions of cyberspace, it must be recognised that the matrix is also an ordered space. Unlike Dos Passos' New York, cyberspace is constructed on a linear grid system, a set of mathematical and geometrical points organised in such a way as to make it accessible and functional to its users. Seen this way, the kinetic style merely represents a speeded-up version of the more sedate movement from point to point.[5]

As a strictly structured grid or matrix, cyberspace is only fantastic because its scale is infinite and the amount of data in it so intricately organised. Scientific realism, in the form of mathematically and geometrically structured space, provides a metaphor for, and a way of controlling, disorienting elements of the fantastic in cyberspace. In fact, in attempting to find a way of making sense of information space, Gibson has undermined its fantastic potential.[6]

However, cyberspace is profoundly ambiguous precisely because the dialogue between realism and the fantastic cannot be finally resolved. The balance between the two discourses varies depending on the conventions and their reading, so that moments of subversion (of realism) and rational ordering (of the fantastic) coexist within the text. In *Neuromancer*, for example, we are confronted by a jumble of ordering and disordering metaphors:

> And in the bloodlit dark behind his eyes, silver phosphenes boiling in from the edge of space, hypnagogic images jerking past like film compiled from random frames. Symbols, figures, faces, a blurred, fragmented mandala of visual information.
> Please, he prayed, *now* –
> A gray disk, the color of Chiba sky.
> *Now* –
> Disk beginning to rotate, faster, becoming a sphere of paler gray. Expanding –
> And flowed, flowered for him, fluid neon origami trick, the unfolding of his distanceless home, his country, transparent 3D chessboard extending to infinity.
>
> (*Neuromancer*, §3: 68, emphasis in original)

Disorientation is textualised by kinetic description, in the literal meaning of the words (boiling, jerking), their alliterative texture (flowed, flowered, fluid), and by fragmentation ('film compiled from random frames'). However, the quote also makes use of a number of geometrical metaphors (mandala, disk, sphere) before describing the 'transparent 3D chessboard' which represents the ordered grid of the matrix. Furthermore, in moving from a state of fragmented experience to one of order, the passage narrates Case's control over the disorder of cyberspace (Bukatman 1993b: 205). This imposition of structure parallels the generic victory of scientific realism over the fantastic.

But the opposite scenario also occurs, though, when moments of fantastic uncertainty enter, however briefly, into the text. The key vehicle for this in Gibson's work is the presence of *vodou* in cyberspace. At the end of *Neuromancer* a number of artificial intelligences (AIs) unite and become fully conscious. They immediately fragment into many smaller intelligences and disperse throughout the matrix. For reasons which cannot easily be explained here, they subsequently take the form of the *loa* (spirits) of Haitian *vodou* in *Count Zero* and *Mona Lisa Overdrive*. I would like to develop Bukatman's argument that this destabilises the order of cyberspace:

> The interface of voodoo superstition with cybernetic certainty has a literally subversive effect upon the rational, geometric perfection of cyberspace. The modernist 'mythology' of rationality, the mechanisms of instrumental reason, are undermined by a new set of postmodern tactical incursions.
>
> (Bukatman 1993b: 214)

In this way the fantastic subverts the rationality of the text and of the represented space. Perhaps the most dramatic example of this comes at the end of *Count Zero*, when one of the *loa* enters a private area of cyberspace which simulates Park Güell in Barcelona. We experience a hesitation in the text, one which is also experienced by other characters, as at first the *loa* cannot be described: 'something plucking at his [Bobby's] sleeve. Not his sleeve, exactly, but part of his mind, something . . .' (*Count Zero*, §32: 318). The *loa* then manifests itself in Virek's park as a wooden cross with all its ritual accoutrements, even though the reader 'knows' that it is an AI operating in a rationally designed computerised dataspace. The tension between the two interpretations – *loa* or AI – is not maintained for long but it is still capable of being powerfully estranging.

A less startling moment of fantasy is experienced by Bobby Newmark at the beginning of *Count Zero*, which acts to introduce (but not to explain) the nature of these fantastic denizens of cyberspace: 'And something *leaned in*, vastness unutterable, from beyond the most distant edge of anything he'd ever known or imagined, and touched him' (*Count Zero*, §3: 32, emphasis in original). Gibson therefore provides a complex and ambiguous fictional space for readers to explore, one which is rationally ordered but also open to fantastic uncertainty. To examine the success of these attempts to convey the experience of cyberspace, we need to turn to the readers.

Reading cyberspace: 'It's real vague'

The discussants' conversations about cyberspace can be read as the identification of a problem with Gibson's depiction and the various responses which can be made.[7] In the first case, they grappled with what they saw as the 'vagueness' of Gibson's descriptions of cyberspace, which I interpret as an anxiety over the lack of detailed 'thick' descriptions of space. Their solutions to this perceived lack are fascinating, as they mobilised different explanations to account for it, and one aspect of this involves reading cyberspace in a dialogue with their own experiences of information technology.

To develop these ideas I wish to talk about the more fantastic or ambiguous aspects of Gibson's representation, and then go on to discuss the various ways in which the readers resolve the textual hesitation which characterises the fantastic.

Cyberspace is an ambiguous place, as the following exchange shows:

MikeR: [. . .] – I think it's somewhat vague how one approaches these things and then there's a sense in which you steal data –

John: Yes. Oh, it's real vague. [. . .]

Jael: It's deliberately vague! [laughs]

(B2)

MikeR's description of cyberspace also emphasises Gibson's lack of explicit description:

> [. . .] – there are solid entities which represent the data and they're virtual objects [agreement from John and Amanda], so you're still – and the impression – I'm not sure it's stated explicitly – is that you fly around in this space [agreement from John] but then you interact in some ill-defined way with the data.
>
> (B2)

The general reaction to this vagueness was frustration or uncertainty:

> [. . .] it's so sort of undefined in that he tells you bits of it, but he doesn't actually say, 'here's what happens, here's what happens', you know. It's weird – you know, if you can go into a place where you have no body that you can see, you look down, nothing.
>
> (Ragnar, C2)

What motivates this response? I suggest that it is rooted in the nature of the genre. When these readers are confronted with a new and estranging space like the matrix, they look for a way to order it. This is the origin both of Gibson's attempt to describe cyberspace and of the desire of many readers for maps and descriptions of this impossible space. The attempt to conceive of the inconceivable is therefore a joint project.

Support for this argument can be found in these readers' discussions of Gibson's writing of spaces. Amanda says that she did not read him for place:

> [. . .] the first time I read any of them, I didn't really think about landscape really, because I was so excited by what was happening, I couldn't take the time to construct it, so I just kind of had an impression in my mind which was enough background really to um, to read the novel perhaps, if you know what I mean. And it's only actually re-reading them [laughs] since we started this discussion group, that I've noticed erm . . . you know, like, finding out points about the landscape and actually where the action is happening, even – you know, like is it in the Sprawl, or is it wherever [agreement from MikeG]. 'Cos you know – which part of the Sprawl? 'Cos you don't really take it – it's so fast-paced, that you don't take the time to construct it, it's too complex to construct quickly, and you get bored with doing that, 'cos you just want to find out what happens next.
>
> (B4)

This inattention to landscape is due to the fact that Amanda was reading the fictions for the first time. Drawing upon Barthes' *S/Z* (1975), Henry Jenkins suggests that the desire to resolve the narrative is strongest on first reading. On subsequent readings, 'Interest shifts elsewhere, on to character relations, onto thematic meanings, onto the social knowledge assumed by the author' (1992: 67). Or, perhaps, on to the fictional landscape. The strategies used to thicken spaces may not be taken up by some readers, or at least not until a second or later encounter with the text. MikeR took up Amanda's point that the pace of the narrative makes thickening space difficult:

> It's like – not having reread them, I – erm to look at how he does it, but I got very much that sense of speed [*agreement from Amanda*] and somehow he implied enormous detail, even though I'm not sure it's really there – [. . .] You cut and paste something in your mind which is an amalgam of things that you're familiar with.
>
> (B4)

In this passage Amanda recognises the thinness of Gibson's represented space.[8] Shortly afterwards, she suggested that thickening space is a style of reading which *may* be used if the reader wishes:

> I think what he's done really is the most any writer can be expected to do, he's described the landscapes up to a point and then it's left to the reader really to fill in the gaps, and to *make the landscape whole if they can be so bothered*. If they don't, well, then they can just enjoy the scenery as they go past, kind of thing.
>
> (B4, emphasis added)

What is so interesting about the readers' discussions of cyberspace is that the vagueness of Gibson's depiction seems to make it hard for them to 'enjoy the scenery'. In fact, they are keen to fill in the gaps and thicken this space. This parallels their discussion of other science fictional texts, where ambiguous or unsettling representations are developed and explained through scientific and rational frameworks. Discussing Brian Aldiss' *Helliconia* series (1982, 1983, 1985), MikeG suggested that he was able to understand the planet's unusual nature because he has a degree in astronomy and physics, and was able to place Helliconia in a plausible scientific framework. Similarly, members of Group A explained their understandings of C.J. Cherryh's *Downbelow Station* (1983) and Frank Herbert's *Dune* (1965/84) in terms of ecological possibilities. So how did these readers respond to Gibson's ordered but uncertain representation?

Rationalising cyberspace

Significantly, cyberspace proves quite manageable for many of the readers. I have already discussed the presence of the fantastic in cyberspace in the form of *vodou*. We might expect the readers to be hesitant about explaining this subversion of rational space. This is not the case, as this (admittedly fragmentary) discussion makes plain:[9]

Alvin: [. . .] – in fact, he [Gibson] makes it seem as if people misconceive technology, like the way all those er – I can't remember the [.?.] – how they saw the AIs as being vodou –

Ragnar: Oh, yeah, gods.

Alvin: Yeah, you know, it's all misunderstanding, and misusing it.

(C2)

Alvin suggested that Gibson's characters have been 'left behind' by technological developments and that the *loa* seem strange only because the AIs are sophisticated enough to fool the ignorant. This is an excellent example of a reading which explains away fantastic elements through the use of a framework of scientific rationalism.

Other examples of this strategy are more creative, as the discussants worked harder to rationalise cyberspace. The first two examples of this display different uses of scientific realism, beginning with Alvin's contribution:

> You never get a clear idea of how – I mean, for example, how Case is manipulating cyberspace in a way, [agreement from Ragnar] [. . .] You're never sure, so I mean cyberspace is very vague. [. . .] I mean, in that way he sort of leaves it up to you, to view it in the way you want [agreement from Ragnar], he sort of leaves it very open-ended, 'cos that's why it's supposed to be an extremely user-friendly computing environment. You can sort of like perceive it the way you want, maybe someone else would actually perceive cyberspace in a completely different way, although functionally it would be the same [agreement from Ragnar]. [. . .]
>
> (C2, emphasis added)

Textual vagueness becomes a kind of 'user-friendly' software when Alvin reads Gibson's writing style through a technological metaphor. Mark added that he saw a parallel to this user-friendly vagueness in his own experience of multi-user games (C2).

A second realist explanation for vagueness also depends upon technological factors. Mark suggests:

> I see them [representations of cyberspace] all minimalist sort of style, because the processor count, the speed the information sort of travel, and obviously the basic [system's?] not gonna have the detail [agreement from Ragnar] – [. . .] – it looks a bit more abstract.
>
> (C4)

Mark explains the 'basic' nature of Gibson's cyberspace in terms of the ease of running this kind of system. Similarly, Ragnar suggests that the determining factor would be 'commercial viability', leading to the standardisation of information (C2), and Simon said 'you can never get a real picture of what it's like . . . because it works at the speed of the computer' (C4). Through these ideas, the readers colonise the blank spaces of Gibson's descriptions of cyberspace, providing realist explanations for vagueness which are consistent with the technology Gibson is describing. These are fundamentally science fictional strategies.

There are other, more general strategies which can be used. Ragnar, frustrated in his attempts to visualise cyberspace, turns to film:

> Like, you know, *Tron*, even before I'd heard of the idea of cyberspace, that you know, [.?.] a good movie, but a very good representation of cyberspace, and the basic idea as well, inside a computer, and um . . . you know, for the time it was really excellent, [. . .]
>
> (C2)

Tron arguably 'set' a powerful representation of cyberspace for many readers before they read Gibson. This use of a visual medium allows the reader to produce the thick space which is lacking from Gibson's descriptions; visual spaces like these are already thicker than literary ones, as filmic images automatically capture the *mise en scène* in a way which has no parallel in written texts.

Cyberspace, the Internet and virtual reality

The second main way in which readers make sense of cyberspace is through their personal experience of the spaces of information technology. This process is clearly dialogical: reading Gibson makes sense of these technologies but using them shapes reading Gibson, as we can see in these two examples. Rob, who likes to 'wander around America' on the Internet (A2), described the links he had made between reading *Neuromancer* and his job working with computer networks:

> [. . .] – when I read *Neuromancer* and then started at this place [his work], I could hack out to the Internet, stuff like that, it's almost like – obviously you don't plug it into your head [laughter from James], but I'm wandering around, you know er, computer networks all around the world, so I could be sort of er, sort of talking directly to a computer in Houston, Texas, and at the same time I could be getting stuff back from one in Washington, and it's all instant, it's all happening right there on my screen, but I can do the two things at the same time, or more. So it is almost like, you know, *you're actually physically there*, somewhere in Houston there'd be a hard disk that's turning because it's getting information and it's like porting it through the network back to me –
>
> (A1, emphasis added)

For Rob it is 'almost like' being simultaneously in Houston, in Washington and in London looking at his screen, able to cause physical motion in Houston. This captures something of the placelessness of Gibson's cyberspace.

John's description of his experience of computing focuses on the sense of speed associated with cyberspace:

> – the interesting thing about the perceived feeling of working with cyberspace, is the absorption, the *tremendous perceived speed* of doing everything is that working with computers with a screen and a keyboard or mouse can be like that now, if you're sufficiently well-practised in what you're doing, and the equipment is reasonably fast. I have – I basically spend my entire working day either writing programmes, writing about programmes or doing desktop publishing, and it is frequently the case *that I disappear entirely*, I'm just about consciously perceiving the screen but I'm not really looking at it, if you want to attract my attention you have to touch me [sounds of agreement]. It's somewhat of the same absorption as getting stuck into a very good piece of reading, or really being carried away with an idea – he took the same experience and he gets it over very well.
>
> (B2, emphases added)

The sense of transcendence, of being elsewhere (or nowhere) when reading or thinking is quite a common one (de Certeau 1984), and here it is extended to the interface between cyberspace and the human: John is absorbed *into* his work.[10]

Beyond these comments – which find parallels between Gibson's imagined space and experiences of information technology – the readers also developed an understanding of cyberspace which compares it to other ideas of virtual reality. Their discussions are significant for two reasons: first, they draw upon their familiarity with computers to establish the nature of Gibson's dataspace, and second, in doing so the very rationality of cyberspace itself becomes fantastic. John described the *forms* of VR and cyberspace in terms of their *functions*:

> [. . .] you're trying to provide a way of talking to something, a way of perceiving something that's efficient for the job you're trying to do [agreement from Jael], and that suits the ways you're trying to think.
>
> (B2)

Thinking about cyberspace, John suggested, involved a new interface and therefore a new way of working with computers which he described as 'unwinding the desktop paradigm' (B2) – in other words, finding a different metaphor for computing as *work*. This could be related back to Alvin's reading of Gibson's textual vagueness as the equivalent of a 'user-friendly' technology which makes it easy for the reader to understand. In defining and discussing cyberspace many of the discussants were careful to differentiate it from the virtual reality technologies described in cyberpunk or experienced in real life:

> [. . .] there is a difference between the virtual reality that Gibson offers and virtual reality that they're predicting, which is a complete – you know, the idea is that the graphics are so good that it will be

> indistinguishable from reality, erm, whereas Gibson's world is very much made up of computer lines – [. . .] – and grids – [. . .] – it's obviously a computer world [agreement from Jason], you know, he doesn't try to make it like reality. [. . .]
>
> (James, A2)

This recognises that the ordered, geometrical nature of cyberspace is in many ways the antithesis of reality. MikeR develops this in a very interesting way:

> [. . .] *we have gone to the computer*, we've not made the computer manifest itself in a form which we are familiar with, we have gone into another world, which is one which is more familiar, which – in some ways you imagine it as a natural state of the computer.
>
> (B2, emphasis added)

Cyberspace does not simulate the real world for our benefit; it simulates the 'natural state of the computer'. This is a fundamental change in our constructions of information technology; a 'user-unfriendly' environment. The readers seem to suggest that this most rationally ordered of worlds possesses an estranging quality *because* it is so unnatural.

These comments add a further twist to ideas of cyberspace, suggesting that in its geometrical perfection it is potentially alien and disorienting – we have come through scientifically realist explanations and out into the fantastic again. However, these ideas must be treated with caution and a concern for the narratives in which they are found. Depending upon their presentation by the author, these VR technologies can be more fantastic or more structured than cyberspace. This is further complicated at the moment of reading: 'the world of the computer' is read both as a user-friendly interface and as a new and inhuman place in the readers' discussions reproduced above.

Conclusions

In each of the readings presented here – Gibson's writing of cyberspace and the interpretations of readers, critics, and my own suggestions – there exists an element of ambiguity. Cyberspace is a highly polysemic representation; it invites, but does not demand, readers to work at 'conceiving the inconceivable'. However, it would have been impossible to develop this insight without recognition of the reader's role in using the conventions of science fiction to think about this space. While generic rules may 'fix' the practices of interpretation which are used to make sense of these technologies, these rules are flexible enough that readers may be able to resist, tactically, and to make their own kinds of sense based upon personal experiences of information technology, or of related texts. However, while readers are immensely creative (de Certeau 1984), we should not confuse this activity with resistance.

Indeed, returning to the immensely complex issue of the textual mediation of ideology, it would seem that the readers I interviewed have merely reproduced the image of cyberspace as 'the heady cartographic fantasy of the powerful' (Ross 1991: 148), with all the trappings of a 'masculinist' space (Rose 1993).[11] As I have already said, this critical argument rests largely upon the *form* of cyberspace: its geometry and order. The form of cyberspace thus becomes like gazing upon New York from the heights of the World Trade Center: 'It transforms the bewitching world by which one was "possessed" into a text that lies before one's eyes. It allows one to read it, to be a solar Eye, looking down like a god' (de Certeau 1984: 92).

To some extent, the problems of naming and knowing 'new' spaces – the gap between knowledge and language – have been faced before, by explorers and colonists of Europe's others (Davis 1987; Carter 1987). It might be possible to see in the ambiguity of these readings of Gibson's cyberspace some echoes of these earlier struggles to impose order upon the unknown: to force places to make sense, and to make them *work* for the reader.

But such a critical account also represents a flattening of the complexity of these interpretations, a generalising tendency which has its own logics of abstraction and control. It is important to remember that this is a representation embedded within the genre of science fiction. The form of cyberspace is not simply a consequence of patriarchy, colonialism, or global capital. Rather, the politics of these readings exists only at the moment of performance. Readers create; they do not simply consume and reproduce. The ambiguity of cyberspace, between scientific rationality and the fantastic, also invites us to keep our interpretations open.

For example, since cyberspace seems to be a rather ambiguously ordered space, how can it be simply and unproblematically gendered? I am wary of the suggestion that constructions of cyberspace simply *reflect* the subject position of the 'typical' masculine science fiction reader, a figure who has been extensively mythologised by popular discourse and media (Jenkins 1992). This argument requires us to see the text as effectively 'transparent', and to agree with the idea of the 'autonomous and self-celebrating reader who transforms the text into a mere pre-text' (Brosseau 1995: 91). Indeed, the fact that the two women readers also interpreted cyberspace as 'ordered' suggests that we should begin to turn our attention to the role of reading practices, and to examine the extent to which they are themselves already gendered (Flynn and Schweickart 1986).

This is not to say that there are no political ramifications of their readings. I would have preferred the discussants to challenge the orderly world of Gibson's cyberspace, and see a certain conservatism in their failure to do so. However, unlike the critics discussed above, I am not prepared to apportion blame to Gibson or the readers. Instead I am more interested in the way that the use of these *conventions* embodies a particular politics. In this way the conventions can be seen as the technologies of power which are expressed in writing and reading practice, the technologies which transform subversive fantastic elements into conservative realist understandings.

Notes

1 SF books go through many reprints. To aid readers, I have given two dates for novels cited in the text; the first refers to the original date of publication, and the second to the edition I have used myself. In addition, the chapter where the quoted passage appears is shown after the symbol §, so that (§4: 45) refers to page 45, Chapter 4 of the edition cited.

2 Significantly, these interpretations of cyberpunk are often based upon different readings of cyberspace.

3 In-depth group interviews differ from focus groups by their longevity and loose structure. Within geography their use was pioneered by Burgess and Harrison at University College London in research conducted during the mid 1980s. This work rejected the market research tradition, adopting the principles and practices of group-analytic psychotherapy to explore the environmental discourses of lay people (Burgess *et al.* 1988a, 1988b; Burgess *et al.* 1990). Key features of this method include the development of a *group identity* which situates dialogue and argument within group relationships, and a freer discussion than is found in the directed interviews of focus groups.

4 Key metaphors for the appearance of data in this space are stars in the night sky and city lights; both appear in the excerpt reproduced above.

5 'Case punched again, once; they jumped forward by a single grid point' (*Neuromancer*, §9: 140).

6 In this sense it is also significant that the 'cyber-' prefix is derived from cybernetics, the study of *control systems*.

7 In-depth group interviews with three sets of readers of Gibson's SF were carried out in October 1992 and April/May 1993. Transcriptions are identified by the group (A, B, C) and session number (1–4), so that (B4) refers to the transcript of Group B's fourth session. The members of the groups are as follows: Group A: James, Rob, Jason, Chris, Piers, and Maria. All were in their early twenties, and all except Maria had degrees or higher qualifications in a science subject. At the time of interview, Rob was a computer network systems designer, Jason, Chris and Piers were students, and James and Maria were looking for jobs. Group B: Jael, MikeG, Amanda, MikeR, and John. All were in their early thirties except Jael, who was twenty-one. John, Amanda, and MikeG worked with computers or technical support, MikeR was a research chemist, and Jael was a student. Group C: Alvin, Simon, Ragnar, Steve, and Mark. All students in their late teens apart from Steve, who was a staff nurse in his early twenties.

8 Amanda and the other members of Group B also explicitly contrasted Gibson's textual spaces with the thicker, more detailed descriptions of Brian Aldiss, John Crowley, and others.

9 The sound quality of the recording of this session was very poor.

10 MikeR suggested that John's experience was very much like Gibson's description of kids playing arcade games where 'you could see that they wanted to be *in* the game' (B2, emphasis added).

11 In general, critical interpretations of cyberspace take one of two positions. Either cyberspace is read as a masculinist conception and that the geometrically ordered and modernist *form* of cyberspace serves to constrain the transformations of gender and identity that might be possible in a new space; 'Cyberspace is a vehicle for allowing the fluidity of social and sexual relations to be confined within the

rationalist configurations of information technology' (Wolmark 1993: 118; see also Springer 1991). The second position argues that cyberspace is originally feminine (Stone 1991); (male) users masculinise it and impose order on it by force as 'metaphoric rapists' (Nixon 1992: 229). Unfortunately, these readings, along with variations on these themes by Ross (1991) and Bukatman (1993b), ignore both the role of generic conventions and the creativity of the reader. I hope I have made clear just how difficult it is to consider the gendering of *texts* in the light of their readers' interpretations.

13 On boundfulness

The space of hypertext bodies

Michael Joyce

The chapter I am writing: heterotopic dwelling or I am here aren't I

> The things he showed were primarily traces; the photograph alluded to knowledge that it would scarcely show. It took the visible tip, a detail or a surface, and found other ways to indicate what was not there, what had been relegated to the distance, when not altogether cut. This was the thrust of emptiness in the third city.
>
> ('In the absence of the parisienne . . . ', Molly Nesbitt 1992)

Some time ago I was sent the text of the proposal for this collection which included, among other chapter abstracts, the following:

> 15) Ethereal Texts: words in the web (*) Michael Joyce, Vassar College, or Mike Crang, University of Durham
>
> The possibilities of virtual geographies are not simply 'out there' to be commented upon, they also inform the nature and possibilities of those commentaries. Developing issues raised in the previous chapter, this essay will explore the ways in which recent developments in electronic media might be affecting the ways in which these virtual geographies are represented. Looking specifically at the world of hypertext, the chapter takes up the argument that hypertext offers a microcosm of the web as a whole – that is, that it takes the form of links between disparate fields of knowledge in an electronic space. It asks how hypertext shapes these links, about the new constellations of knowledge that its geography makes possible and, more tentatively, about the power it makes available to the reader to radically re-order the text to create new, perhaps more open forms of knowledge. But it sets its enquiries in a historical perspective that questions how far such developments do indeed move beyond older forms of text and modes of textual representation and reading.

This is, of course, the chapter I am writing, although at the time it was proposed to me it could have been someone else who was writing it (the someone who had written the abstract, one supposes, perhaps the editor, Mike Crang, who is named here, although someone else could have likewise given its injunction to him). A year or so ago I received a similar solicitation from an editor to contribute a chapter regarding hypertext for a proposed collection on re-reading. That solicitation was addressed to Michael Moulthrop, a conflation of my name and that of Stuart Moulthrop.

It has ever been thus for editors and authors and proposals and collections and so I don't suppose to suggest that these anecdotes necessarily limit 'the possibilities of virtual geographies . . . [which] inform the nature and possibilities of those commentaries'.

Yet the plausibility of interchangeable authorship is also the plausibility of interchangeable identity.

Well, I'm here, aren't I?

The strategy of making myself the centre of a text about boundfulness is suspect. However, we may ask to whom the identity of this 'I' matters (and likewise who is this 'we' that may ask when she's at home). As Habermas says, 'Even collective identities dance back and forth in the flux of interpretations' (1992: 359) A chapter on hypertextuality is wanted and whomever (preferably someone with at least one name beginning with M) may write it.

A chapter on hypertextuality is found wanting. Hesiod saw the world as founded (or found the world he saw) upon wanting.

At the level of style, it may be that a certain readership exists who finds this particular kind of lyrically self-reflexive playfulness identifies a certain 'author', namely me. Without doubt, I can number myself among such readers. The number is either one or the zeroth.

At the level of genre this one is a variety of 'My-story', Gregory Ulmer's (1989) intentionally artificial, proto-hypertextual genre of successively (and heuristically) generated spaces.

At the level of general organisation (we could say argument or organic unity were we more old-fashioned; we could say apparent organisation were we more postmodern), any of us would do. Although it must be supposed that I have done as well or better than at least one other in this instance since the editor, one such other, has seemingly allowed this entry (or so I surmise, however projectively, as I write this).

I can find some consolation in set theory I suppose, since for each of the two collections, I am the one recurring element in the set of possible writers. But in fact my putative authorship (insofar as it is not included in the otherwise seemingly plentitudinous authorships which open out from the questions at hand) merely marks the bounds of the questions which are examined here, something Mike Crang's abstract did long before me and which in some sense I therefore always fulfil if only in the suggestion of its absence, a presence which my authorship metonymically enacts.

The reader functions likewise, here as well as in hypertext, although there is less chance that the text here will slip from its putative authorship within the confines of the chapter.

Ethereal Texts: words in the web, by Michael Moulthrop or Hesiod.

I'm here, aren't I, at least until page ____ (the last word I am incapable of writing since the page itself does not exist at the time of this writing and so therefore is tokenised, marked by underlined blanks as a sign of some future authorship, which may or may not be filled).

The author serves as the singular. *The* space of hypertextual bodies in the platted (plaited, plaintive) provocation of the singular title of the chapter, 'On boundfulness: the space of hypertext bodies', within which this chapter, 'Ethereal Texts: words in the web', does and does not appear.

> The section you are reading, or have read, depending on how you constitute this coda, originally appeared at the end of the chapter I am writing, until the editors intervened with the congenial suggestion that 'if you moved the last section about your authorship to the front it might prepare the reader about how to interpret the essay . . . [since] unless they read the editorial (pious hopes) they won't know in advance this is about hypertext'.

This is about hypertext.

Thinking one self else where

> Seasoned and bendy
> it convinces the hand
> that what you have you hold
> to play with and pose with
>
> ('A Hazel Stick for Catherine Ann', Seamus Heaney)

> This particular Irish literary genre, the immram, deals with high-spirited ventures into an unknown ocean . . . However fantastic the adventures and imaginary the geography, the immram record experiences that were real.
>
> ('Irish Seafaring', Carl O. Sauer)

Dieberger doesn't think he's here – what could this mean? – he thinks he's somewhere else. It doesn't do to ask him, I can tell you. He's not alone in thinking this, wherever he is. Let's ask him.

He wants to be everywhere, he wants to be nowhere. Utopia=pantopia. He has no leg to stand on. He's ectopic, 'beside himself' as the saying used to go. He's used to it, he's a juggler.

He's the new wave, it says so on the programme. Likewise, likeable, he waves to the crowd as he, slightly Teutonic (i.e., 'I dun't read anysing . . .) – Austrian actually, says:

> I don't read anything on the web anymore. I just check out the links and mark the ones I want to come back to later. Though I never do really.

The betweenness which is not strictly nomadic nor instrumental nor algebraic but rather volitional, constitutive. The break of a link in a web conceived as everywhere impinging surfaces, a topological space, skin, surface of surfaces without surface, amen.

It is important to ask ourselves how we constitute the body which sees itself in boundfulness, permeable though not less permanent than the rest of the impermanent world, wholly interstitial and yet no less whole.

Ectopic{≠}osmotic.

Not by negation. It is something other than the nomadic body which, however deterritorialised, occupies its own space.

Marcos Novak thinks not:

> I coin the word pantopicon, pan+topos, to describe the condition of being in all places at one time, as opposed to seeing all places from one place. The pantopicon can only be achieved through disembodiment, and so, though it too speaks of being, it is being via dis-integration, via subatomization of the consciousness, rather than by concentration or condensation.
>
> (1996, on-line unpaginated)

It is something more than a Berkeleyean turn (the bishop not the campus) to say that Dieberger (the mountains) means differently. He's no places at many times and though nowhere all at once elsewhere here.

Hear hear.

His pronouncement is met with great applause (a great many boys in the audience claim via huzzahs that they too no longer read links, just swing from url to url like Tarzan; meanwhile back at the main camp Jayne Loader, a WebWench but no Jane, builds a literature of links upon the boyish schema – not Schama, no Simon upon which is built a church – of proto-(pre-frame)-Suck.

There are certain kinds of literacies which can unpack the previous sentence, though they are not co-terminal merely coincidental. (Tarzan and Jane are, of course, the pre-post-colonial icons of certain pre-post-feminist baby boomers; Jayne Loader's WebWench site (http://www.publicshelter.com/wench/) exploits a contrapuntal aesthetic first popularised by the webzine *Suck*, viz. www.suck.com, in which embedded links smartly play off a smartass text, the pleasure being in the simultaneous recognition of the implicit disjuncture and plenitude of the web. (Modernity being, as Edward Soja puts it, both 'context and conjuncture', (1993: 147), of which disjuncture is but a special case; though when I first read the phrase I confess

I thought he said conjecture.) Loader's film, *Atomic Café*, surely is among pre-webbed predecessors of this aesthetic of interwoven disjuncture and plenitude – Novak calls this shift of consciousness 'from the society of the centripetal panopticon to the society of the centrifugal pantopicon', a Centrifug(u)e.

The apposition of the initially innocent sentence and the long, dense and overdetermined parenthetical paragraph is not merely a stratagem of modernism, duly (and dully) carried forth through postmodernism and hypertextuality but also itself an instance of boundfulness.

To the extent the pantoptic young men on the flying trapezoids move from link to link without alighting there may, of course, be reasons to suppose that the centrifugue is the same old song. What, we could ask, is the meaning of the swan song of the one who is moving on? What production, mea culpa mea marxisme, is involved in passing over, links or space? In a proto-post-Marxist melody of his own, David Harvey suggests that 'denigration of others' places provides a way to assert the viability and incipient power of one's own place'. Young anthropoids make such calls. All this frantic swinging from branch to branch may merely be place-marking in what Harvey calls 'the fierce contest over images and counter-images of place' and where 'the cultural politics of places, the political economy of their development, and the accumulation of a sense of social power in place frequently fuse in indistinguishable ways' (1993: 23). In a consideration of hypertext browsers (here meaning programs like Netscape or Internet Explorer) in which the reader 'cannot change the text but can only navigate among already-configured trajectories', as instances of de Certeau's wanderers (Wandersmänner), Mireille Rosello likewise finds an underlying production in which browsers (that is, readers) 'passively consume what others have produced or written {and} the steps of the walker across the city are disembodied, like weightless information saturating a network . . .' (1994: 136). She evokes de Certeau's assertion that 'Surveys of routes miss what was: the act itself of passing by' (1984: 97) suggesting that 'Like a Derridean trace, such maps keep the memory of an absence . . . The trace left behind is substituted for the practice.' For the pantoptic young men that trace is embodied, is the body in its boundfulness. They are walking (or flying) maps of unexplored meaning.

The gesture of the appositional other-handedness and scholarly stitchery, on the other hand, closes off a dialectical ticking which bounds the feathery space upon which lie kinds of systematic literacies (no less (post)modernist) which can pack and unpack the previous sentences and paragraphs like the suitcases of some wanderer, a foreigner for instance, making repeated journeys elsewhere without settling and, outside our seeing, disappearing without a trace.

I've elsewhere suggested as a figure for hypertext Kristeva's characterisation of the 'otherness of the foreigner' in terms of 'the harmonious repetition of the difference it implies and spreads' in a fashion which she links to Bach's Tocattas and Fugues: 'an acknowledged and harrowing otherness . . . brought up, relieved, disseminated, inscribed in an original play being developed, without goal, without boundary, without end. An otherness barely touched upon and that already moves away' (1991: 14).

Though the otherness may be seen as without boundary, the foreigner is the image of boundfulness as well. Persistence of vision, for instance, or simply simple repetition. Foreigner or fugue, especially as these are repeated, make space that is both within and somehow simultaneously outside the space of the text. The gesture of the parenthetical, the dialectic, the thematic, the rhythmic, the fugal, the isobaric, the metonymic, the list, the link, the litany as well as any and all other – whether en-dashed or no – appositional stitchery constitute the space of hypertextuality. Boundfulness, in this sense, is space that ever makes itself, slice by slice, section by section, contour by contour; never getting anywhere is Dieberger on the links (his hole-in-one a torus, ever a single surface).

> [Internote: in the course of looking up Zeno of Elea in order to make the link between this space of ever opening boundfulness and his famous paradox, I discover that when text is copied from the Microsoft Bookshelf 1996–97 hypertext version of *The Concise Columbia Encyclopedia* into a document, at least within Microsoft products, the program automatically (and invisibly in most views) generates a footnote containing copyright information. It is of course possible to imagine that someone copying terms from the Microsoft Bookshelf into the span of terms previously copied from that source could create a fully boundful text of successive citational edges, not unlike topographic contours. Quod erat demonstrandum.]

Dieberger is Andreas Dieberger, whose screen name (and avocation) is juggler, author of 'Browsing the WWW by interacting with a technical virtual environment – A framework for experimenting with navigational metaphors' (1996: 170). Here (or there) is the ACM Hypertext '96 panel on Future (Hyper)Spaces.

Dieberger filled in at the last moment for Mark Pesce, co-creator of VRML, virtual reality modelling language, (a language as yet apparently unable to put him in two places at once) and so Dieberger is not listed as a participant in the program. His comments are not printed or posted on the web. I may have made them up. You had to be there.

The phantom limb and metonymic imagination

> It is not a matter of deciding to go into cyberspace. We are always already in it, before the literal condition.
>
> 'B e i n g @ H o m e . . . a s B e c o m i n g I n f o r m a t i o n a n d *H y p e r s u r f a c e*', Stephen Perrella

I just dropped in to see
What condition my condition was in

(Kenny Rogers and the First Edition)

Even so we could consider the post-prior experience of the current condition.

The space of the node – not properly a 'screen' or a 'page' – a meta-element yet computationally exact, at least to the extent that it is locatable, is thus metonymic. It summons the whole of the web of its relations by means of the disclosure of its failure to participate within that whole, which of course does not exist in any coherent state beyond the suggestion of its absence, a presence which the partiality of the metonymic element enacts.

Whoa. Woe. Woebegone. Wo bin ich?

Let us go through this more simply. A web page currently is represented as a window with some method of indicating depth (scrolling, etc.) which also suggests a measure of local closure or boundedness. Yet that measure (the scroll-bar for instance) is not so much meta-textual as extra-textual. One is not apt to say or think 'I found a picture of a pig seven ninths of the way down the scroll bar on a fully opened window upon a 640×480 monitor'. The window itself is merely an artefact of a data structure within the operating system, or the browser, or the html structure (which may itself be a formal hierarchy or a happenchance and opportunistic network as yet unformally represented).

Yet each of these spaces – the proprioceptive measure of a pig seven ninths of the way down or the un-formal abstraction of the containing container – can be said not merely to embody but to constitute the hypertextuality, especially of a random site which for argument's sake I am only presenting through the single discrete element (itself 'located' elsewhere, in a gif file for instance, i.e, 'This little piggie went to mark-up . . .').

The experience of this space within the node stands metonymically both for the space of the abstract structures of its representation (window, system, browser, frame) and for the composite space (the site, the web, the story, the reading) within which we experience it.

The insistence upon 'click-through' as the measure of advertising effectiveness on the web inverts a rhetoric established by the advertising industry (conventionalised as 'the advertising world': the whole earth became a hippie brandname before Stuart Brand became a technological guru) wherein advertising was purported as neutral and extra-textual. Thus *Wired* magazine proposes that advertising ought perhaps to enter content (as if product placement remained to be invented). Actually the columnist (I know I should footnote this but, even if this attribution is a lie, the argument, if such, doesn't suffer and the writing in that magazine isn't meant to persist only exist) suggests that advertising be woven through content (as if *textus* meant anything else; as if a weaving were not a penetration; as if a parenthesis were not a lung).

The placement of this particular argument (if argument it be; click-through not yet determined) is a version of the same inversion. Assigning boundaries to this discourse depends upon an invented perception of its meaningfulness. Locality, Arjun Appardurai suggests, appropriating Raymond Williams term, is 'a structure of feeling' (1996: 199). The reader must suppose that the indulgence of the spatialised voice here represents (enters) the content of the argument which is itself represented by the perception of that interwovenness (representation, entrance). Perhaps 'I' am simply mad or a sloppy thinker.

Following a talk in Hamburg (53° 33′N 10° 0′E) for instance a woman comes forward to ask if I am a Buddhist, saying 'I had to listen very closely because I could not be certain from moment to moment whether you were discussing the poem by Milocz, or the scientific concept of catastrophic singularity, or hypertext theory.'

'That is the scientific concept of singularity,' the sensei replied.

The modernist notion that meaning presents itself in (or après) fragmentation is ceded (seeded) in postmodernism to a present-tense fragmentation of meaning as a constant stream. Is it fair to ask where this stream is situated? This whole earth? This seven ninths piglet? The space within a node? The Powerbook in the cinematic twister?

The world of tokenised representations does not exist in any coherent state beyond the suggestion of its absence, a presence which the metonymic element enacts.

The disembodied juggler, all boundful, die Berger, exists in absence. He's on his way not nowhere (the nomos) but elsewhere (osmosis). The endless sectioning of the boundfulness of the web turns Deleuze and Guattari's speculation on the smooth and striated to pure reportage. Osmosis is the liquid equilibrium along a semi-permeable barrier.

The web transcends the inevitable spatiality of other hypertexts by becoming primarily osmotic and ephemeral. (Before the hyperfusty bookishness of *Myst*, Rand and Robyn Miller created a transcendently hyperfictional happy-ever-otherworld in *Cosmic Osmo*.) By circling round our senses of confirmation, disclosure, and contiguity, we find ourselves falling into sense. A recognition of traversal prompts my student Samantha Chaitkin to offer 'a brand-new metaphor' in a critique of spatial hypertext representations:

> I'd rather . . . jump up into the air and let the ground rearrange itself so that I, falling on to the same spot, find myself somewhere different. Where am *I* going as I read? No, more where is the Text itself going, that I may find myself there.
>
> (1996: unpaginated)

The 'I' is considered constant (as here) and avoids the conventionalised gesture which blurs I/eye and thus either objectifies the world within a gaze or subjectifies the world within a discourse. The ground is given its own authorship when the link is seen as *tour en l'air. En bas*, however, the self stands as metonym on terra firma, part not for but *as* the whole.

The world (ground) (re)assumes its agency in the elsewhereness of our boundfulness (a phrase that in its comic compounding of -nesses summons a lost sense of utopianism, Shaker hymnal phrases, the sussurus of – sweet Jesus – praises). The question at hand is resurrection: what body?

It is difficult not to imagine that something goes on (this ground) elsewhere as we land again and again on a turning world. The most common anxiety which my students report as they come to read hypertext fictions is a feeling that the story goes on elsewhere either despite their choices or unmindful of them.

Hypertext novelist and theorist Stuart Moulthrop early on in the recent history (to make a space) of hypertext literary theory foresaw this metonymic quality of hyperfictional space: 'To conceive of a text as a navigable space is not the same thing as seeing it in terms of a single, predetermined course of reading.' Moulthrop contrasts 'the early intimations of wholeness provided by conventional fiction [which] necessitate and authorise the chain of particulars out of which the telling is constituted' to the distinctive 'metaphor of the map' in hypertext fiction which rather than preferring 'any one metonymic system . . . enables the reader to construct a large number of such systems, even when . . . these constructions have not been foreseen by the text's designer' (1991: 129).

I've suggested elsewhere that this sort of claim for hypertext fiction is beyond, if nonetheless beholden to, Umberto's Eco's literally transformative formulation of the 'open work' which offers the reader an 'oriented insertion into something which always remains the world intended by the author' (1989: 19). Yet even Moulthrop (however necessarily, since at its earliest stages – a mere decade ago – we all conceived a single reader and writer in a shifting dance on a single web) argues from the perspective of the single reader, herself unaware of her compeers' actual or imagined progress elsewhere through the shifting text:

> Metonymy does not simply serve metaphor in hypertextual fiction, rather it coexists with metaphor in a complex dialectical relationship. The reader discovers pathways through the textual labyrinth, and these pathways may constitute coherent and closural narrative lines. But each of these traversals from metonymy to metaphor is itself contained within the larger structure of the hypertext, and cannot itself exhaust that structure's possibilities.
>
> (1991: 129)

My students' awareness of the simultaneous elsewhere which they construct from their boundfulness is heightened because they are aware their choices shape the story, or better still their continued presence embodies it.

'Hypertext navigation', as Terry Harpold notes, 'means not only traversing a space between two points in the narrative; it means as well electing to diverge from a predetermined course' (1991: 129).

The phantom limb is the story that goes on elsewhere while we experience the story of the repeated here. We have the feeling that we are elsewhere, on another terrain. Yet the elsewhere is also here, what differentiates metonymy from metaphor is its suborbital flight, its semantic shift is lateral, a spatialising displacement within the space of the literal without the orbital escape velocity of metaphor. 'The possibility that the reader may choose to digress from a path of the narrative, and remain within a field/terrain that is still identifiably that of the text she is reading,' says Harpold, 'greatly complicates metaphors of intentional movement that may be applied to the act of reading' (1991: 129). Simultaneously within and without, the reader doesn't feel her loss as much as the loss of embodied presence of the body's boundfulness. Terrain and body alike are systematic longings.

'The body', as Donald Kunze notes, 'is not an abstract idea but a living entity. The body is what the body does, and the body is thus allied with the process of enactment that fleshes out the "4th dimension" reinserted into its new position between representation and world.' Kunze locates this enactment in 'a gap in the system of Cartesian dimensions . . . between dimensions #2 and #3' where '"spatialised time and temporalised space" has to do with the muscularity, enaction, emplotment, and dynamics of moving from images to solid realities, i.e. the world in which human action becomes actual, the real and sharable world' (1995: on-line unpaginated).

Act II, scene iv: *The Real enters, accompanied by various lords and ladies, the Frenchman and others:*

> There are also probably in every culture, in every civilisation, real places – places that do exist and that are formed in the very founding of our society – which are something like counter-sites, a kind of effectively enacted utopia in which the real sites, all the other real sites that are found within the culture are simultaneously represented, contested, and inverted. Places of this kind are outside of all places, even though it might be possible to indicate their location in reality. Because these places are absolutely different from all the sites that they reflect and speak about, I shall call them, by way of contrast to utopias, heterotopias.
>
> (Foucault 1986: 24)

The speaker is the Frenchman, Foucault, and he wrestles with an angel (although on-lookers may perhaps mistake this for a dance). In his so-called real heterotopic space I want to situate a little dwelling, a ritual shed perhaps, in which, for purposes of this drama, we may distinguish between the flying young men on their trapezoidal machines and the young woman who spins in the air.

Heterotopic≠ectopic

For Samantha's longing is not, I think, for the body outside the body, the ectopic self, outside temporality and thus beyond mortality. Not a longing for self-evident and embodied truth, a proof. For, as Kunze notes, 'by definition, proofs of the body are refused the possibility of detachment; they commit the logical fallacy of self-reference. . . . That is to say, by attempting to step outside the human condition in order to describe any element of it, the mind must falsify its status as body, as member of the subject in question.'

Foucault has wrestled with this angel as well, looking for/at the spaces where self-reference can be seen. 'I believe that between utopias and these other sites':

> there might be a sort of mixed joint experience, which would be the mirror. I see myself there where I am not, in an unreal virtual space that opens up behind the surface; I am over there, there where I am not, in an unreal virtual space that opens up behind the surface; I am over there, there where I am not, a sort of shadow that gives my own visibility to myself, that enables me to see myself there where I am absent
>
> (1986: 24)

Although perhaps there is nothing to look at (or through). Instead (in the place of) of the ectopic unreality of present absence perhaps what we see is the heteroptopic flicker, the here-there of the woman in the air, the 'epistemic shift toward pattern/randomness and away from presence/absence' of what N. Katherine Hayles characterises as 'flickering signifiers' (1993: 73ff).

Heterotopic=osmotic

'Interacting with electronic images rather than materially resistant text', says Hayles,

> I absorb through my fingers as well as my mind a model of signification in which no simple one-to-one correspondence exists between signifier and signified. I know kinesthetically as well as conceptually that the text can be manipulated in ways that would be impossible if it existed as a material object rather than a visual display. As I work with the text-as-image, I instantiate within my body the habitual patterns of movement that make pattern and randomness more real, more relevant, and more powerful than presence and absence.
>
> (1993: 71)

Exeunt The Real, the Frenchman, the spinning girl, et alia.

LEGEND

In an attempt to approximate the space of hypertextuality in this linear form this essayis written in a series of overlays, viz. Lippard 1983, unmarked by typography or, for that matter, any but the most rudimentary narrative or syntactic markers, save perhaps the section conventions of which this is an – anomalous – instance. Thus only the temporal marks this stratigraphy and the claim here, as in any legend – cartographic, folkloric, or mythic, – is susceptible to physical observation and verification. Even so I intend a perceptible differentiation of the spaces here, not unlike a mapping, or better still something akin to the 'global cultural flows' of Appadurai (1996: 33), i.e., '(a) ethnoscapes, (b) mediascapes, (c) technoscapes, (d) financescapes, and (e) ideoscapes'. In fact, my own affectation (my own private Idaho one might say) is that the various sections here constitute imaginary physical elements of a virtual landscape. Thus one section in my mind is 'the desert' while another is 'the city of fractal towers' and so on. These remain unmarked (although – to summon the inevitable Derridean iterative – not unremarked).

Hypertextual contours and heterotopic proprioception

> 'AKiddleedivytoo', *i.e, a kid'll eat ivy too*
>
> From a children's song

> The imagistic approach begins with image but ends in translation. The metonymical approach would begin with translation – actually, the failure of translation – and end in image, which is the only means of sustaining an ambiguous relation of polyvalent meanings.
>
> 'The Thickness of the Past: The Metonymy of Possession,' Donald Kunze

The feeling of being beyond oneself must of course reside within one's self or else there would be no sense that one had projected it beyond the self.

Everything depends upon that "it," although the allusive structure of phrasing this sentence so in certain literary circles marks it as a static modernism (viz. William Carlos Williams).

I have long ago accepted the fact that I am post-priorily a retro modernist. (Viz Soja supra re con -text & -juncture.)

The first generation of hypertext fiction writers, pedagogues, and literary theorists, as already noted above, were practised in systems where writers and readers of relatively bounded texts ('not infinite but very large', as Jay Bolter termed them) enjoyed relatively rich interactive environments. With the emergence of the image-driven Web following the development of Mosaic, readers and writers take their place in a network of relatively unbounded texts (where texts are understood in the pomo sense which includes image – both dynamic and static, as well as sound and collaborative interaction) which paradoxically only afford relatively sparse interactive environments. A third generation (java-pushed, one might say) is likely to trade off access to unbounded space for enriched experience within commercially circumscribed (intranetted or infotained) environments. These latter network spaces might be not unlike glass-bottomed touring boats moving soundlessly and invisibly above the sprawl and noise of the richly populated but sparsely interactive net beneath, searching out the neon tetra, Atlantis, and debris of fallen flights or provocateur submarines.

The image works only if one imagines the glass-bottomed boat outfitted with amenities: cocktail bar, avatar, shuffleboard, search engine, captain's table, real audio.

The image can never work if it must be explained appositionally.

It is crucial to make the distinction between 'work' and 'doing work' as the third-generation, Berlin hypertheorist pointed out when the plump, cognitive scientist at the meeting in Hamburg (15: 11: 13 7Jan97) reported (regretfully, regretful he) upon the tenth anniversary of the publication of my hyperfiction *afternoon* that his data showed it did not work. ('He proves by algebra', Buck Mulligan said, 'that Hamlet's grandson is Shakespeare's grandfather and that he himself is the ghost of his own father.')

The image, in hypertext, always works appositionally in the biological sense of the growth of successive cellular layers.

'When I first spoke of living systems as autopoietic systems', Maturana says,

> I was speaking of molecular systems. Later, when I made a computer system to generate an autopoietic system . . . I realized it was necessary to make the molecularity of living systems explicit in order to avoid confusions. . . . a computer model takes place in a GRAPHIC SPACE generated by the computer, and this is why we did not claim to have a living system. . . . Yet it could have been proper to call all autopoietic systems regardless of the space in which they occur, living systems.
>
> (1991: 375–376)

In a usage of that term which most likely would set a geographer's teeth to grating, I have tried in a series of essays spanning nearly a decade to explore a notion of what I called hypertextual contour (I won't cite them all here given recent press reports that there are software agents which, speaking of contours, can generate a citational index, a sort of GIS view of the intellectual landscape, which is sensitive to self-citing and flattens the cumuli of self-referential plateaus).

I meant nothing more (or less) than to describe the readers' (now the placement of this apostrophe, the surface structure, is significant) sense of changing change across the surface of a text. I had in mind something less isobaric than erotic, the sense of a lover's caress in which the form expresses itself in successivenesses without necessarily any fixation.

My most discrete formulation of this notion:

> Contour, in my sense, is one expression of the perceptible form of a constantly changing text, made by any of its readers or writers at a given point in its reading or writing. Its constituent elements include the current state of the text at hand, the perceived intentions and interactions of previous writers and readers which led to the text at hand, and those interactions with the text that the current reader or writer sees as leading from it. And which are] most often . . . read in the visual form of the verbal, graphical or moving text. These visual forms may include the apparent content of the text at hand; its explicit and available design; or implicit and dynamic designs which the current reader or writer perceives either as patterns, juxtapositions or recurrences within the text or as abstractions situated outside the text.
>
> (Joyce 1996: 280)

suffers beyond its seriosity (to use Woody Allen's term) from its fixity. In the first ellipsis in the block quote above I've excised a claim (here restored) that 'Contours are represented by the current reader or writer as a narrative'. The same claim is restored in action by Heather Malin:

> when i'm writing a ht [*hypertext*] strange things happen. i am writing and creating, and i am fairly sure nothing coherent is going on. i am losing it. perspectives multiply as i realise the possibilities of my discourse. i see what is evolving, and i see contours, shapes, movements . . . i was not planning them, they were not of my authorial consciousness.
>
> i try to follow the emerging currents; i start riding them and bringing them to some completion or exhaustion. but i am not sure that they are mine, although they are of my making. unless the text is writing itself, i am making it happen. somehow. . . .
>
> i end up stuck in the middle of my own movement.
>
> (1998: forthcoming)

This stuck in the middle of movement is a giant step (albeit one suspended in stop-time like the children's game of 'Mother May I?') away from beginning-middle-end. In the space of what I've called the hypertextual 'story that changes each time you read it', story becomes a matter of where you've been, de Certeau's already-mentioned 'act itself of passing by'. In seeking to describe a notion of hypertextual contour I meant nothing more than what Malin suggests here: readers and writers do report the recognition of a form perceived outward from the middle of their own movement. This is the proprioceptive measure (proprioception is gut surveying, in which the surveyors' levels spin around an inner sense of space *and* a sense of inner space alike) which literally embodies our on-board and in-born (and thus the first) global information system. A similar measure (in the musical sense of the Centrifug(u)e) seems to inform what John Pickles sees as an explicit connection between ht and GIS, i.e., that:

> With the emergence of spatial digital data, computer graphic representation, and virtual reality . . . [t]he principle of intertextuality common to both hypertext and GIS directs our attention to the multiple fragments, multiple views, and layers that are assembled under new laws of ordering and reordering made possible by the microprocessor.
>
> (1995: 9)

Which is to say (again) that the gesture of the parenthetical, the dialectic, the thematic, the rhythmic, the fugal, the isobaric, the metonymic, the list, the link, the litany, as well as any and all other – whether en-dashed or no – appositional stitchery constitute the space of hypertextuality.

> More importantly this is space lived *in*, in the double sense of 'in the body' and 'habitat', both body and space being the arena of habitual, which by dictionary definition, means 'established by long use'. In hypertext the forms of stories are quite literally established by use, not unlike the way that Sauer, in a gentler but no less threatened era for the science of geography, characterised the geographer's art as 'distinctly anthropocentric, in the sense of the value or use of the earth to man' and, not unlike notions of hypertext contour, again and again located the actual (areal) for of the world in the act itself of passing by:
>
> > We are interested in that part of the areal scene that concerns us as human beings because we are part of it, live with it, are limited by it, and modify it. Thus we select those qualities of landscape in particular that are or may be of use to us . . . The physical qualities of landscape are those that have habitat value, present or potential.
> >
> > (Sauer 1963: 393)

Edge wise

> Even the lemellae of the desert slide over each other, producing an inimitable sound.
>
> (Gilles Deleuze and Félix Guattari, 'Treatise on Nomadology – The War Machine')

> the critique of the 'instrumentality' model doesn't explicitly take into account the aporetic, inconsistent effects of desire. It would be interesting to pursue the model of the pervert's reading of the link, to determine how it might cast instrumentality in ways that exceed mere utility.
>
> (Terry Harpold, 'Author's Note' 1996)

A knife is the figure of the inside out. I don't like to think about them. A well sharpened blade repeats, and in the repetition anticipates, the contour of the inside out.

Where has the knife come from? The gesture is that of the *policier*. Its mere existence makes a mystery of known space. The knife is likewise history. The spatialisation of history (its cut) makes a place for the play of reading and writing (which is of course the/a play of/on words) and the cultural margins (whose knife? the murderer or the emperor? do they differ?) which contain and represent them.

> If hypertext is constituted by boundfulness–space that ever makes itself, slice by slice, section by section, contour by contour, never getting anywhere; then, like the spaces of our perception and occupation newly opened by GIS, it always both contains and simultaneously escapes the new orders to which it is subject. Each new slice of Zeno's GIS offers real estate for imperial economies yet likewise also opens a space of contestation for the foreigner armed now with the blade of the heretofore unexpected extents he natively inhabits.

In a section called 'the silenced spatiality of historicism' within a chapter theorising from the perspective of Foucault's heterotopia (and to which I am obviously indebted) Edward Soja defines historicism as 'an overdeveloped historical contextualisation of social life and social theory that actively submerges and peripheralises the geographical or spatial imagination' (1993: 140). Indeed in a dazzle of boundfulness and a fearless succession of (self- and othering) citational edges, Soja in his chapter 'the Trialectics of Spatiality' in *Thirdspace*, cites himself citing, in *Postmodern Geographies*, the proto-hypertextual ur-text of '[the Argentinian writer Jorge Luis] Borges' brilliant evocation of the Aleph as the place "where all places are"' (1996: 56) as a locale where '*[e]verything* [his italics] comes together' including:

> subjectivity and objectivity, the abstract and the concrete, the real and the imagined, the knowable and the unimaginable, the repetitive and the differential, structure and agency, mind and body, consciousness and the unconscious, the disciplined and the transdisciplinary, everyday life and unending history.
>
> (Soja 1996: 57)

Hypertextuality considered in its widest aspect (whether Web or 3space) has already fallen prey to an historicism of Soja's sort, ranging from the overdeveloped contextualisation of so called cyberspace by webs of interlinked advertisements for other adverts and search-engines to the submersion and peripheralisation of locality and embodiment alike – which thus far constitutes virtual reality. Of the latter, it is enough to say for now that VR perhaps provides the most obvious instance of how cyberspace exhibits what David Harvey calls 'speculative place construction'.

'Profitable projects to absorb excess capital have been hard to find in these last two decades,' says Harvey, 'and a considerable proportion of the surplus has found its way into speculative place construction.' (1993: 8).

VR aside (and where else, pray tell?) there is likewise a kind of Cliff Notes cartography of certain human interface specialists which too 'submerges and peripheralises the geographical or spatial imagination'. Consider for instance the successive taxonomic slices of Fabrice's (1990: 27–49) obviously earnest, if ultimately muddled, metaphors for data-types in an essay titled 'Information Landscapes'. Florin suggests five categories (one hesitates to say spaces or entities); the following is my annotation of Dieberger's summary of them:

- collections of data which are represented as (literal) fields in the landscape. That is, fields with older data vanish to the horizon like certain kinds of legal testimony and lovers' excuses;
- interactive documentaries are visualised as a kind of village in which, one supposes, people come and go talking of Michelangelo. These are differentiated from the following;
- annotated movies which are characterised by their linear structure and so are represented (Heraclitus not withstanding) by rivers and (Al Gore and Bill Gates standing by) highways and so on;
- networks of guides which are visualised as other persons in the landscape each one one supposes garrulous as streams and various as trees and always ready to help a stranger, Stranger; and
- hands-on activities which range from simple games to complex simulations and which perhaps we ought to imagine as playgrounds, amusement parks, or legos, i.e., little recursive versions of the same five data-types.

This schema is of course merely an exteriorised version of the briefly popular 'social interface' wherein we move from a desktop metaphor to include the kitchen sink. Here both desk and sink are leashed and taken outside for a walk. Because most HCI (human computer interface) specialists are innocent of history and naive as dollhouses or miniature railroads, the problems which accompany slicing the world into landscapes and which are the grist of post-colonial *policiers* don't bear rehearsing to them. Yet 'reading "the iconography of landscape"', as Peter Jackson notes (appropriating the quoted phrase from Cosgrove and Daniels 1988) means 'arguing from a world of exterior surfaces and appearances to an inner world of meaning and experience' (1989: 177). To be sure one might long for an information landscape even half as rich in meaning and experience as Breughel's or at least one which could represent as much as Breughel's poet, the old master Auden, did of suffering and 'its human position: how it takes place/While someone else is eating or opening a window or just walking dully along' ('Musée des Beaux Arts'). All this is, however, cut out of the world when taken as interface. Disembodied interface puts the world on edge, contourless. It is necropsy.

> Against it there stands a different kind of cutting, the biopsic boundfulness of what Soja vis-à-vis the Aleph calls 'all-inclusive simultaneity' which
>
> > 'opens up endless worlds to explore and, at the same time, . . . invokes an immediate sense of impossibility, a despair that the sequentiality of language and writing, of the narrative form and history-telling, can never do more than scratch the surface of [its] extraordinary simultaneities'.
> >
> > (Soja 1996: 57)

Florin's information landscape would seem at first heterogeneous enough to suit a contemporary view of both simultaneities and socially constructed space, given its mix of landform, social scape and potential human interaction. However it is not coincidental that Florin comes to HCI from the world of television. This peopled landscape is sliced by the relentless transits of the surveying beam which sections scape and self and site alike.

Because the Web links edgewise, I have said elsewhere, it suggests that every screen is linked to another; thus on the Web the true hypertext is the severing of one screen from another. Exclusion and inclusion interact, the outside defines the centre (in the body this is called proprioception, that is how the body perceives depth by its own depth, surface by its own: is the printed whorl where the finger ends or the world begins?).

There is a story in every slice. This is the story of contour, at least if taken as something other than a metric.

Joseph Paul Jernigan of Waco, Texas, convicted murderer, http://www.nim.nih.gov, the male Visible Human, comprises 1,878 slices for a total raw data of fifteen gigabytes, including MRIs, CAT scans, and photographs. The anonymous female Visible Human comprises 5,189 cross-sections. Jernigan's last meal comprised two cheeseburgers, fries, and iced tea. Jernigan was missing one tooth, his appendix and a testicle at the time of his death.

Because of the computational power and dataspace he requires, the Visible Human exists only in networked iterations and dynamic representations. Because he is dead, he cannot be said to exist. Because he exists

Space in the Singular(ity)

> I placed a jar in Tennessee
> And round it was, upon a hill.
>
> ('Anecdote of the Jar', Wallace Stevens)

> I take SPACE to be the central fact to man born in America, from Folsom cave to now. I spell it large because it comes large here. Large and without mercy.
>
> (*Call me Ishmael*, Charles Olson)

Here is the parable of singularity, as told by the Boddhisatva of Ice:

> My wife's son-in-law tells a story which for reason's of parabolical compression I usually retell as if it happened to him. He is driving back to his home in winter (for purposes of the parable I make this a cabin in the Sierra foothills reachable only by four wheel drive along a logging road which is often closed; outside the parable he lives in Michigan, his name is Joe Moon) carting a large bottle of drinking water, the kind of oversized thick plastic jug that you sit inside a water cooler. He has been doing errands all day, it is a long drive from the cabin to town. At the cabin he puts the water jug up on his shoulder and begins to cart it up the path to the cabin when, suddenly, the sloshing stops and he feels it turn to ice in an instant, a solid block on his shoulder.

This is the singularity, the point where a system shifts states in perturbation (this sentence once had the word "causes" in it, once its subject was its object). When I tell the parable I often say there is no single point, no zero Centigrade, where the water becomes ice. Instead imagine the slow jostle of the Subaru (let's say) along the mountain roads, the rhythmic lift and dip of the jug on my wife's son-in-law's shoulder, the sudden shift when sloshing sleet is solid.

Yet the same is true of the parable, whether told here or in whatever there (the Philosophy Building at the University of Hamburg, for instance: you could mark one such actual telling so with coordinates of a GPS or the actual date of one of its tellings – 8 January 1997 – yet there is no zero Centigrade which constitutes a telling, no measurable parable, no historical story).

Already, above (where is this 'above' I cite so casually?) the story has threatened to grow interstitially beyond the controlled sense in which I have let its multiplicities and simultaneities express themselves (as if they could be stopped from doing so). Some of these expansions seem as much musical (temporal) as spatial, for instance I was tempted (in fact, actually entered and then cut and repasted here) to add the phrase upside-down to the following phrase now no longer above, now no longer the phrase I was tempted to add to, but another phrase with its own life here: 'carting a large bottle of drinking water, the kind of oversized thick plastic jug that you sit upside-down inside a water cooler'.

Yet even the temporality here is multiple. At very least it is doubled, with the first temporality residing within the time (the prosody) of the sentence in which 'upside-down-inside' takes on a certain pleasant rhythm and a comic (if not cosmic) joy; and the second the temporality (largely spatial) of 'in illo tempore', the vouching for the perceptible truth in a narrative which induces an audience or a reader having experienced an upturned jug of a water cooler to endow this story with plausibility (in fact Joe Moon has such a water jug in Michigan, in fact that is his name, in fact that, rather than Tennessee – or for that matter the California of Sierra foothills – is the state where he lives).

Yet where is the space of this part of the parable? Does it reside in the space of an imaginary foothill? the single topological surface of the torus upon the hero's shoulder? the notion of singularity as expressed by an only half-comprehending fiction writer afraid of being found out by a real physicist? the water cooler a reader imagines? the proprioceptive memory of the heft of an upturned jar?

It doesn't matter of course (it isn't matter of course). The space resides in the web of its tellings. Sometimes the space is literal (an obvious pun compounds into littoral) as in the paragraphical void after the framing device ('above' in illo tempore) of the Boddhisatva of Ice which enables one to occupy several narrative perspectives (where one is two at least: writer and reader); or the packed syntactic syntagma of 'wife's son-in-law' which, almost ideo-grammatically, tells its own story of postmodern succession, a story however 'untold' here and thus one which truly (insofar as it is a story not included in the otherwise seemingly plentitudinous stories which open out from the parable) marks the bounds of the story which is told here.

This meta-telling is its own sort of verification and placement in several senses. You can see the space where the sentence ('above') was not edited for the rhythm of the inside upside down, and where (also 'above' though elsewhere) where it was so emended. You can see the self-reflexiveness of the story begin (parenthetically) to unravel. (A list can be likewise parenthetical, witness the mini-story of the extended phrase which marks my anxiety speaking of singularities among physicists, geography among geographers.)

Some such unravelling marks the space of plausibilities.

For instance in coming to write this story I began to think that the urgency which is gained by endowing the oral telling of the parable of the trip to town with the Sierra locale is somewhat lost, if not threatened altogether, in print where Sierra reads as a metonym for spring or melted snow (albeit romanticised – since a back-packer knows giardia defuses some figures for clear water). I wondered, briefly, whether I should account for this, or rather whether such an accounting would gain in parabolical plausibility what it loses in sluggish scientism. (Words have this same, non- or not merely syntactic, trade-off: does 'parabolical' or 'scientism', each authenticated OED usages, cause the reader to attend to the sentence or to fall from it? and what of extended parenthetical musings? or the tokenised metaphor of defusing as a description of the action of a parasite upon a less-tokenised but conventional metaphor of mountain streams?)

A narratologist in the Hamburg audience was quite angry: 'Space! Space! All you talk about is space! What's happened to time? A story is temporal' (8 January 1997 Temperature 4°C The Alster frozen).

The hypertextual story is the space opened by its telling. And the space that it opens is called . . .

'. . . house again, home again jiggitty . . .'

> This is the lost and found knowledge, the assurance of touch, head to foot. This is buoyancy, hazard, and waywardness – what it is to be at home, unhoused, ongoing . . . Deprived of the elemental; world – and who isn't, with a globe divided, the whole planet sectioned, roofed, cut and pasted – even its waters – what can a body do, if it is a body, but acknowledge, salvage, the elements in its own boundaries. Draw them out. Wring them out. Host. House.
>
> (*The Body in Four Parts*, Janet Kauffman)

This is about hypertext. 'About' as in around&about. Or roundabout. Home&Page. Homepage. The enclosures of form which form us. Jiggity. Our boundfulness. Jag.

All around us, if you haven't noticed, homepages are disappearing, torn down or overwhelmed by on-line shopping malls. All around us, if you haven't noticed, homepages are appearing, persistent as toadstools in the parking lots of shopping malls. All around us home is here and not here, we are bounded by it, bound for it. Home is the heterotopic place 'outside of all places' in Foucault's telling, 'even though it might be possible to indicate their location in reality'. And the name of their location (in reality) is house.

heterotopic=house

The space in which we do and do not appear is the house. Foucault suggests that a history of spaces is a history of powers including 'the little tactics of the habitat' (1980: 149). The heterotopic strategy of making oneself the centre of a text about boundfulness is house-building. 'With the house that has been experienced by a poet', Bachelard says, 'we come to a delicate point in anthropo-cosmology. The house then, really is an instrument of topo-analysis; it is even an efficacious instrument, for the very reason that it is hard to use' (1964: 47).

Bachelard situates the poet's house so that in some fashion it reverses the flow of Kunze's fourth dimension with its 'dynamics of moving from images to solid realities, i.e., the world in which human action becomes actual, the real and sharable world'. For Bachelard the geometrical rationality of housal space 'ought to resist metaphors that welcome the human body and the human soul'. Instead 'independent of all rationality, the dream world beckons' (1964: 47).

The architecture of an argument (a body, a hypertext, a readiness for something to happen) is not the same as the house (the world, a reading, the lines of desire) which sustains it. Kunze writes:

> 'Architecture', while it requires artifacts to sustain it, is more like a readiness for something to happen. It forms the lines of desire and the

> defenses against danger. It crystallises between our hunger for and revulsion with the world. Architecture is not identical with the objects it needs to sustain it. But when does a stone stop being a rock and start being the key of an arch?
>
> (1995: on-line, unpaginated)

One answer is when we need it to be so. 'The body is our general medium for having a world,' Merleau-Ponty argues (1962: 146). Of his three possible worlds, including the biological which the body 'posits around us' in 'actions necessary for the conservation of life'; and the cultural in which 'meaning . . . cannot be achieved by the body's natural means' and so it 'build[s} itself an instrument'; it is the mysterious third which suggests the space of heterotopic hypertext . . . Merleau-Ponty has no convenient container like biology or culture at hand for this sense which he baptises a 'core of new significance'. As the middle world in his original text it is presented as a literal movement from the biological to the cultural world, manifested through 'motor habits such as dancing'.

> And so we come full circle (it runs in the family of unrelated Joyces) to Foucault's mirror of 'mixed joint experience' where 'I see myself there where I am not' and behind me I see Samantha spinning while the young men soar link to link overhead. It is here halfway spinning halfway soaring that I set up my ritual shed (this is about hypertext) and call it
>
> homme sweet home

14 Unthinkable complexity? Cyberspace otherwise

Nick Bingham

> [I]f we had to describe those imbroglios of computer chips, organisations, subjectivity, software, legal requirements, routines, and markets without using modernist or postmodernist idioms, how would we proceed?
>
> (Latour 1996b: 305)

Introduction: everywhere the same

Once upon a time, cyberspace was just a word; assembled, as William Gibson, the writer who coined it, puts it:

> from small and readily available components of language. Neologic spasm: the primal act of pop poetics. Preceded any concept whatsoever. Slick and hollow – awaiting received meaning. All I did: folded words as taught. Now other words accrete in the interstices.
>
> (1991: 27)

No longer merely a word, cyberspace is now a full fledged thing, and that thing is everywhere. We know that it is everywhere because we are told that it is everywhere. It must be everywhere: look what we have now that we did not have then. We have magazines about cyberspace (*Wired* and others), we have newspaper supplements about cyberspace (*The Guardian*'s *Online* etc.), we have how-to guides about cyberspace (*The Rough Guide to the Internet* etc.), we have memoirs of cyberspace (*Surfing the Internet* etc), we have (non-sf) novels about cyberspace (*Email* etc.), we have films about cyberspace (*Johnny Mnemonic* etc.), we have endless television and radio programmes about cyberspace (*The Net* etc.), we have celebrities about cyberspace (Sadie Plant etc.), we have cafés about cyberspace (*The Hub* etc.), and we even have the obligatory media scares about cyberspace ('Free Kiddie Porn Shock' etc.). And now, last and quite possibly least, we have academia about cyberspace: the titles are too many to mention, but suffice it to say that cyber- has replaced post- as most favoured prefix (cyber-space, cyber-bodies, cyber-cities, cyber-sex, cyber-futures (the bracket remains open ended) . . .

On its own, this state of affairs is fairly remarkable. What is even more so is this: with few exceptions, the story being told through each of these diverse media is essentially the same. Over and over again the same. The newspapers the same as the films, the magazines the same as the academics. As Thrift has recently noted of the latter:

> There is a mode of writing about electronic telecommunications technologies which is now becoming ubiquitous. According to this body of literature what we are now seeing is nothing less than a new dimension coming into existence. This new space (and interestingly, it is nearly always a space) goes under many names [. . .] but they all signify the same thing.
>
> (1996a: 1465)

And that is, as Sean Cubitt has recently put it, that 'the technological determinist McLuhanism revived by Jean Baudrillard is almost an orthodoxy' (1996: 832).

It is this orthodoxy which I want, in this chapter, to join Thrift and others in challenging. It is an urgent task: already a cyber-discourse has solidified, and it will take much work to make the already too familiar strange again (Dienst 1994). But above all it is a necessary task if the academy is to take seriously its new found reflexiveness about the sorts of story that it tells and the ways in which it tells them. For as James Carey and John Quirk wrote over 25 years ago: '[t]he promotion of the illusion of an electronic revolution borders on complicity by intellectuals on the myth-making of the electrical complex itself' (Carey 1989: 138).

In terms of structure, I approach this issue by dividing the chapter into two main parts. In the second, I draw on the work of Michel Serres and Bruno Latour (amongst others) in an attempt to flesh out a style of articulating cyberspace that might act as a productive alternative to the rather limited way in which it is currently treated. Before that, however, I want to characterise in more detail, that mode of thinking our virtual geographies which is currently so prevalent. Specifically, I want to argue that what we are offered is cyberspace as a contemporary manifestation of the technological sublime. (It is worth noting here that in both cases I will be largely considering 'cyberspace' in its narrower sense, that is as the 'terrain' produced by and through computer-mediated-communications (CMC) such as e-mail, newsgroups, and on-line chat functions).

Cyberspace as technological sublime

> Cyberspace. A consensual hallucination experienced daily by billions of legitimate operators in every nation . . . A graphic representation of data abstracted from the banks of every computer in the human system. Unthinkable complexity. Lines of light ranged in the non-space of the mind, clusters and constellations of data. Like city lights receding . . .
>
> (Gibson 1984: 67)

The figure of the sublime has had a long and complex history. It has been taken up and (re)defined by a number of different authors writing in a number of different locations and periods – Longinus in the sixth century, and Immanuel Kant, Joseph Addison, and Edmund Burke in the eighteenth being perhaps the most notable – and has undergone something of a revival in our own. It has slipped back and forth between describing a rhetorical strategy and a category of aesthetic experience, and has been applied to an almost endless list of objects. What has remained relatively stable throughout these metamorphoses, however, are the sorts of qualities that can be said to characterise the sublime. As catalogued by Burke, these include 'power, deprivation, vacuity, solitude, silence, great dimensions (particularly vastness in depth), infinity, magnificence, and finally obscurity (because mystery and uncertainty arouse awe and dread)' (Williams 1990: 85–86): all traits, as Scott Bukatman puts it, that 'suggest realms beyond human articulation and comprehension' (1995: 266). The same author summarises the feelings aroused by such sensations thus:

> The sublime initiates a crisis in the subject by disrupting the customary cognized relationship between subject and external reality. It threatens human thought, habitual signifying systems, and, finally, human prowess: the mind is hurried out of itself by a crowd of great and confused images, which affect because they are crowded and confused. The final effect is not a negative one, however, because it is almost immediately accompanied by a process of, and identification with, the infinite powers on display. The phenomenal world is transcended as the mind moves to encompass what cannot be contained.
>
> (ibid.: 266–267)

Now, as I have already signalled, the precise origin of that which 'cannot be contained' has varied over time. Originally associated with poetic language and spoken rhetoric in particular, by the eighteenth century it was accepted that the sublime might be evoked by the visual representation of the grandeur of the natural world. Features such as mountains, deserts, oceans, and so forth became a staple of a tradition of landscape painting that, as Bukatman writes, was less concerned with mimetic accuracy than with encouraging 'specific spectatorial behaviours', notably that of mediation upon the magnificence of the Creator (ibid.: 271–272). However, after Burke shifted attention away from the natural by suggesting that feelings of sublimity might be aroused by human constructions of great dimensions – what he called an 'artificial infinite' – and as the nineteenth century heralded what was widely perceived as an industrial 'second nature', the sublime became to be employed more and more with reference to an increasingly technological environment (Williams 1990: 88).

Like its father-figure, the technological sublime also has a history. As David Nye has extensively shown, successive developments – from the railroads, through telephone networks and skyscrapers, to electrification and the atomic bomb – have all, as they gained widespread usage, been thought through in terms of their sublimity (1994; see also Marx 1964 and Carey 1989). In each instance, the same notion has been turned to in order to ground an understanding of an ostensibly novel phenomenon (Bukatman 1995: 284). The technological sublime then, has become a way of 'making sense' of what we might call 'the shock of the new'. Given this background, it is hardly surprising that much of the current thinking around our 'virtual geographies' has followed the same route.

For, as Bukatman puts it:

> The startling rise of mediating electronic technologies has precipitated a crisis of visibility and control. If cultural power now seems to have passed beyond the scales of human activity and perception, then culture has responded by producing a set of visualisations – or allegorisations – of the new 'spaces' of technological activity.
>
> (ibid.: 281)

These new 'spaces' (or, more usually, space singular as we shall see), have been referred to by various terms, from (to pick some of the more influential) 'space of flows' (Castells 1996), through 'postmodern hyperspace' (Jameson 1991), to the most common of all, 'cyberspace' (Gibson 1984). Various terms, but the same rhetoric: the vocabulary of the technological sublime is mobilised by each of these writers. Not (as with the natural sublime) in order to conceptualise the encounter with an object of physically overwhelming size or power, but rather as a means by which to 'get a grip on' the complexity of the processes and relations which, while increasingly facilitating the activities of everyday life, seem to be beyond rational comprehension (cf. Crowther 1993: 164). The frequency with which this move is now made has meant that the technological sublime has become the preferred trope by which to represent our virtual geographies. What it also means is that at least three interrelated tendencies, internal, I would argue, to that figure, are reproduced in an ever-increasing number of locations.

tendency #1: virtual totality

The first is that of thinking the world (only) in terms of totalities. As Joseph Tabbi has recently summarised, our contemporary 'crisscrossing networks of computers, transportation systems, and communications media, successors to the omnipotent 'nature' of nineteenth-century romanticism, have come to represent a magnitude that at once attracts and repels' (1995: 16). Faced with such, the imagination can react either passively or actively: either it

> wishes to be inundated in the network, and thus risks experiencing a loss of identity or 'anxiety of incorporation,' or it desires to oppose or replace

> the sublime appearance with a linguistic construction of its own, to 'possess' verbally the object of its anxieties'.
>
> (ibid.: 16–17)

By choosing the second option, thinking through the technological sublime allows the mind to 'encompass what cannot be contained' as we saw earlier. In doing so, the vast external unintelligibility is subsumed into a single image of homogeneous infinity. The most famous articulation of this process regarding the electronic space(s) with which we are concerned here, is provided by William Gibsons's famous lines from *Neuromancer* which open this section. What we have in that quote, as elsewhere in Gibson's work – most notably perhaps (and certainly most visually) in the film adaptation of one of his early short stories; *Johnnny Mnemonic* (for which the author wrote the screenplay) – is cyberspace depicted as a gridded, Euclidean world, stretching uniformly and endlessly in all directions: a geo-metric totality.

Now, if *Neuromancer* and the rest of the 'Sprawl' trilogy were just the latest in a long line of science fiction novels drawing upon the language of the technological sublime in order to communicate the experience of yet another futuristic landscape, they would be of little more than passing interest here. But the fact is, as David Tomas has documented, Gibson's 'powerful vision' has become much more than that, influencing 'the way that virtual reality and cyberspace researchers are structuring their research agenda and problematics' (1991: 46). For Sandy Stone, its effect has been even more widespread: she writes that by articulating a new 'technological and social imaginary' and crystallising a new research community from a number of disparate fields, *Neuromancer* in particular acted as

> a massive intertextual presence not only in other literary productions of the 1980s, but in technical publications, conference topics, hardware design, and scientific and technological discourse in the large.
>
> (1991: 95, 99)

That 'large' has included the social sciences, and it is easy to see why. For the sort of unified vision of cyberspace that Gibson offers fits in very nicely with a strand of the thinking of society in general and technology in particular that has been dominant for at least the last century (Thrift 1996a). In its quest for the 'big picture' this tradition – epitomised by the less modest forms of Marxism – has, as Paul Crowther has noted (1989: 163–165), often invoked the rhetoric of the sublime, and today writers such as Fredric Jameson, David Harvey, and Manuel Castells all write of the need to conceptualise the 'global social totality' (Thrift 1995). The epic accounts of 'globalisation', and 'time-space compression' produced by this style seem to lead inexorably to the sorts of commentaries on electronic spaces that have provoked Bryan Winston to talk of the emergence of an 'unthinking "basic litany"' regarding computer-mediated-communication technologies. A variety of assumptions – that they can be thought of as unproblematically 'new', as producing a 'crisis' of representation, and as acting as the interface to an-other, somehow 'unreal' world (ibid.: 228–232) – have become cyber-clichés, and consequently discussion of our virtual geographies has become constrained by a very limited frame and the need and/or desire for easy stories. For as Jim Collins puts it, positing a 'systemacity to information production, circulation, and repetition' results in the construction of a 'matrix that makes everything far more manageable, narratively, and ideologically'. Any emphasis on exceptions to such a formulation, he continues, 'would, of course, undermine the binary opposition between cowboys and corporations' (1995: 14, 15).

tendency #2: virtually deterministic

Collins' last sentence not only serves as a neat introduction to the second tendency that I want to identify as internal to the figure of the technological sublime and the writings that draw upon it, but also signals how it is in many ways a consequence of the first. For by thinking in terms of totalities, as we have seen much of the social science literature that I am concerned with here does, such work also presupposes a stable and unproblematic distinction between the social and the technical. Or more precisely here, between the self and cyberspace. The encounter between the 'cowboy' and the 'corporation' that Collins mentions, may be a reference to the narrative structure of Gibson's novels, but it also stands as an illustration of a wider inclination in the non-fictional work which has employed the same rhetorical devices.

The reason why this should be so is not difficult to understand. As we have already noted, the sublime sets itself up as an encounter between an individual and an external object – in this case technological – of literally awe-inspiring dimensions. Even when the mind moves to grasp this 'artificial infinity', the observer is left with a feeling of helplessness in the face of an autonomous non-self. As Rosalind Williams puts it more concisely, 'the aesthetic of sublimity implies technological determinism' (1990: 89), in that it attributes (or perhaps better distributes) agency such that the non-human part of the 'equation' is credited with a significantly greater capacity to act, to exert force. Technological determinism, of course, is something which from which, almost without exception, social scientists wish to dissociate themselves. Nevertheless, it remains the most powerful (and hence the most popular) discourse of material–social change currently in circulation, and as such is probably the best way to describe much recent theorising about cyberspace and associated phenomena (Bingham 1996; Thrift 1996a). While the traditional 'billiard-ball' model, according to which a technological innovation 'rolls in from outside and "impacts" elements of society' (Fischer 1992: 8) is eschewed, the newer 'impact-imprint' model is prevalent in discussion of the 'effects' of computer-mediated-communication:

> According to this school of thought, new technologies alter history, not by their economic logic, but by the cultural and psychological transfer of

> their essential qualities to their users. A technology 'imprints' itself on personal and collective psyches.
>
> (ibid.: 10)

By remaining wedded to the traditional modern agenda that insists a priori that the social and the technical are separate and to be treated as such, the sort of commentary that Claude Fischer is referring to repeatedly ignores the findings of a heterogeneous but coherent body of work that reasons not from 'the properties of tools' but 'what people do with the tools' (ibid.: 11; Chartier 1997). The notion that the spread of electronic media might best be expressed as histories of the practices by which such technologies become part of the fabric of the day-to-day lives of all sorts of groups and individuals, and that these processes might vary spatially and temporally, seems totally antithetical to the grand ambitions of prevailing social science orthodoxy regarding such matters. Or as Collins puts it:

> The one position that remains unarticulated [. . .] is that this cyberspace, rather than being imagined as a sum total or a shape, might be more productively thought of in terms of discontinuity, a dissonant cacophony resulting from different technologies being put to radically different uses.
>
> (1995: 15)

tendency #3: virtual mastery

Once again, then, we are reminded that, by making certain things more thinkable than others, the technological sublime, like all forms of rhetoric, is not a 'disinterested' aesthetic strategy but caught up with discourse as a form of power (Williams 1990: 88). This point is underlined by the last of three tendencies through which I want to characterise it. By now, it should be no surprise that I wish to draw attention to what underpins the vocabulary of 'confrontation and mastery' that typifies the sorts of accounts of cyberspace that I have been considering here. For, as Bukatman notes, 'one must acknowledge (at least briefly) the recurrent fantasies of sexuality and power at work within many of these texts' (1995: 287).

And they are not hard to find: as the initially destabilising moment of being faced with the 'unthinkably complex' is transcended, the position of the observer and the observed are reversed, leading to a 'renewed and newly strengthened experience of the self' (ibid.: 284) which is now 'free' to apprehend 'the whole' all at once. This, of course, is the masculinist 'god-trick' (Haraway 1991b) par excellence: the dream of a disembodied viewpoint that yields an (imaginary) totalisation (Deutsche 1991: 9 after de Certeau), of an (impossible) 'august position' – the place of Critique – in which 'one is always in the right, the most knowledgeable and strongest' (Gibson 1996: 112 after Serres).

Even this 'privilege', however, is only granted to the few. For the most part in that social theory which relies on the device of the (technological) sublime, the 'masses' continue to be regarded as more or less helpless. When that work concerns cyberspace, individuals tend to be portrayed as drowning in a sea of information, unable to form a coherent identity in a world in which an excess of signs has swept away established measures of space and time (Collins 1995: 31–32). Only those who have done the detective work and have discovered the key that unlocks a 'secret' reality (Perniola 1995) can possibly have the perspective from which to tell us what's *really* going on. Now although such an argument is patently untenable – for basic semiotic reasons on top of everything else (Collins 1995: 32) – it is religiously retained because the other option is held to be descent into the 'random and undecidable world of microgroups' (Jameson quoted in ibid.: 37). Once again an ill-conceived dichotomy, a false problem: there are other options.

Cyberspace as message-bearing system

> I will begin by telling you an ancient legend.
>
> Late in life the emperor Charlemagne fell in love with a German girl. The barons at his court were extremely worried when they saw that the sovereign, wholly taken up with his amorous passion and unmindful of his regal dignity, was neglecting the affairs of state. When the girl suddenly died, the courtiers were greatly relieved – but not for long, because Charlemagne's love did not die with her. The emperor had the embalmed body carried to his bedchamber, where he refused to be parted from it. The Archbishop Turpin, alarmed by this macabre passion, suspected an enchantment and insisted on examining the corpse. Hidden under the girl's dead tongue he found a ring with a precious stone set in it. As soon as the ring was in Turpin's hands, Charlemagne fell passionately in love with the archbishop and hurriedly had the girl buried. In order to escape the embarrassing situation, Turpin flung the ring into Lake Constance. Charlemagne thereupon fell in love with the lake and would not leave its shores.
>
> (Calvino 1992: 31)

A medieval French myth may seem an unpromising location from which to begin proposing alternative techniques of articulating cyberspace, but by examining a bit more closely that which acts as both 'narrative link' and 'real protagonist' of this wonderful story, an-other way of telling begins to reveal itself. The entity in question, of course, is the magic ring because, as Calvino notes:

> it is the movements of the ring that determine those of the characters and because it is the ring that establishes the relationships between them. Around the magic object there forms a kind of force field that is in fact

> the territory of the story itself. We might say that the magic object is an outward and visible sign that reveals the connection between people or events.
>
> (ibid.: 32)

What I want to suggest is that the magic ring of Calvino's legend operates as an imperfect – but in this context highly suggestive – exemplar of what the philosopher Michel Serres has christened 'quasi-objects'. As I have shown at greater length previously (Bingham 1996), disillusioned by the impoverished position that 'things' have traditionally held in philosophy and social science (to summarise: very strong or very weak) Serres has sought to produce a 'decent philosophy of the object' that grants all sorts of entities their due role in the (co)construction of the world. Re-visioned as quasi-objects – 'multiple in space and mobile in time, unstable and fluctuating like a flame, relational' (Serres 1995a: 91) – the active role that non-organic actants play in constituting 'the group that thinks, that remembers, that expresses itself and, sometimes, invents' (1995b: 50) can finally be re-cognised. Like the magic ring above, they accomplish this function by circulating amongst – and hence linking – some (but not others) of the variety of local spatialities on which Serres' notion of the social is premised:

> my body lives in as many spaces as the society, the group, or the collectivity have formed: the Euclidean house, the street and its network, the open and closed garden, the church or the enclosed spaces of the sacred, the school and its spatial varieties containing fixed points, and the complex ensemble of flow-charts, those of language, of the factory, of the family, of the political party, and so forth. Consequently, my body is not plunged into one space, but into the intersection of the junctions of this multiplicity.
>
> (1982: 44–45)

Each culture, for Serres, may be described by the way it 'constructs in and by its history an original intersection between such spatial varieties' (ibid.: 45). Both this intersection and that which it holds together, of course, alter with time: societies metamorphosise as certain morphologies mutate or fade away while others are born or gain in importance: 'the conception, the construction, the production of rapports, of relations, of transports – communication in general – evolve so fast that they continually construct a new world, in real-time' (Serres with Latour 1995: 114). This is the process – or the processes – that Serres has attempted to comprehend throughout his work via the legend (in both senses of that word) of the Greek god Hermes (see especially 1982) who:

> by renewing himself, becomes continuously our new god, for as long as we've been humans – not only the god of our ideas, of our behaviour, of

> our theoretical abstractions, but also the god of our works, of our technology, of our experiments, of our experimental sciences. Indeed, he is the god of our laboratories, where as you [Latour] have pointed out, everything functions through networks of complex relations between messages and people. He is the god of our biology, which describes messages transmitted by the central nervous system or by genetics. He is the god of computer science, of rapid finance and volatile money, of commerce, of information, of the medias.
>
> (Serres with Latour 1995: 114)

Latterly, however, the extent of this last set of developments has prompted Serres to classify his project as 'a general theory of relations, like a theology in which the important thing would be an angelology – a turbulent array of messengers' (ibid.: 108). For '[t]oday what comes up even more than the figure of Hermes is the figure that he will take on his death [. . .] that of the multiplicity of angels' (ibid.: 118). In his most recent work (particularly Angels (1995b)) and consistently in the work of Bruno Latour who has, in many senses, translated Serres' project from philosophy to the social sciences, a certain importance has been granted to the sorts of relations traced by a subgroup of the category quasi-objects that Latour has termed 'immutable mobiles'. Immutable mobiles are 'materials which can easily be carried about and tend to retain their shape' (Law 1994: 102) and, as Robert Cooper illustrates, provide the curious possibility of allowing things to be simultaneously far and close:

> Administrators and managers, for example, do not work directly on the environment, but on models, maps, numbers, and formulae which represent that environment; in this way they can control complex and heterogeneous activities at a distance and in the relative convenience of a centralised work station. Events that are remote (that is, distant and heterogeneous) in space and time can be instantly collated in paper form on the desk of a central controller. This has the paradoxical effect of bringing remote events near while, at the same time, keeping them at a remove through the intervention of representations. In other words, the power of representation to control an event remotely is a form of displacement in which representation is always a substitution for or re-presentation of the event and never the event itself.
>
> (1992: 257)

Hence, (and to simplify), we might say that, historically, a number of sociotechnical developments including writing, print, paper, money, a postal system, cartography, navigation, and telephony have all generated both new forms of immutable mobiles and (consequently) the potential for new configurations of centres where they may be combined and peripheries from where they may be gathered (Law 1994: 103–104). I have argued elsewhere that the technologies that form the conditions of possibility for computer-mediated-communication might fruitfully be treated as the most recent of such 'innovations', and it is that line of thought that I want to continue (in a slightly different direction) below. Specifically, I want to propose three affordances in particular that such a move offers as compared with the style of writing which I identified in the first half of the chapter, as (sadly) hegemonic in this area.

affordance #1: from surfaces to networks

First, then, what I shall call for convenience the 'amodern' style of thinking through (cyber)space suggests that we might be better off conceiving of the social in terms of networks rather than totalities; as having a 'fibrous, thread-like, wiry stringy, ropy, capillary character that is never captured by the notions of levels, layers, territories, spheres, categories, structure, systems' (Latour 1997: 2). Ironically one of the examples by which Latour demonstrates how such a change in metaphor can lead to accounts that are at once more subtle, more historical, and more empirical, is a technological network:

> Is a railroad local or global? Neither. It is local at all points, since you find sleepers and railroad workers and you have stations and automatic ticket machines scattered along the way. Yet it is global, since it takes you from Madrid to Berlin or from Brest to Vladivostok. However, it is not universal enough to be able to take you just anywhere. It is impossible to reach the Auvergnat village of Malpy by train, or the little Staffordshire town of Market Drayton. There are continuous paths that lead from the local to the global, from the circumstantial to the universal, only as the branch lines are paid for.
>
> (1993: 117)

Ironically, because Latour's point is that we should apply the lessons learnt from technological networks (where we usually have 'no difficulty' reconciling their 'local aspect' and 'global dimension' (p. 118)) to those other elements of our societies such as organisations, markets, and institutions whose size and solidity we are only to ready to exaggerate (pp. 120–121):

> They are composed of particular places, aligned by a series of branchings that cross other places and require other branchings in order to spread. Between the lines of the network there is, strictly speaking, nothing at all: no train, no telephone, no intake pipe, no television set.

In the case of the technological networks that constitute cyberspace, however, being told of in terms of the sublime has meant that here too the space between the connections has been filled in, and we are too often left with yet another sleek surface of the sort from which Latour is striving to depart. Thus, although it may seem obvious, it is necessary to assert that 'the Net' is neither local nor global. It is local at all points since you always find terminals and modems. And yet it is global since it connects Sheffield and Sydney. However, it is not universal enough to take you just anywhere. After all, I cannot e-mail my next-door-neighbour and between a third and a half of the world's population still lives more than two hours away from the nearest telephone.

However, contra the rhetoric of the 'incredible shrinking world' (Kirsch 1995) upon which cyberspace as technological sublime is based, the argument here is not merely that networks have the ability to cross (read transcend) space and time considered as 'an unshakeable frame of reference inside which events and place would occur' (Latour 1987: 228 emphasis in original). Rather, the claim is (much) stronger:

> Space and time are not, contrary to Kant's demonstrations, the *a priori* categories of our sensibility. Gods, angels, spheres, doves, plants, steam engines, are not *in* space and do not age *in* time. On the contrary, spaces and times are traced by reversible or irreversible displacements of many types of mobiles. They are generated by the movements of mobiles, they do not frame these movements.
>
> ((Latour 1988b: 25 emphasis in original)

Just as we saw that Calvino's magic ring formed a 'kind of force field that is in fact the territory of the story itself', so too – as Pierre Lévy (one of Serres' students) has shown (1996) – immutable mobiles such as e-mails, newsgroups postings, and IRC messages trace by their circulation at once new (cyber)-space-times and – which amounts to the same thing – new sorts of collectivities. It is to these last that I now want to turn.

affordance #2: from technical vs social to technosocial

In the re-visioning of cyberspace that I have begun to outline, then, 'heres' and 'theres' do not pre-exist, but are created by the connections that bind them: the social fabric is not self-subsistent, but relational and in the process of constant evolution. One way to consider this in terms of the regional spaces with which we are more familiar, is to think of networks as 'folding' together what are metrically distant points (Serres with Latour 1995: 60, Mol and Law 1994: 649–650). The fold 'locates where and how the world has become compressed' as Tom Conley (1993: xvi) puts it in his preface to Deleuze's *Le Pli* (not 'whether or not' note, but 'where and how') and usefully expresses the effect that networks have of connecting the disconnected. New elements can be mixed together, new juxta-positions created we might say. The outcome of such bringing close is always uncertain: 'the third', as Jeanette Winterson remarks in her *Sexing the Cherry* (1992), 'is not given'. It might be, as Michel Maffesoli has postulated, that new 'tribes' might be generated:

> computer bulletin boards (for amusement, erotic or functional purposes) may create a communicational matrix in which groups with various goals will appear, gain strength and die; groups that which recall somewhat the archaic structures of village clans or tribes. The only notable difference which characterises the electronic nebula is of course the very temporality of these tribes. Indeed, as opposed to what is usually meant by this notion, the tribalism we are exploring here can be completely ephemeral, organised as the occasion arises. To return to an old philosophical term, it is exhausted in the act. As has become clear in many statistical reports, more and more people are living as 'singles'; but the fact of living *alone* does not mean living *in isolation*. According to the occasion – especially thanks to the computer services of the Minitel [a French domestic fore-runner of the services now available internationally through the Internet, nb] – the 'singles' can join a given group or activity. The 'tribes' based on sports, friendships, sex, religion, and other interests are constituted in many ways (the Minitel is just one), all of them having various lifespans according to the degree of investment of the protagonists.
>
> (1996: 1839–1840 emphasis in original)

Most treatments of the emerging new(s)groups associated with computer-mediated-communication (Maffesoli's included) are happy to talk of them as the result of the 'impact' of technology on society or, symmetrically, as the outcome of society 'shaping' technology according to its needs and desires. In both scenarios, the resulting 'tribes' tend to be treated as if they were 'suspended in a vacuum' (Serres with Latour 1995: 142). What the conceptual framework elaborated by Serres and Latour and selectively summarised here affords (in the second place), by contrast, is a way of elaborating human groups as always already thoroughly technosocial (to use an awkward word), always already bound together by quasi-objects. As Lévy remarks:

> One could recount the history of humanity from its very beginnings as a succession of emerging objects, each of them inseparably linked to a particular form of social dynamics. One would then observe that every new type of object induces a particular brand of collective intelligence and that every truly important social change necessarily involves the invention of an object. In the span of anthropological time, collectives and their objects take shape within the same general movement.
>
> (1996: 6–7)

The new collectivities of cyberspace (if that is how we choose to refer to them), then, represent nothing more and nothing less than the latest instalment in a long series of co-productions between people and things. Whether the entities in question have been tools, stories, corpses, or bulletin board contributions, the quality which identifies them remains the same: 'the power to catalyse social relationships' (Lévy 1996: 5; Serres 1995a: 87; Strum and Latour 1987).

affordance #3: from explanation to adequate stories

In principle then quasi-objects always serve the same function (and we must recognise this in our stories). In practice, of course, they will always act differently: the paths traced, the relations made, the groups formed, the spaces linked and created will vary according to context (and we must recognise this in our stories as well). The means to fulfil both these criteria at once represent the final affordance that I want to identify the amodern move as offering. In many ways it is the most difficult chance to take up and operationalise, difficult because we – as social scientists – are so wedded to another way of 'doing things'. For taking seriously the project that Serres and Latour are following through ultimately requires that we abandon our most cherished weapon: the position of domination, as I have already mentioned, that is Critique.

Too often the accounts of computer-mediated-communication we are offered are steeped in what Latour calls the 'politics of explanation' (1988b): they wish to explain cyberspace. What I mean by this is that they seek – as any good Critique work does – to divide the world into two packs: 'a little one that is sure and certain, the immense rest which is simply believed and in dire need of being criticised, founded, re-educated, straightened up . . .' (Latour 1990: 85). By holding on to the short list, whether it contains (for example) 'the cogito', 'the transcendental', 'the class struggle', 'discourse' (ibid.) or – in the case of cyberspace – (for example) '(late-) capitalism', 'masculinity', 'the culture of narcissism', the 'unthinkable complexity' of the rest may be reduced to manageable proportions: explained.

But why seek to explain at all? And why is a powerful explanation – where the elements of the short list are 'correlated' with more of the elements of the long list – seen as inherently better than a weak one? Because what a strong explanation allows us to and a weak one does not is 'act at a distance' (Latour 1988b: 159). If one can remain in (short list) A and still act on (long list) B then what one enjoys is *power*: being able to explain cyberspace or anything else through a simple formula is the first step to gaining authority. And this, argues Latour, is exactly what much social science, as conventionally organised, is about: empire building, the will to recognition. We want to be like the 'real' scientists who we see as representing an ideal because they have 'mastered' explanation.

As Robert Koch has written, '[f]ollowing Serres, who is in turn indebted to the work of René Girard, Latour suggests that the course of academic and political debate is really a *tribunal* in which responsibility for effects is determined and allocated' (1995: 344 emphasis in original):

> It means that a cause (factor, determinant, pattern, or correlate) is the outcome of a trial of responsibility through which a few elements of the network are taken to be the impetus behind the whole business. It is, in practice, very much an election of representatives or, depending upon the

> outcome, an accusation made against a scapegoat. The belief in cause and effect is always, in some sense, the admiration of a chain of command or the hatred of the mob looking for someone to stone.
>
> (Latour 1988b: 162)

Instead of repeating this 'regime of accusation' (Koch 1995: 345) and offering us yet another short, power-full list or 'metalanguage', what Serres and Latour provide is better described as an 'infralanguage' (Latour 1988b, 1997). A set of tools, that is to say, that opens up the possibility of following the actors and the series of transformations, substitutions, and delegations, through which they pass in any given setting, while at the same time holding on to the 'sturdy theoretical commitments' (Latour 1997: 6) that I have sketched out above. The challenge, then, is to produce accounts of the world in which the big categories to which we remain so attached are not assumed in advance but result, if at all, from the actions of the various quasi-subjects or -objects involved and invoked.

The difference that computer-mediated-communication makes . . .

. . . then, cannot be specified in advance. The difference that computer-mediated-communication makes is always different. As is evidenced by some recent interdisciplinary work on the subject, what we might be able to offer, however, by bringing the insights of Serres and Latour to bear on the problematic of cyberspace, are some tentative statements concerning what Latour has recently called the 'regimes of delegation' (1996b: 304) involved in information architectures of various kinds. These are designed to 'follow at once the dissemination of an indefinite number of entities and the limited number of ways in which they grasp one another' (ibid.: 304), which is as good a way as any of describing the efforts of Susan Leigh Star and Karen Ruhleder to reformulate the notion of 'infrastructure' in a way which is appropriate to a world of electronic media.

Asking the question 'when' rather than 'what' is an infrastructure to emphasise that it is always something that 'emerges for people in practice, connected to activities and structures' as opposed to 'a thing with pregiven attributes frozen in time' (Star and Ruhleder 1996: 112), Star and Ruhleder identify eight dimensions to the concept. First, infrastructure is *embedded*, '"sunk" into, inside of, other structures, social arrangements and technologies'; second, it is *transparent* to use, 'in the sense that it does not have to be reinvented each time or assembled for each task', acting instead as an invisible support; third, infrastructure has '*reach or scope*' beyond 'a single event or one-site practice'; fourth, it is '*learned as part of membership*' of a 'community of practice' which acquires a state of 'taken for grantedness' with respect to the 'artefacts and organisational arrangements' of a particular infrastructure; fifth, 'infrastructure both shapes and is shaped by' the '*conventions*' of a community of practice; sixth, infrastructure acts as an '*embodiment of standards*' by 'plugging into other infrastructures and tools in a standardised fashion'; seventh, it 'does not grow *de novo*', but rather is '*built on an installed base*', and thus 'inherits strengths and limitations from that base'; finally, a normally backgrounded infrastructure '*becomes visible upon breakdown*' (all ibid.: 113).

In much greater detail than I have space to enter into here, Star and Ruhleder proceed to employ these eight precepts in an illuminating analysis of the ways in which the work patterns of a geographically-dispersed group of geneticists were altered by their introduction to both a large-scale, custom-built collaborative software tool, and (coincidentally in terms of timing) the Internet with its various utilities. One short passage is certainly worth highlighting, however, if only for the way in which it presents so emphatically the reason why the sort of symmetrical vocabulary developed by Serres and Latour to allow us to speak in the same breath and the same register of both (quasi-) subjects and (quasi-)objects, is so necessary if we are to 'do justice' to the sort of imbroglios of which the milieus of computer-mediated-communication are but a particularly 'recent' manifestation:

> Scientists do not 'live on the net'. They do make increasingly heavy use of it; participation is increasingly mandatory for professional advancement or even participation, with a rapidly changing set of information resources radically altering the landscape of information 'user' and 'provider'; and the density of interconnections and infrastructural development is proceeding at a dizzying rate. That development is uneven; is an interesting mixture of local politics and practices, on-line and off-line interactions, and filled with constantly shifting boundaries between lines of work, cohorts and career stages, physical, virtual and material culture, and increasingly urgent and interesting problems of scale.
>
> (ibid.: 131)

From this perspective, as Latour himself has put it:

> Organisations, finally, no longer look the same now that to their local interactions, and to their dispatchers, has been added so many computers and data banks, so many artefacts and intellectual technologies, so many stories, so many centres of calculation and information processing rooms, so much distributed and situated cognition. It is no longer clear if a computer system is a limited form of organisation or if an organisation is an expanded form of computer system. Not because, as in the engineering dreams and the sociologist's nightmares, complete rationalisation would have taken place, but because, on the opposite, the two monstrous hybrids are now coextensive.
>
> (1996b: 302)

Conclusion: beyond emblems of modernity

> Visions of new technologies revolutionising the way that we live are often bold, sweeping, and millenarian. They are exciting to hear; they sell books; they can earn one a good living on the corporate lecture circuit. But their shelf life is roughly equivalent to that of a Big Mac.
>
> (Fischer 1997: 113)

As in the case of many technological networks before (Buck-Morss 1989: 90–91), a potent brew of equally zealous boosterism and scepticism has elevated cyberspace to the position of an 'emblem of modernity' (to use Fischer's felicitous phrase): truly a sign of the times regardless of whether one views it with hope or despair. What I have argued in this chapter is that taking our virtual geographies seriously must mean moving beyond the myriad temptations such fast food accounts offer towards something a bit more nourishing. By drawing on the work of Serres and Latour, what I have attempted to show – hopefully convincingly – is that there are other ways of telling of computer-mediated-communication than the ones currently on offer, ways that are perhaps more adequate to the task of learning and mapping precisely why, when, and where it truly makes a difference. Let us ask not what cyberspace is or what it stands for, but what it binds together . . . and what it holds apart.

15 Virtual worlds

Simulation, suppletion, s(ed)uction and simulacra

Marcus A. Doel and David B. Clarke

Plato blushes for shame

(Nietzsche 1968: 41)

Introduction

Any old object, individual or situation is today a virtual ready-made

(Baudrillard 1996: 28)

If we let slip a yawn at the mere mention of virtual reality, cyberspace, and embodied virtuality, or roll our eyes at the naming of telepresence, teletopia, and electronic cloning, it is because something has been missed in the headlong rush to exit the common-or-garden experience of everyday life for the apparent wonderment of the latest technologies. According to Robins (1991: 64), the 'new technologies promise . . . nothing less than the "re-enchantment" of our mundane existence', acting as timely antidotes to the long-standing disenchantment and alienation wrought on humanity by abstract codes and machines. More often than not, however, what accompanies discourses on both these latest technologies and the new articulations of space-time that they express – whether utopian, dystopian, or measured – is an impoverished understanding of the real and the virtual; that is to say, an impoverished understanding of space-time. What is striking about this impoverishment is that it infiltrates both the virtual *and* the real, to the point where the virtual is invariably collapsed into a badly analysed version of the real (its degraded or resolved double) – wherein the real and the virtual are no longer distinguishable according to *qualities* (powers and affects), but only according to *quantities* (more or less). This 'collapse' in the configuration of space-time typically takes one of two paths: the virtual as a 'false approximation' of the real (a mere duplicate); or the virtual as a 'resolution' or 'hyperrealisation' of the real – the extremums being occupied by technophobes and technophiles, respectively. In the following two sections of the chapter we consider in detail this bifid collapse of the virtual, before turning to consider more directly the implications of the 'virtual illusion' in

terms of an insatiable desire for the hyperrealisation of the world's possibilities; and, in the final section, the virtual character of the real itself. What is at stake here is not just how we think about reality, virtuality, and virtual reality; it is also how we figure space-time itself. Perhaps the most important error that we wish to highlight is the reduction of reality to actuality and virtuality to possibility: as if the actual and the virtual were the *given* and the *pre-given*, respectively. It is the need to re-think space-time, rather than any new-fangled technologies, which poses the most pressing challenge.

Virtual reality 1: Simulation. Or, the false approximation of the real

> In those days the world of mirrors and the world of men were not, as they are now, cut off from each other. They were, besides, quite different; neither beings nor colours nor shapes were the same. Both kingdoms, the specular and the human, lived in harmony; you could come and go through mirrors. One night the mirror people invaded the earth. Their power was great, but at the end of bloody warfare the magic arts of the Yellow Emperor prevailed. He repulsed the invaders, imprisoned them in their mirrors, and forced on them the task of repeating, as though in a kind of dream, all the actions of men. He stripped them of their power and of their forms, and reduced them to mere slavish reflections. Nonetheless, a day will come when the magic spell will be shaken off . . . shapes will begin to stir. Little by little they will differ from us; little by little they will not imitate us. They will break through the barriers of glass or metal and this time will not be defeated
>
> (Borges 1974: 67–68)

In commonsensical terms, the virtual is to the real as the copy is to the original: it is nothing more than a reflection, a representation, and a reproduction – the 're' signifying just another of the same (the good and dutiful copy), rather than a transformation or differentiation (the bad and perverted copy). The virtual-as-copy is a late-comer and add-on, which paradoxically attaches itself to, whilst differentiating itself from, the original thing. The virtual-as-copy is secondary, derivative, and supplemental. Superficial, decorative, and accessorised, it has nothing essential about it. It participates without belonging. Yet one should not forget that every duplicate is duplicitous, and every supplement dangerous. As a stand-in or double for the real-original, the virtual-copy may come to efface and occlude it, along with the discernibility of the difference between the one and the other: money, the fetishism of commodities, and the play of seduction being obvious examples.

The foregoing conceptualisation is dominated by a correspondence theory of re-presentation, in which the (virtual) image/imaginary is subordinated to the original self-identity of the real. Exchange-value and sign-value should reflect real value (labour time, libidinal investment, utility, etc.), just as writing should mirror speech. As a superficial effect the virtual should not 'float' freely: it should be anchored on to something substantial, which it dutifully expresses. In harmony with this enslaving image of thought, Robins (1991: 61; citing Stone 1990: 32) argues that 'The terms "virtual" and "artificial" reality "refer to the computer generation of realistic three-dimensional visual worlds in which an appropriately equipped human operator can explore and interact with graphical (virtual) objects in much the same way as one might in the real world."' Thus, the virtual can never be anything more than a pale imitation of the real: a mere simulation. At best, then, 'Telepresence is the extent to which one feels present in the mediated environment' (Steuer 1992: 76). Yet this subordination of the virtual to the real does not depend on referentiality; it does not require the virtual to re-present an actually existing fragment of reality.[1] Rather, it rests upon a strict separation of the real and the virtual, such that there is an immutability of essential forms. For whilst there can be a localised transfer of affects between them (as in the visions and spells of diabolism and witchcraft); there can be no becoming-other (as in the transmutations of alchemy and the transmogrifications of lycanthropy) (Deleuze and Guattari 1987; Doel 1996). In short, and for the sake of appearances and experiences, the virtual may seem real and vice versa, but in their respective essences, never the twain shall meet. Their parallelism is evidently laid out in Euclidean space. (In a curved space, however, such as that of the earth, parallels may criss-cross and interlace.) Hence the fact that so much of the literature confines itself to hunting down superficial and ultimately illusory effects: Are you deceiving us? And for whose benefit?

As an analogue of the real, the virtual necessarily degrades it. No analogical stand-in can reproduce the resolution of an original; likewise for a copy of a copy. This is why virtual reality de-realises the world. It is a xerography – a dry inscription – that fades to grey. Hence so much of the agitation concerning the emergence of virtual communities, virtual politics, virtual companies, virtual wars, virtual sex, etc., and the increasing fascination and queasiness with regard to apparent boundary disputes between the real and the virtual (cf. Bauer 1995; Brook and Boal 1995; Pepperell 1995; Shields 1996). These concerns are underwritten by an inkling that 'what is significant' in 'the virtual microworld' 'is that the user is removed from the fullness of "real" human existence' (Robins 1991: 66). Or as Hayles (1993: 91) pleas in her parting shot on *Virtual bodies and flickering signifiers*: 'As we rush to explore the new vistas that cyberspace has made available for colonisation, let us also remember the fragility of a material world that cannot be replaced.'[2]

Almost inexorably, then, the badly analysed concept of 'false approximation' leads to an onto-theological cult of authenticity, in which the real is figured as a mundane, fragile and passive victim of a virtual seduction.[3] Or rather: the real is a victim of a virtual *s(ed)uction*; for 'the fantasy is structured round the evacuation of the real world. . . . The real world that was once beyond is now effaced: there is no need to negotiate that messy and intractable reality' (Robins 1991: 65–66). Thus, it is tempting to 'no longer see any need whatever for this residue which has become an encumberance', suggests Baudrillard (1996: 42), somewhat sadly. 'A crucial philosophical problem, that of the real which has been laid off.' Little wonder, then, that virtuality in the discourses of false approximation should be so easily conflated with parasitism, vampirism, and autism. The virtual is a dangerous supplement of the most orthodox kind. Like writing or drawing, each virtual-reality technology is a *pharmakon* – cure, remedy, potion, drug, poison, magic, etc. They reli(e)ve memory and enhance experience and sensation, but in so doing they facilitate forgetfulness, dis-ablement, and sensory deprivation (cf. Derrida 1981). On the basis of such dangerous supplements human beings may attempt to relieve themselves of some of their amassing possibilities and responsibilities. For example, Baudrillard (1996: 40) notes how 'the video plugged into the TV takes over the job of watching the film for you', and suggests that 'the idea that there is a machine to store [one's repressed possibilities] and filter them, into which they go to die away quietly, is a profoundly comforting one'. And so it is that 3M's *Scotch*™ videotape has as its pop refrain: 'Re-record – don't fade away!', which in the TV advertisement is sung by a skeleton in an endlessly replayed loop of film, just to underscore the fact that the virtual technology is supposedly there to preserve the humanity in you, even though it is doing so by de-realising the universe. The said product carries a 'lifetime re-record guarantee,' but whose life is underwritten is far from clear: yours, the tape's, the virtual's, or the real's? (Ominously, the graphic depicting 3M's test of the four-hour tape's 'luminance quality' only goes up to the 2000th recording; almost as if to acknowledge the fact that, in a certain sense, the year 2000 will not have taken place: Baudrillard 1986.)

Evidently, the privilege granted to the real *vis-à-vis* the virtual as a 'false approximation' is not so much temporal as ontological. But as we shall see, this ontology will have always already been witness – in the spectral transpearance of a veritable hauntology (Derrida 1994) – to a ghostly, ghastly revenge.[4] Suffice it to note, for now, that there is nothing new in this prioritisation of the so-called original over *its* copy. Think of Platonism, iconoclasm and the BBC's *Antiques Roadshow*. As Abrioux (1995: 61) stresses, 'Western culture remains largely under the cruelly watchful eye of its Platonic superego.' Accordingly, the virtual – like all images, concepts, and ideas – must be kept in its place; it must be anchored in and be subservient to that which it re-presents and/or dis-places. The real must not be eclipsed by *its* shadows or *its* alibis (elsewheres). Duplicity must be to the advantage of the same, which passes through its others only for the benefit of expanded reproduction. Thus, whilst the latest image technologies are frequently defined as 'post-photographic' – a claim that betrays a very poor understanding of photography –, in that they carry the inscription of light beyond the mere re-presentation of the visible spectrum into pure simulacra; to the point where the digital manipulation of an anterior reality passes over into a self-sufficient, although obviously intertextual, generation of digital images;[5] they are invariably characterised in Platonic terms. For example, Mitchell (1992: 225) closes his study of *visual truth in the post-photographic era* with a reassurance, a warning and a return: 'For a century and a half photographic evidence seemed unassailably probative,' he recalls. But today, 'we must face once again the ineradicable fragility of our ontological distinctions between the imaginary and the real, and the tragic elusiveness of the Cartesian dream. We have indeed learned to fix shadows, but not to secure their meanings or to stabilize their truth values; they still flicker on the wall of Plato's cave.'

Now, not only do post-photographic, computer-generated 'images' embrace a 'Platonic ideal,' as Youngblood (1989: 15) puts it – 'They refer to nothing outside themselves except the pure, "ideal" laws of nature they embody' –, but for many the computer itself aspires to becoming a 'universal machine,' to the extent that it can simulate, encompass, and ape every other medium. In such a universal machine, not only would the medium become the message, but the medium would also become a purely transparent screen. 'The "virtual reality effect" is the denial of the role of signs (bits, pixels, and binary codes) in the production of what the user experiences as unmediated presence. . . . As in the *trompe l'oeil* of illusionist art, the medium must become transparent for the represented world to become real. . . . "with a VR system you don't see the computer any more – it's gone. All that's there is you"' (Ryan 1994: §8; embedded quotation from Lanier and Biocca 1992: 166). That is to say, the universal machine would no longer be modelled on the virtual depth of a mirror (perspectival reflection), since it would be instantaneously self-present, re-transmitting itself without *différance* or degradation (cloning, ghosting). Indeed, the very term 'cyberspace', with its etymological link through 'cyber' to the Greek *kybernan* (meaning to control or steer), displays a certain fixation on the construction of a 'control-space' that would enable one to exorcise, or at least contain, the ghosting of presence that would accompany such a universal machine.[6] It betrays a neo-Platonic yearning to reduce the virtual to a mere facsimile of the real. For whilst Platonists constrain the virtual image to a re-presentation of the timeless and spaceless Ideal forms choreographed into reality,[7] neo-Platonists invariably bracket out such idealisation in order to leave us with a world and its perfect copies: computer-generated autoreferentiality becomes perfection itself. Yet the real never holds sway over images and virtual realities. 'Whereas representation attempts to absorb simulation by interpreting it as a false representation, simulation envelops the whole edifice of representation itself as a simulacrum' (Baudrillard 1994a: 6). In short, the virtual is not cast – thrown off, shaped, projected, conjured – by the real. It comes forth from an altogether different dimension. The attempt to reduce virtuality to an occult and xerographic de-realisation of the world is not even skiamachy (imaginary or futile combat; fighting with shadows). The virtual is not shadowy; it is not an oblique skiagraphy of the real and the ideal: a shading of each into the other. To the contrary: the virtual is perfectly real. But note the polysemy: is the virtual perfectly real in the sense of completely, flawlessly, faultlessly, entirely, absolutely and exactly real; or is it perfectly real in the sense of quite real or just (as) real? In the final section of the chapter we will pick up on the latter sense in which the virtual is *(only) just* real. Meanwhile, the next two sections will pursue the former sense in which the virtual is a *perfection* of the real: the first of which outlines the banal version (the [hyper]realisation of the real); and the second follows such banality into fatality (the ex-termination of the real). And whilst both versions of the virtual turn out to be somewhat 'more' than the real, they are fundamentally different in their effects: the excess of perfection seeks to annihilate the real; the excess of real virtuality merely opens actuality to other, unforeseen events. One leads from being-in-the-world to nothingness; the other from being-in-the-world to the vi(r)t(u)al life of becoming-only-to-fade.

We will begin, then, by simply casting off the denigration of virtuality as a dubious and duplicitous xerography. Hereinafter, it is acceptable to love the simulacrum; and to love it all the more for departing from re-presentation – for turning the 're' of representation from just another one of the same to a wholly transformative production of something that is other than the same (cf. Deleuze 1994). In fact, simulacra do not re-present anything (not even themselves); they are acts, events, and happenings that take place on the surface of things with a consistency all of their own. Given the fall of the onto-logical Iron Curtain that forcefully segregated mind and matter – a theatrical curtain which was always already falling from the moment that it was raised – the simulacrum has only itself: although the simulacrum may, through a certain *trompe l'oeil*, give the material illusion of a resemblance, reproduction, and representation – so that this photograph of me is not me (do not confuse image and thing), nor is it like me (do not confuse the event that it is with the resemblance that it is not). Borges' allegory of the 'mirror people' foregrounds the violence of reducing otherness to a ventriloquy of the same, of forcing virtuality to shadow reality, and of making the virtual a duplicate of the world. In commenting upon Borges' tale, Baudrillard (1996: 149) notes that: 'Behind every reflection, every resemblance, every representation, a defeated enemy lies concealed. The Other vanquished, and condemned merely to be the Same' – 'every representation is a servile image, the ghost of a once sovereign being whose singularity has been obliterated.' Yet precisely because the Same thinks that it has the Other pinned down, that the virtual is merely a (dutiful or rebellious) representation of the real, it misses otherness, difference, alterity and virtuality in and of themselves, and it even misses the missing of them. Virtual realists do not look for reflections of themselves in others. Rather, they strain to feel and sense the approach of the mirror people, who may or may not be hiding in the shadows, like a frightened child or a patient assassin, and who may or may not be clinging to the surface, like a character from a slap-stick comedy film who fortuitously avoids detection by inadvertently holding on to doors as they are opened.

Furthermore, virtual realists, unlike critical realists, shun all notions of a depth that would claim some ontological privilege 'over and above' the superficial effects of simulation and simulacra. Whereas the critical realist goes 'deep, deep, deep undercover', as Eddie Murphy once put it, diving down, down, down for the essentials, like a pearl snatcher, virtual realists are content to glide across the surface, like a thrown pebble skimming the surface of a lake; or like a reader's sideways glance darting over the page. But we are not yet able to say, quite simply, that we no longer sense a world *and its double*: that there is (only) just one world, which is immanent to itself and autopoietic. We are not yet able to return to the surface, having discovered, like Alice in Wonderland, that the old 'depths' are themselves nothing but a certain folding, unfolding, and refolding of the surface (cf. Deleuze 1990b; Lyotard 1994). And as we shall see, this virtually wafer-thin world is perfectly real; it is a real virtuality. Here and now, the world is haunted: not by ghostly apparations coming from another world or dimension, such as transcendent Ideas and Forms that must be reproduced here on earth; but by the fact that the 'one' world is folded in many ways. Real virtuality is not duplicitous, but multiplicitous – it is an immanent manifold, the consistency of which depends, precisely, upon one's point of view (cf. Deleuze 1992).

As we float on the surface, having freed ourselves from enslavement to the depths, we must therefore negotiate another version of virtuality. This version loves the virtual, but it is still too devoted to 'lack'. Specifically, instead of the virtual lacking reality, authenticity, etc. (as a mere copy), it is now the real that lacks virtuality. If one looks at the world from the point of view of this latter, one will find that the world is not full; it is impoverished. But it is not a world of scarcity that is brought about through depletion – as if a given stock were being run down. The real world has always been marked by scarcity. On this basis, evolution figures as a painfully slow attempt to 'fill out' the world a little, and to realise a few more of its possibilities. Rather than a society of preservation and conservation, or a society of production and reproduction, we perhaps need a society of creation and invention: there is so much to conjure up in the world; so many experiments and mutations to carry out. Every conceivable possibility should be realised, if only in thought. To speed things up, perhaps one should suspend the selective principle of the survival of the fittest: it is too limiting and filters out too many possibilities. (After all, the old, modernist formulae of 'Form following Function' and 'Fit for purpose' betray a depressing lack of imagination, which only really fit the imperative of satisfying the demands of given environments.) In this regard, one could engage in all manner of transgenic mutations – and if the resulting life-forms cannot live long and prosper, or find a suitable purpose and motivation, in any of our existing environments, then one will have to create special reserves for them, as happens for the invaginated dog and the offspring of the genetically spliced Human-Fly, 'Brundlefly', in *The Fly II* (directed by Chris Walas, 1989). But of course, we already have such reserves for our products, our ideas, our desires, our cars, and even ourselves. On the basis of this image of thought, the world that is abandoned to us should be multiplied and mutated to infinity. Welcome to *Virtual Reality II*: relief at last for a world of lack, for a world that lacks the realisation of infinite difference. . . .We virtual realists are no longer deceiving you, to be sure; but are we making you queasy? (As with daytime TV talk shows, we only ask because it is just one more possibility to realise and tick off – check off and tell off, that is.)

Virtual reality 2: Suppletion. Or, the resolution of the real

> Aristotle thought that the goal of techne was to create what nature found impossible to accomplish
>
> (Guattari 1995: 33)

By contrast with characterisations of the virtual as a degraded, dangerous, and duplicitous *copy* of the real, as a practice of occult occlusion, of xerography and skiagraphy, versions (mis)taking the virtual for a *resolution* of the real work through an inversion of the discourse of approximation, falling short of any deconstruction, which would *pervert* it – 'in the literal sense of "turning away from"' (Massumi 1992: 145), or of twisting it out of shape (see Patton 1994; Wigley 1993). In this case, the virtual is to the real as the perfect is to the imperfect. Here, it is the real that is figured as partial, flawed, and lacking; whilst the virtual promises a rectification and final resolution to come. Priority is not given to the fallen first; it is withheld for the full second. The virtual exudes wonderment by 'correcting' defects in the real; by surpassing the constraints, inadequacies, and limitations of the real – particularly the drag of (real) space-time. Witness, for example, the currency of terms such as distanciation, disembedding, and space-time convergence, compression, and annihilation. Or again: think of poor eyesight and spectacles, bedazzlement and sunglasses. In this way, the virtual no longer degrades the fullness of some original position by appending some occlusive contingency and arbitrariness to it (as in the discourses of false approximation); rather, it adds to it, comple(men)ting and supplementing it.[8] The given real always already has room for the supplement within itself; it always already has an opening or fissure that anticipates what is coming. Thus, the real ordinarily lacks what the virtual will have come to furnish. Sadly, reality rarely suffices. Fortunately, virtuality invariably reli(e)ves. Baudrillard (1995a: 106) refers to this repletion-to-come as the 'virtual realisation of the world'. Yet such a full realisation of the real – an hyper-realisation – is not an *ex nihilo* gift of nature; it must be produced as a special effect. Hence the fact that privilege and priority is accorded to the full supplement, as it comes to reli(e)ve and fulfil the lack in the original. So, one should forget the onto-theological quest for an authentic representation. For there are no Second Comings in this or any other world. It is difference that returns. As with the endless remixing of musical tracks, repetition is a transformer and dissimilator (cf. Lyotard 1990).

Nevertheless, there is a duty in the forced hyperrealisation of the world. Consider, for example, the hyperbole surrounding genetic engineering and the Human Genome Project. An obsessive hygeia insists that everything must be debugged: from outer space to inner space; from computer programs to DNA. Such is the purity of codes. Yet, once again, there is nothing new in rendering the real as a flawed version of some ideal. One finds it in all manner of idealist, materialist and utopian thought, not to mention in landscape gardening and flower arranging. Nor is there anything new in discerning something in the real that nevertheless gives (assured?) access to such an ideal. It is the stock-in-trade of all forms of immanent critique. And whether this rendition is given an idealist or materialist slant, our relation to the virtual is always framed in terms of realisation. By way of the virtual, one is able to realise (actualise) the latent potential of the world. (Hence its *hyper*reality effects.) One secures access to what appears more real than reality. (Hence the experience of transcendence acted out in the exhaustion and ex-termination of the world.) Thus, 'when we finally come to be immersed in this "cyberspace" we should be able to realize our true and full potential' (Robins 1991: 59). And then it will be over – and out. In short, 'by wishing the world ever more real, we are devitalising it. The real is growing and growing; one day everything will be real; and when the real is universal, that will spell death' (Baudrillard 1996: 46).

One cannot, however, escape the anthropocentrism of this version of the relationship between the real and the virtual, which evidently requires the presence of a subject to *embody* the difference (Braidotti 1994; Grosz 1990, 1994; Massumi 1996). For the wonderment of the virtual is not just confined to its *immersive* reality-effects (simulations and simulacra); it also extends to its *interactivity*: 'Whereas immersion may be a response to a basically static representation, interactivity requires a dynamic simulation,' and 'Virtual reality . . . reconciles immersion and interactivity through the mediation of the body' (Ryan 1994: §25 and §39). This is obvious in the context of digital pornography. In the virtual realisation of the world, the subject acts as a pivot between the flaws of the real (especially the drag of space-time) and the perfection of the virtual (a universal telepresence, from which the sentient and sensuous body can withdraw). Furthermore, the products and apparatuses of technoscience increasingly allow the real (that is to say, flawed) user to function in the virtual (that is to say, perfected) register. For example, prosthetic virtual-reality body-suits and head-sets do not simply help realise the latent potential of the body by sublating the all-too-real limitations of the user's space-time embedment; they simultaneously *produce* this embedment/embodiment as a limitation and a flaw – just as normalisation engenders deviancy, memory creates the possibility of forgetting, order produces disorder, reason evokes madness – and the invention of the car was simultaneously the invention of the car crash. And so in one of *Aura*'s advertisements for their *Interactor Cushion* and *Backpack* – 'the virtual reality games wear that really packs a punch!' – the consumer is promised an 'Experience [of] games, films and CDs as they were meant to be' by 'turning all the on-screen action into thumping vibrations'. Apparently, one can even 'discover what music feels like' (*Littlewoods* 1997: 987) – as if music were not ordinarily felt.

Paradoxically, since ability, competence, and performance are context-dependent rather than given in advance, virtual-reality technologies dis-able; or more accurately, 'differently enable'; the body in order to sublate the flaws they retroactively generate (Virilio 1992).[9] Sometimes, through fortuity, a given body will fit snuggly into the new constellation of intensities, powers, and affects: a home from home. Sometimes it will need to get into shape in order to become one of the 'in crowd'. And sometimes it will remain a misfit, be excluded, or take flight. To that extent, virtual-reality technologies and their social apparatuses function rather like the criminal justice system or the urban environment: they all process heterogeneous bodies into recognizable and manageable forms. There can be no flaw without perfection, no original repletion without originary suppletion, and therefore no reality without virtuality. Such is the 'revenge of the crystal'. The 'object strikes back'.

In contrast to the discourses of false approximation, then, there is no radical disjunction of immutable and inalienable forms between the real and the virtual, because the latter is but a dilated version of the former. The parallelism of the two series no longer maintains their separation through relations of mimesis, analogy, and resemblance. Since the parallel series criss-cross, interlace, and become indiscernible, each deterritorialises the other in a mutual becoming or aparallel evolution. In this way, the real and the virtual are no longer opposed but denote different degrees of potentialisation and actualisation along a simulacral continuum from the real to the hyperreal. Differences of degree – the more and the less: speed and slowness, intensities and affects, etc. – overcode the previous differences of kind. And with this change of emphasis, we are in a better position to understand the limitations of Wark's (1994b: vii) distinction, not between the real and the virtual – 'virtual geography is no more or less "real"' he insists; 'It is a different kind of perception, of things not bounded by rules of proximity, of "being there"' –, but between virtual *reality* and virtual *geography*: 'If virtual reality is about technologies which increase the "bandwidth" of our sensory experience . . . then virtual geography is the dialectically opposite pole of the process. It is about the expanded terrain from which experience may be instantly drawn.' In a similar, although more rigorous manner, Virilio (1992: 79) avers that '"real-time" is not . . . to be contrasted with "recorded time", but rather with present time alone'; 'teletechnologies of "real-time" . . . kill present time by isolating it from its here and now, to the benefit of a commutative elsewhere which is no longer that of . . . our "concrete presence" in the world but of a discontinuous "remote presence".' Witness, for instance, the diminishing legitimacy of the football referee in the age of the instant replay. What hitherto stood in for the real is increasingly subjugated to the hyperreality of the real-time recording. At this point we find ourselves in an all too familiar situation – of the disembedding and distanciation of social interaction through space and time; of space-time convergence and compression; and of the annihilation of space through time.[10] Hereinafter, the real and the virtual are no longer distinguishable according to qualities (powers and affects), but only according to quantities (more or less).[11] Henceforth, the difference between them will be measured in terms of the range and the territory that they can command, and it will be exposed in terms of instantaneity and (tele)presence: 'the notion of exposure supplants . . . that of succession in the measurement of present duration and that of extension in the area of immediate expanse' (Virilio 1992: 81. See also Virilio 1994b, 1995). On this account, then, the virtual is more, not less, (real) than the real. It is hyperreal.

So much for both the false approximation and the resolution of the world. Without further ado, let us simply wrap things up by noting how this twofold collapse of space-time into either a degraded copying and distantiation of an original position or a final resolution of a retroflex incapacity turns out to entail a mutual and reversible predication: no virtuality without reality and no reality without virtuality. In short, there can be no 'simple' presence (being) without a 'remote' (tele)presence (an alibi, a being elsewhere). Yet in this reversible becoming-real of the virtual and becoming-virtual of the real – a double-take on the system's degree of realisation –, what remains of their respective specificity? (Almost) nothing. For what is hinted at in this bifid collapse of space-time is a badly analysed composite: presence itself (or if you prefer: 'actual reality', although as we will argue in due course, such a phrase barely hangs together). It is as if the real were collapsed into a succession of integral points, such that each 'point' in real space-time can be occupied by only one Thing and that these points are mediated through an invariable contiguity. Hence the striated drag of real (i.e. extensive) space-time, which is perhaps the most important disability retroactively engendered in the 'real' through the process of hyperrealisation. Likewise, it is as if the virtual were collapsed into an array of differential points, such that each 'point' in virtual space-time can be occupied by an infinite number of Things and that these points are immediated through a supple discontinuity. Hence the absolute speed of virtual (i.e. intensive) space-time. Such a collapse gives the real its physical security and the virtual its ghostly telepathy. (Parenthetically, one could say that if the real is the fold between here and there, then the virtual is its un-fold or re-fold: NowHere, NoWhere). Henceforth, the real will connote the resistance and drag of matter, whilst the virtual will connote the flight of spirit: arduous voyages versus motionless trips; ontological fixity versus hauntological drift; real bodies versus virtual ghosts (Derrida 1994). But as you may have already gleaned from our tone and phraseology, there are problems with this characterisation of the real and the virtual in terms of integral and differential instants (presence versus telepresence). Moreover, we need to pursue in a little more breadth the passage from a banal hyperrealisation to a fatal ex-termination of the imperfections in any given – that is to say actual – reality.

Virtual reality 3: S(ed)uction. Or, the virtual illusion of a final solution

> Year Zero of Virtual Reality
>
> A community of Tibetan monks have for centuries devoted themselves to transcribing [the] nine billion names of God, and once they have accomplished this the purpose of the world will be achieved, and it will come to an end. The task is a tiresome one and the weary monks call in technicians from IBM, whose computers do the job in a few months. In a sense, the history of the world is completed in real time by the workings of virtual technology. Unfortunately this also means the disappearance of the world in real time. For suddenly, the promise of the end is fulfilled and, as they walk back down into the valley, the technicians, who did not really believe in the prophesy, are aghast to see the stars going out one by one
>
> (Baudrillard 1996: 43, 25)

As always, we have been far too restrained in our consideration of virtual reality as a final resolution of the fault-ridden world that is given to us. In this section we will endeavour to go all the way with this illusory image of thought. 'What is the idea of the Virtual?', asks Baudrillard (1995a: 101): 'the unconditional realisation of the world.' It is the desire for a 'resolution of the world ahead of time by the cloning of reality and the extermination of the real by its double' (Baudrillard 1996: 25). In this respect, the so-called 'virtual technologies' are simply the latest in a long line of apparatuses designed to accomplish the impossible dream of actualising in real-time the entire set of the world's possibilities. This is much more than mere affine redundancy, since the recurrent dreamwork of modernity in general, and technoscience in particular, is to decode and debug the world fully; to enact the final (re)solution to the 'problem' of the world; to exhaust the entire matrix of possibilities; and, therefore, to have done with the world. The world itself is a(t) fault. 'This is perhaps the fate . . . of the world: its accelerated end, its immediate resolution . . . though with no hope of salvation, apocalypse or revolution. Merely hastening the final term, accelerating the movement towards disppearance pure and simple' (Baudrillard 1996: 26). And so there remains 'the vitally urgent need to stay this side of the running of the programme, to de-programme the end', continues Baudrillard (1996: 48), but 'the aim of our system is precisely the opposite: to drive right through to the end, to exhaust all the possibilities.' Little wonder, then, that Baudrillard should call the yearning for such a final solution the programmation of a 'perfect crime'. For were it to be carried off, it would leave no traces. Needless to say, this dreamwork lives on in many of the more technophile and technophobic versions of postmodernity (post-industrialism, infantile capitalism, etc.). Modernity has always mobilised Reason in the vain attempt to annihilate the necessary ambivalence of the world (Bauman 1991). It tried to destroy the world as appearance and enigma, but only succeeded in amplifying it.

Fortunately, the perfect crime of the exhaustion of the world's stock of possibilities, finishing everything off without leaving any remainder or trace, is triply impossible. First, it is impossible because the stock of possibilities is not only infinite, it is also an infinity that ceaselessly changes owing to the fact that it is open and folded according to a variable consistency (it is a chaosmotic infinity rather than a given infinity). Second, it is impossible because there are always remainders, traces and excesses. Third, it is impossible because 'reality' evades both the hypothesis of its existence and the hypothesis of its non-existence: 'The point is not, then, to assert that the real does or does not exist,' says Baudrillard (1996: 46). The 'presence' and 'absence' of 'reality' is always both undecidable and ghostly. Hereinafter, 'the body of the real [will] never [have been] recovered. In the shroud of the virtual, the corpse of the real is forever unfindable.' All in all, then, it is undecidable whether or not there will have been a perfect crime: there is no definitive evidence one way or the other. For the 'reality principle' itself, which concerns the existence and co-presence of different terms and events in a single space-time, is an 'objective illusion', says Baudrillard (1996: 52). For reasons that will become clear in the following section, 'integral presence is only ever virtual. . . . "Real" time does not, therefore, exist; no one exists in real time; nothing takes place in real time – and the misunderstanding is total.' For something to cast off the objective, material illusion of being self-present and fully integrated, it must differ and defer a manifold of relations to what is different, other, and absent; and it must simultaneously dissimulate such an act of differing and deferring in order to appear to stand all by itself without need of support. The objective illusion is to simulate presence and identity in the here and now, when in actuality – or rather, virtuality – it is nothing but an infinitely disadjusted dissimilation. (For example, even tautologically, 'a' is not 'a' – it is only ever asymptotically 'itself', and only then to the extent that 'it' is the set of negative differences established between itself and every other: not-b, not-c, not-d . . . not-z, not-aa, not-bb, not-cc, not-dd . . . not-zz, etc. Needless to say, this set is infinite and untotalisable. Moreover, since the indeterminancy cuts all ways, none of the other terms against which the integrity of 'a' is calculated are integral in and of themselves. Everything is vaguely precise. But one need not generally worry too much about this infinite disjointure, since it usually suffices to fold it all up into one singular point: '*it is "a"*. . .') Baudrillard gives the example of looking at a star. Owing to the relative speed of light, what one sees may have already disappeared. 'By the fact of dispersal and the relative speed of light, all things exist only in a recorded version, in an utterable disorder of time-scales, at an inescapable distance from each other. And so they are never truly present to each other, nor are they, therefore, "real" for each other' (Baudrillard 1996: 52). Perhaps the only event that can be said to have truly existed in real time was the creation of the universe itself. 'Once that initial (and perfectly hypothetical) state came to an end, the illusion of the world began.'

So, what seems to be a gesture of self-definition is necessarily a boundary-drawing act. To erect an identity, one must draw a limit between what is inside and what is outside. In this way, the manifold is folded in two, giving rise to the material illusion of a 'this' and a 'that'. Be that as it may, each one is irreducibly twofold, to the extent that it is halved together as the inside of the outside. This twofold is not an arbitrary identification of given forms, as nominalists would have us believe; it is a cutting out of relational figures. In short, the play of folding is dehiscent without being discerptible: the fold gapes open but what it unfolds can neither be divided up nor plucked out; the fold unfolds – and reciprocally. As with origami, the event of folding is never given in and of itself; it never appears as such (Doel 1996, forthcoming). Transfixed on only one side of the fold, the purported act of self-definition invariably misses its other or double. Hence the fondness in Western metaphysics for binary and mutually exclusive oppositions. This kind of act of self-definition, which modernity made its own, invariably casts the other, upon which it depends, as a *flawed version of itself*. The bad is a flawed or incomplete version of the good; the illusory a deficient version of the real; writing a death mask for living speech; and so on. Hence the modern dream of a perfect order creates disorder as its own alibi (elsewhere, 'excuse'). The will to order is the screen through which multiplicity is filtrated into both order and disorder. Disorder and chaos are special effects projected by a certain regulated play of folding. It is in this sense that 'Nothing is perfect, because it is opposed to nothing' (Baudrillard 1996: 75). Blindly ignorant of its own impossibility, however, the hyperreal dream of 'virtual reality' steers itself in vain towards the exhaustion of all the world's possibilities, through the execution of what Baudrillard (1995a: 102) refers to as 'the code *for the automatic disappearance of the world*'.

This fetishised ideal of the virtual would amounts to living in the (tele)presence of a full realisation of the world's possibilities. The sought-for accomplishment of the virtual in real space-time would entail forcing the world to confront the limits of its possibilities. Such is the specificity of this most s(ed)uctive version of the virtual. And it is with regard to such a situation that the real is increasingly being seen as a (real) drag. This drag should be taken literally: it is the friction, extension, and duration of matter in extensive space-time. Conversely, the world of cyberspace, telepresence, and virtual reality would be 'an ideal world, a world beyond gravity and friction' (Robins 1991: 60). It would actualise, in the here and now, an expenditureless superconductivity that knows no bounds: 'at once *NowHere*, simultaneously nowhere and everywhere' (Friedland and Boden 1994: 45). Such is the dream of a literal freedom from the drag of space-time, which conveniently forgets the enormous array of socio-material apparatuses needed to sustain it. The virtual illusion yearns for an intensive space-time, in which the whole would be folded into the punctum, without a trace of extension: the One-All of unmediated presence.

'Fortunately,' says Baudrillard (1995a: 106), 'all this is impossible. . . . There is no place for both the world and for its double.' Accordingly, it must be recognised that the virtual-reality technologies are anything but virtual. Indeed, they have sought to hijack the idea of the virtual, since they are dedicated to the unlimited realisation of the world (which is to say, to the total annihilation of illusion; to radical and absolute dis-illusion). Whereas the real is always already predicated in terms of its opposition to the illusory, which it renders as a flawed, derivative, and subordinate version of itself that is thereby amenable to sublation by the real; the illusory rests on the non-opposition of the real and the illusory, on their reversibility and indiscernibility, and on their duality/duelity and irreconcilable antagonism.[12] Since virtual-reality technologies are dedicated to the forced realisation of the world, and are thus pitched against the world as illusion, they belong to modernity. Yet they are also implicated in a more fundamental shift. This situation, suggests Baudrillard (1995b: 94), 'is perhaps the only case in which we can take the term "postmodern" seriously' – inasmuch as the modern world 'of terms and the opposition of terms' (1995b: 93) has reached its end. Or, more precisely, it has already passed *beyond* its end (Baudrillard 1994b; Clarke and Doel 1994).

The hypothesis of a postmodern condition does not, of course, suggest the simple removal of modernity and all its accomplishments from the world (as if it ever could). If modernity is equated with 'the immense process of the destruction of appearances . . . in the service of meaning, . . . the disenchantment of the world and its abandonment to the violence of interpretation and history' (Baudrillard 1994a: 160), then the postmodern amounts to 'the immense process of the destruction of meaning, equal to the earlier destruction of appearances' (Baudrillard 1994a: 161). Thus, whilst the modern zeal for imposing order and discernibility instigated and required a process of boundary-drawing and boundary-maintenance, the very existence of the boundaries it erected established the conditions of possibility for their straddling, and thus the subsequent proliferation of undecidability, indeterminacy, and indiscernibility. Hence Baudrillard's (1994b) equation of the postmodern with the process of ex-termination.[13] For example, we are beyond the order-words of 'good' and 'evil' owing to their short-circuiting in absolute indifference and undecidability; no more real and illusory, but their short-circuiting in simulation and simulacra. This sketch is a useful outline of the postmodern condition. However, one additional feature is important to recognise. Against all those depictions of the postmodern as idealistic, metaphysical, and inconsequential, it must be recognised that the postmodern itself scrambles the integrity, coherence, and exactitude of both these terms and their alibis.

Accordingly, whilst one may counter that the postmodern is material rather than idealistic, this would surely be a materialism of an 'incorporeal' kind (Deleuze 1990b; Foucault 1982); a 'material illusion' (Baudrillard 1996). In short, therefore, there is nothing *unworldly* about the postmodern. Indifference, indiscernibility, and undecidability are in the world and of the world. Thus, whilst modernity aimed at achieving a social *order* (a principle giving rise to certain self-propelling, if contingent, forms of social arrangement), postmodernity bears witness to the decomposition of that principle – and to the irruption of new, surprising forms of social and technological arrangement. And so Baudrillard (1995a: 101) speaks not only of an ex-termination, but also of a final (re)solution[14] implied by this ideal of radical disillusion. For this passing beyond the end itself indicates a certain casting adrift from the irreversibility of the reality principle in its pure form – not at all in the sense of a return to the reversibility and reciprocity of symbolic exchange (Baudrillard 1993), which would be the pure form of reversibility characterising the world as illusion – ; but in the form of a general *indeterminacy* characteristic of a world with no terms at all. The world is not so much realised as volatilised and vapourised: 'we are in the fractal, the molecular, the plural, the random, the chaotic' suggests Baudrillard (1995b: 93). Perhaps, therefore, at a certain point, the level of reality-effects cast off by the virtual illusion attained a critical mass, or else the process of realisation reached a certain limit, which we have since passed beyond: 'And having to make sense of a world where the end is not ahead of us but behind us and already realised, changes everything' (Baudrillard 1995b: 95).

To sum up, the unlimited realisation and the virtual programmation of the world would amount to the perfect crime (Baudrillard 1996): not only would it simultaneously conjure and exhaust in 'real time' a perfect world – from which we would be expunged –, it would annihilate all traces of its production and seduction. One could not even sense the criminality. '*In the perfect crime it is the perfection itself which is the crime*' (Baudrillard 1996: no pagination). Hence, the perfect crime always already amounts to the perfect conspiracy. There would be no redress. And the 'absolute real', which amounts to the technological dream of virtual reality, would leave no room at all for the dangerous imperfections of humanity, extensive space-time, or mundane reality. For the conflation of the virtual and the real would amount to a state of *aphasis*, a condition in which the subject is coextensive with a state of affairs, tied to actuality, and paralysed by (tele)presence: a subject *possessed* (Grosz 1994). Environmental determinism *par excellence*; technological determinism *par excellence*. Everything would have been accomplished in the degree-zero punctum of intensive space-time. UTOPIA ACHIEVED: the final (re)solution of a faulty world without trace. (No more shadows. No more [ex]tension.) So, while 'the "original" crime is never perfect, and always leaves traces – we as living and mortal beings are a living trace of this criminal imperfection – the future extermination, which would be the result [of] the absolute determination of the world and all its elements, would leave no traces at all' (Baudrillard 1995a: 106). And yet, it would be undecidable whether or not this perfect crime would have automatically generated the perfect alibi. Since an alibi is a 'being-elsewhere', the very condition of possibility of such an alibi would have been short-circuited by the abolition of the drag of space-time. In effect, the perfect crime is absolutely impossible.

There are always traces – and nothing but traces. Yet we still 'dream of perfect computers', notes Baudrillard (1995a: 102). 'But . . . we don't allow them to have their own will. . . . No liberty, no will, no desire, no sexuality. We want them complex, creative, interactive, but without spirit.' Nevertheless, 'It seems that they have an evil genius for dysfunctions . . . which save them, and us in the same way, from perfection and from reaching the end of their possibilities' (Baudrillard 1995a: 102–103). For if the absolute realisation of virtual reality were finally accomplished, as so many technoscientists and their hyperbolists seem to desire, we would be obliged to step out of the world without leaving a trace. We would never have been (t)here.

Virtual reality 4: Simulacra. Or, real virtual(real)ity

> We opposed the virtual and the real: although it could not have been more precise before now, this terminology must be corrected. The virtual is opposed not to the real but to the actual. *The virtual is fully real in so far as it is virtual.* Exactly what Proust said of states of resonance must be said of the virtual: 'Real without being actual; ideal without being abstract'; and symbolic without being fictional.
>
> (Deleuze 1994: 208)

The discourses of hyperrealisation and ex-termination make the mistake of confusing the virtual with the possible: the latter pre-empts, opposes, and resolves the real. The problem of possibility is realisation and de-realisation: that of virtuality is actualisation and counter-actualisation. Now, what can we deduce from the three versions of virtuality sketched out and deconstructed above? That nothing has been resolved – and so, the simulacrum, despite 'having been banished ontologically to the margins' (Genosko 1994: 28), continues to haunt Western culture's Platonic superego. 'The platonic dialectic commands that one distinguish between the model and its copies and, among them, the correctly copied from the flawed' (Alliez and Feher 1989: 55). Where the good copy (*eikon*) is endowed with *resemblance* (inasmuch as it belongs to the ideal or model; that is, it participates in the Idea of the thing), the bad copy (*phantasma*) is marked by a mere *semblance* (which only appears – with bad will – as a likeness, having no proper entitlement to a filiation with the ideal). The simulacrum (*eidôlon*), however, is a bad copy that produces 'an effect of *resemblance*' (Deleuze 1983: 49) – and in so doing it establishes itself as the corner-stone that would bring down the entire Platonic edifice. For the simulacrum is a category with neither fixed identity nor essential form, operating within 'a dimension in which objects may be said to be simultaneously both hotter and colder, bigger and smaller, younger and older' (Bogue 1989: 56). This is not, as Deleuze's (1993: 39) remarks on Lewis Carroll's *Alice in Wonderland* make clear, a simultaneity of being(s), but an unhinged and reversible becoming:

> When I say 'Alice becomes larger,' I mean that she becomes larger than she was. By the same token, she becomes smaller than she is now. Certainly, she is not bigger and smaller at the same time. She is larger now; she was smaller before. But it is at the same moment that one becomes larger than one was and smaller than one becomes. This is the simultaneity of a becoming whose characteristic is to elude the present. . . . It pertains to the essence of becoming to move and to pull in both directions at once: Alice does not grow without shrinking, and vice versa.

The logic of becoming heralded by the ambivalent category of the simulacrum discloses a virtual world that eludes the present – a world characterised not by the full completion of presence usually associated with the real, but by a ghostly presence; a haunting (Derrida 1994). The point is (s)played out. It is not given in and of itself; it is relayed or refolded. Space-time is always already (s)played out. But let us be clear: this is a world where the virtual is disclosed as being more, not less, than the actual. Yet by the same token, the *hyper*reality of the virtual is not one of full realisation or complete actualisation; it is one through which the Event eludes the present and evades every actual state of affairs: to grow, to cut, to live, to die. The virtual is the becoming of the Event: *this* is what happens. There is no reality effect without virtuality, no being without becoming, and no living presence without a ghostly repetition.

Accordingly, to the subject who would wish to embody the difference between the drag of real space-time on the one hand, and the absolute speed of a virtual false approximation and/or hyperrealisation of the world on the other hand, one should recall that a ghostly or virtual presence always already animates this subject, inasmuch as 'There is no subject without, somewhere, *aphanisis* of the subject' (Lacan, cited in Borch-Jacobsen 1994: 80). This aphanisis refers to the perpetual process of becoming-only-to-fade that characterises subjectivity, in terms irreducible to the completed self-presence of the Cartesian *cogito* (Lacan 1979: 207). For the subject is always marked by a wholly imaginary promise of complete self-presence, which issues from the sense of lack stemming from an entirely mythical originary completeness. This sense of lack results from the subject being afforded a position by the Other (the Symbolic) that cannot contain it; that it necessarily exceeds. And, moreover, it is a constitutive condition of subjectivity that it incurs this cost of division, because subjectivity is necessarily split between the conscious of intended meaning and the unmotivated Other of the (collective) unconscious.[15] The fact that the Other is also figured by lack – in accordance with the patriarchal Law structuring the Symbolic order, whereby Oedipalisation operates in terms of the Name-of-the-Father (that is, in terms of a signifier standing in for an absent Father and therefore a lack) – entails that the subject's desire (which amounts to the desire for complete self-presence) is necessarily insatiable; that the subject can never attain a full self-presence, except in the virtual dimension of a retroactive mo(ve)ment of folded punctuation or quilting.[16] To the extent that such a dislocated self-presentation marks out a disadjusted temporality – since it is discontinuous with the present –, the subject is itself untimely: it is, in short, a virtual, ghostly presence. As Lacan (1977: 304) remarks, as a subject, I am always already 'the future anterior of what I shall have been for what I am in the process of becoming'. Thus, there is no complete, authentic identity for the subject as such. Or again: space-time, the world, and the subject are all distended and (s)played out from the middle.

In line with this portrayal of a world always already subject to virtuality, it should be recognised that the virtual is most definitely not separable from the real, either as a false approximation (mere simulation) or else as a forced realisation (sublation of real limitations; actualisation of latent potential). The real is always already virtual; that is to say, disadjusted and untimely. Reality evades (its reduction to) actuality. Hereinafter, the wondrous power to annihilate the (extensive) drag of (real) space-time promised by all of those virtual-reality technologies will never have taken place. And this insistence hangs not on so-many new-fangled technologies – which pin the subject between a full/flawed reality and a degraded/perfected virtuality –, but on the ghostly (s)play of presence itself. Thus, even before the embodied subject dons a body-suit or helmet, plugs itself into the Internet, recodes its genetic makeup, or sets out for a stroll, it is always already caught up in the disadjusted play of differentiation and differenciation. The subject does not so much exist as transpear through the suturing of this disadjustment. Hereinafter, if the common-or-garden technologies of virtual reality have little to do with (real) virtuality, then this is because everyday life itself is always already a virtual reality, as are the most mundane of tools, instruments, and machines (Guattari 1995).

The comeuppance of our deconstruction of the major discourses on 'virtual reality' is this: reality is not the actualisation of a set of possibilities in a given time and space, an actualisation that would unfold a serial exhaustion of the world's possibilities – reality equals actuality, or, if you prefer, a *given* reality is *only* one of the world's stock of possibilities. For example, think of the commonsensical view that paper or electronic databanks are a kind of 'stock', patiently awaiting their 'real-time' enactment and emplotment by users, who may themselves be machinic rather than human operators. Instead of reducing reality to a tangible actuality, and expelling virtuality in the process, we wish to argue that reality is the actual and the virtual. This 'and' is not a relation of the more and the less, *n* possibilities in addition to the actual *one* that is directly lived. Rather, it is the 'and' of folding, unfolding, and refolding. Reality is the immanent twofold of actuality–virtuality. Such a twofold is never given in advance, like the matrix of possibility is supposed to be; it always has to be created and worked over in situ. The badly thought-out notion that the real realises a pre-formed possibility rests on a retroflex movement according to which the existent is assumed to precede itself and the creative act that constitutes it. Possibility is not just a combinatoric. It is also a retroflex projection. Deleuze (1991: 97) puts it like this:

> the real is supposed to be in the image of the possible that it realizes. (It simply has existence or reality added to it, which is translated by saying that from the point of view of the concept, there is no difference between the possible and the real.) And, every possible is not realized, realization involves a limitation by which some possibles are supposed to be repulsed or thwarted, while others 'pass' into the real.

The virtual illusion, then, amounts to the desire for an unlimited realisation of the possible, to its delimitation and exhaustion in the real. But 'the possible is a false notion, the source of false problems', continues Deleuze (1991: 98):

> The real is supposed to resemble [the possible]. That is to say, we give ourselves a real that is ready-made, preformed, pre-existent to itself, and that will pass into existence according to an order of successive limitations. Everything is already *completely given*: all of the real in the image, in the pseudo-actuality of the possible. Then the sleight of hand becomes obvious.

By contrast, the actual does not resemble the virtual. To the contrary, actualisation is *creation*. The whole is no longer given, as a combinatory matrix of possibility; it is always already open and in endless (de)construction, like the cinematic practice of splicing irrational cuts. 'The Whole must *create* the divergent lines according to which it is actualized and the dissimilar means that it utilizes on each line. There is finality because life does not operate without directions; but there is no "goal," because these directions do not pre-exist ready-made, and are themselves created "along with" the act that runs through them' (Deleuze 1991: 106). In short, real virtuality has nothing to do with resemblance. Still less does it concern a false approximation or a final resolution. (Virtual) reality is nothing but immanent creation and experimentation.

If all of the above has become too abstract and disconnected for your liking, then pose yourself the following problem: what can a body do? (And one should resist the temptation to read into this a human body, which is itself an heterogeneous and variable form rather than a secure essence.) A body is not given in advance, nor does it resemble anything. Rather, a 'body' is a composition and articulation of relations, social as well as object relations: of speeds and slownesses, and of the power to affect and to be affected by other bodies. And such 'a' body changes with context, ceaselessly creating itself in situ; it has to be composed, orchestrated, and performed (Doel 1995). There is, therefore, no solution to the problem of what a body can do. To the contrary, the task for any and every body is to pose itself (as) a problem; to continuously reinvent itself (as) a problem. For instance, in a consideration of the performance artist Stelarc's bodily experimentations – hanging his body from hooks, attaching the limbs to electronic prostheses, rendering audible the fluxion of his body, visualising the interior surfaces and cavities, etc. –, Massumi (1996, no pagination) notes how: 'The problem posed by a force cannot be "solved" – only exhausted.' Not exhausted in the sense of realising a matrix of given possibilities; exhausted in the sense of reworking a problem until it mutates into a qualitatively different problem. Or again: 'The usual mode in which the body functions as a sensible concept – possibility – is radically suspended. The body is placed at the limits of its functionality. The answer to what is being suspended is: embodied human possibility.' Quite literally, the hooks upon which the fleshy body hangs do not only render the force of gravity on the comportment of the body visible – on this 'gravitational landscape' 'ripples and hills . . . form on the hook-stretched skin' – ; they also serve to enable the suspended body to engage in a transformative series of irrational cuts. In other words, the hung body, de-limited and counter-actualised to the point of 'suspended animation', knows only of immanence (what to do?), and nothing of logical possibility (the body is/ought/should). It no longer re(as)sembles a pre-given form that is posed for it in advance of its (dis)connection and experimentation. In shaking off a certain kind of organization, the hung body has opened on to what Deleuze and Guattari (1984, 1987) call the 'Body without Organs' – the fully deterritorialised plane of immanence and collective consistency. Indeed, the problem for a Body (without Organs) is that of counter-acting or counter-actualising the forces that currently organise and constrict it. Yet even here it is not necessary to fret too much about virtual technologies in relation to the body. If truth be told, we know next to nothing about what a body can do in the most banal of respects: eating, loving, walking, fighting, falling, dreaming. . . .What to do with a rope, or a pen, or an egg? Likewise with a body of thought: 'We simply do not know what thought is capable of' (Ansell Pearson 1997: 4). A matrix of possibilities may spring to mind, but that is not the solution. So, 'There is no need for apocalyptic imagery of the cyborg,' avers Massumi (1996, no pagination). 'What is called for is experimental connection to and continuation of the inhumanity in all of us.' In sum, the real is virtually inexhaustible, even though it is bereft of possibilities. Hereinafter, there will have been traces of life after the final (re)solution. For beyond the illusory hyperrealisation and ex-termination of the world's possibilities is the difference-producing repetition of dissimilating and counter-actualising the givens.

Notes

1 In virtual reality 'The question isn't whether the created world is as real as the physical world, but whether the created world is real enough for you to suspend your disbelief' (Pimentel and Texeira 1993: 15).

2 Bradley (1995: 10) distinguishes two discourses on cyberspace. 'The first portrays

cyberspace as a new frontier, an empty and/or formless space 'discovered' in the interstices of information and communication technologies: a new frontier which awaits socialization. The second raises cyberspace to the status of a mission to be carried out according to the "inevitability" of human, social and technological development.'

3 One should note the importance of how the discourses of 'false approximation' gender the real and the virtual. This gendering is not confined, however, to the discourses of 'false approximation', as we shall see in the other major discourses on virtual reality under consideration here: the supplemental rectification of retroflex flaws in an original (motifs of lack), and the seductive desire for a final (re)solution to the 'real drag' of actualising the world's matrix of possibilities (motifs of decorporealisation). Moreover, our three preferred instances of real virtuality – Deleuze and Guattari's notions of 'becoming' and 'Bodies without Organs', Derrida's duplicitous logic of 'hauntology', and Lacan's 'aphanisis of the subject' – are themselves intimately imbricated with notions of sexual difference. However, given the brevity of this chapter we have decided not to give these concerns the consideration they deserve, partly because they are already the subject of voluminous debate.
4 'If . . . tangible certainty and solidity corresponds to ontology, then . . . how to describe what literally undermines it and shakes our belief?' asks Jameson (1995: 86). 'Derrida's mocking answer – hauntology – is a ghostly echo if there ever was one, . . . which promises nothing tangible in return; on which you cannot build; which cannot even be counted on to materialize when you want it to. . . . all it says . . . is that the living present is scarcely as self-sufficient as it claims to be; that we would do well not to count on its density and solidity.'
5 Strictly speaking, such digital generation does not produce an 'image', since it no longer reflects and resembles anything. The digital product is without original, archetype or prototype. It amounts to a simulacrum rather than a simulation.
6 One should not overlook the fact that 'The frequent use of the city as a model for cyberspace suggests that discipline – as a set of power relations between subjects, space and visibility – is transferred, as a "matrix of regulations," from the state to cyberspace' (Bradley 1995: 14. See also Crary and Kwinter 1992; Alliez and Feher 1989). Nor should one overlook the various social apparatuses that regulate the production, mercantilisation, and dissemination of data (Lyon 1994; Lyotard 1984; Poster 1990, 1995). But suffice it to say that these property and control relations no longer hold – if indeed they ever did (Deleuze and Guattari 1987).
7 The resonance of choreography and *chora* are manifold, and are especially charged with respect to sexual difference and the play of folding (see for example: Derrida 1982, 1995; Grosz 1995b).
8 The difference between these two versions of suppletion – the progressive degradation of an original (negative evaluation) versus its perfectible rendering (positive evaluation) – is replayed in the difference between analog and digital reproduction (Mitchell 1992).
9 Virtual-reality technologies succinctly encapsulate this paradoxical reworking of the body: 'Far from being left behind,' notes Richards (1995: 35), 'the flesh forms the essential site of VR.' The sensors and effectors that are attached to the user's flesh not only enable 'an idealised, bodiless "experience", they also place the body in a state of sensory-deprivation with respect to its immediate environment'. Moreover, this interface between machinic sensors and effectors on the one hand,

and bodily receptors and actions on the other, attests to the fact that 'no matter how imperceptible it becomes. . . . [i]t is an intimate layer always watching' (Richards 1995: 36).

10 A different spin is put upon such notions if it is recognised that postmodern space-time is characterised, above all, by a certain 'detemporalisation of space'. For Bauman (1997: 86), 'The projection of spatial, contemporaneous difference upon the continuum of time, re-presentation of [spatial] heterogeneity as [an] ascending series of time stages, was perhaps the most salient, and also the most seminal, feature of the modern mind.' There is, however, always a two-way transference involved in any metaphor: 'The projection of space upon time furnished time with certain traits that only space possesses "naturally": modern time had *direction*, just like any itinerary in space' (Bauman, 1997: 86). The space-time of modernity thus possessed all those qualities necessary for planning out routes, for making one's way in the world, and for guaranteeing *arrival*. It is the loss of faith in any certainty of arrival that marks out our postmodern condition. The rhetoric surrounding virtual technology speaks, perhaps above all else, of the postmodern shock of the detemporalisation of social space, of the need for motionless voyaging in place (motifs of schizophrenia and nomadology come most readily to mind).

11 Cf. Lukás (1971: 90), on the transformation by capitalism of time and the subject: 'time sheds its qualitative, variable, flowing nature; it freezes into an exactly delimited, quantifiable continuum filled with quantifiable "things" (the reified, mechanically objectified "performance" of the worker, wholly separated from his total human personality): in short, it becomes space. In this environment where time is transformed into abstract, exactly measurable, physical space, an environment at once the cause and effect of the scientifically and mechanically fragmented and specialised production of the object of labour, the subject of labour must likewise be rationally fragmented.'

12 Similarly, whilst good is predicated on its opposition to evil, the principle of evil amounts to their non-opposition, on the insistence of a duel between indiscernible and reversible terms. (Accordingly, evil is more, not less than good; the illusory more, not less than the real.)

13 '*Ex-terminus*: what has passed beyond the end, so to speak' (Baudrillard 1995b: 95).

14 Resolve derives from the Latin *re-solvere*: re-release; unfasten again.

15 The unconscious is famously described by Lacan (1979: 20) as being 'structured like a language'.

16 The interminable yet temporary suturing – stitching or joining – of the Imaginary register (characterised by mythical completion) and the Symbolic register (marked by lack) is the mechanism whereby the lack in the subject and its Other is perpetually differed and deferred through dissimulation, dissimilation, and dissemination.

References

Abler, R. (1993) Everything in its place: GPS, GIS, and geography in the 1990s, *Professional Geographer* 45(2): 131–139.

Abrioux, Y. (1995) The (mis)adventures of photographic memory, *Art and Design* 10 (9/10): 61–67.

Acción Zapatista de Austin (1997) *Zapatistas in Cyberspace*, Internet, http://www.eco.utexas.edu/faculty/Cleaver/zapsincyber.htm.

Adam, A. (1997) What should we do with cyberfeminism?, in Lander, R. and Adam, A. (eds) *Women into Computing: Progress from Where to What?*, Exeter: Intellect.

Adam, A. (1998) *Artificial Knowing: gender and the thinking machine*, London: Routledge.

Adam, B. (1990) *Time and Social Theory*, Cambridge: Polity Press.

Adam, B. (1995) *Timewatch*, Cambridge: Polity Press.

Adam, B. (1997) *Timescapes of Modernity*, London: Routledge.

Adams, P. (1996). Protest and the scale politics of telecommunications, *Political Geography* 15(5): 419–441.

Adas, M. (1989) *Machines as the Measure of Men: Science, Technology and Ideologies of Western Dominance*, Ithaca, NY: Cornell University Press.

Adorno, T. (1945) A Social Critique of Radio Music, *Kenyon Review* 7: 208–217.

Adorno, T. (1990) [1934] The Form of the Phonograph Record (trans. T. Levin), in T. Levin, For the Record: Adorno on Music in the Age of Its Technological Reproducibility, *October* 55: 23–66. Originally appeared in German in 1934.

Al-Hindi, K. F. and C. Staddon (1997) The hidden histories and geographies of neo-traditional town planning: the case of Seaside, Florida, *Environment and Planning D: Society and Space* 15(3): 349–372.

Alcoff, L. (1988) Cultural Feminism versus Poststructuralism: The identity crisis in feminist theory, *Signs* 13(3): 405–436.

Aldiss, B. (1982) *Helliconia Spring*, London: Triad Grafton.

Aldiss, B. (1983) *Helliconia Summer*, London: Triad Grafton.

Aldiss, B. (1985) *Helliconia Winter*, London: Triad Grafton.

Alliez, E. and Feher, M. (1989) Notes on the sophisticated city, in Feher, M. and Kwinter, S. (eds) *The Contemporary City. Zone 1 & 2*, New York: Zone Books: 41–55.

Alvarez, J.L. (1996) The international popularisation of entrepreneurial ideas, in Clegg, S. and Palmer, G. (eds) *The Politics of Management Knowledge*, London: Sage.

Amin, A. and Graham, S. (1997) The ordinary city, *Transactions of the Institute of British Geographers* (forthcoming).

Amin, A. and Thrift, N.J. (1992) Neo-Marshallian nodes in global networks, *International Journal of Urban and Regional Research* 16: 571–587.

Anderson, N. (1992) The Telecom New Zealand Story: The Impact of Deregulation and Privatization of the New Zealand Telecommunications Industry on the Industrial Relations Environment and the Conditions of Employment of the Employees of Telecom (NZ) Ltd. Diploma of Industrial Relations, Victoria University of Wellington.

Andrews, G. (1994) Interactive marketing, *IEEE Technology and Society Magazine* July: 12–13.

Ansell Pearson, K. (1997) Deleuze outside/outside Deleuze: on the difference engineer, in Ansell Pearson, K. (ed.) *Deleuze and philosophy: the difference engineer*, London: Routledge: 1–22.

Appadurai, A. (1990) Disjuncture and Difference in the Global Political Economy, *Public Culture* 2(2): 1–24.

Appadurai, A. (1996) *Modernity at Large: Cultural Dimensions of Globalisation*, Minneapolis: University of Minnesota Press.

Aronson, S.H. (1971) The sociology of the telephone, *International Journal of Comparative Sociology* 12: 153–167.

Arthur, W. B. (1994a) On the Evolution of Complexity, in Cowan, G., Pines, D. and Meltzer, D. (eds) *Complexity: Metaphors, Models and Reality*, Reading, MA: Addison-Wesley: 65–77.

Arthur, W.B. (1994b) *Increasing Returns and Path Dependence in the Economy*, Ann Arbor: University of Michigan Press.

Arthur, W. B. (1996) Increasing Returns and the Two Worlds of Business, *Harvard Business Review* July: 100–109.

Atkins, P.J. (1993) How the West End was won: the struggle to remove street barriers in Victorian London, *Journal of Historical Geography* 19: 265–277.

Audirac, I. and Shermyen, A. (1994) An Evaluation of Neotraditional Design's Social Prescription: Postmodern Placebo or Remedy for Suburban Malaise, *Journal of Planning Education and Research* 13(3): 161–173.

Austin, J.L. (1955) *How to Do Things with Words*, Cambridge, MA: Harvard University Press.

Bachelard, G. (1964) *The Poetics of Space* (trans. Maria Jolas), New York: Orion Press.

Bakhtin, M. M. (1984) *Problems of Dostoevsky's Poetics* (ed. and trans. C. Emerson), Manchester: Manchester University Press.

Bal, M. and Bryson, N. (1994) Semiotics and Art History, in Bal, M. (ed.) *On Making Meaning: Essays in Semiotics*, Sonoma, CA: Polebridge Press.

Baldwin, F.G.C. (1925) *The History of the Telephone in the United Kingdom*, London: Chapman and Hall.

Bangemann, M. (1994) *Europe and the Global Information Society: recommendations to the European Council*, Brussels.

Bannister, N. (1994) Networks tap into low wages, *The Guardian*, 15 October: 40.

Barry, A., Osborne T. and Rose, N. (eds) (1996) *Foucault and Political Reason: Liberalism, neo-liberalism and rationalities of government*, Chicago: University of Chicago Press.

Barthes, R. (1975) *S/Z* (trans. R. Miller), London: Cape.

Bataille, G. (1988) *The Accursed Share, Vol. 1* (trans. Robert Hurley), NY: Zone Books.

Batty, M. (1996) Simulating reality, *Environment and Planning B: Planning and Design*, 23: 253–254.

Batty, M. (1997a) Urban systems as cellular automata, *Environment and Planning B: Planning and Design*, 24: 159–164.

Batty, M. (1997b) Virtual Geography, *Futures* 29: 337–352.

Batty, M. and Longley, P. (1995) *Fractal Cities*, London: Academic Press.

Baudrillard, J. (1983) *Simulations*, New York: Semiotexte.

Baudrillard, J. (1986) The year 2000 will not take place, in Grosz, E. A., Threadgold, T., Kelly, D., Cholodenko, A. and Colles, E. (eds) *Future*fall: excursions into post-modernity*, Annandale: Meglamedia, Annandale: 18–28.

Baudrillard, J. (1990) *Revenge of the Crystal: selected writings on the object and its destiny, 1968–1983*, Foss, P. and Pefanis, J. (eds) London: Pluto Press.

Baudrillard, J. (1993) *Symbolic Exchange and Death*, London: Sage.

Baudrillard, J. (1994a) *Simulacra and Simulation*, Ann Arbor: University of Michigan Press.

Baudrillard, J. (1994b) *The Illusion of the End*, Cambridge: Polity.

Baudrillard, J. (1995a) The virtual illusion: or the automatic writing of the world, *Theory, Culture & Society* 12: 97–107.

Baudrillard, J. (1995b) Symbolic exchange: taking theory seriously. An interview with Jean Baudrillard by Roy Boyne and Scott Lash, *Theory, Culture & Society* 12: 79–95.

Baudrillard, J. (1996) *The Perfect Crime*, London: Verso.

Bauer, M. (ed.) (1995) *Resistance to the New Technology: nuclear power, information technology, and biotechnology*, Cambridge: Cambridge University Press.

Bauman, Z. (1991) *Modernity and Ambivalence*, Cambridge: Polity.

Bauman, Z. (1993) *Postmodern Ethics*, Oxford: Blackwell.

Bauman, Z. (1997) *Postmodernity and its Discontents*, Cambridge: Polity.

Beamish, A. (1996) *Communities on Line: community-based computer networks*, (anneb@mit.edu).

Beauregard, R. (1993) *Voices of Decline: The Postwar Fate of US Cities*, Cambridge, MA: Basil Blackwell.

Beck, U. (1992) *Risk Society*, London: Sage.

Belgrave, J. (1993) Public Utilities and Light Handed Regulation. Speech to Institute of Policy Studies Symposium, Wellington, 6 October.

Bellah, R. (1985) *Habits of the Heart*, Berkeley, CA: University of California Press.

Bellinghausen, H. (1997) El domingo, movilización internacional en apoyo a Zapatistas, *La Jornada* 15 February.

Benedikt, M. (1991) Cyberspace: Some Proposals, in Benedikt, M. (ed.) *Cyberspace, First Steps*, Cambridge, MA: MIT Press.

Beniger, J. (1986) *The Control Revolution: Technological and Economic Origins of the Information Society*, Cambridge, MA: Harvard University Press.

Benjamin, A. (ed.) (1995) Complexity. Architecture/Art/Philosophy, *Journal of Philosophy and the Visual Arts* no. 6.

Benjamin. T. (1989) *A Rich Land, a Poor People: politics and society in modern Chiapas*, Albuquerque: University of New Mexico Press.

Benjamin, W. (1968) The Work of Art in the Age of Mechanical Reproduction, *Illuminations* (trans. H. Zohn), New York: Harcourt, Brace and World.

Benjamin, W. (1978) Paris, Capital of the Nineteenth Century, *Reflections* (trans. E. Jephcott), New York: Schocken Books.

Benjamin, W. (1979) *One-Way Street and Other Writings* (trans. E. Jephcott and K. Shorter), London: NLB.

Berger, P. and Luckmann, T. (1966) *The Social Construction of Reality: A Treatise in the Sociology of Knowledge*, Garden City, NY: Doubleday.

Berman, M. (1991) *All That is Solid Melts into Air: The Experience of Modernity*, London: Verso.

Bernstein, M.A. (1994) *Foregone Conclusions. Against Apocalyptic History*, Berkeley, CA: University of California Press.

Bersani, L. (1995) *Homos*, Cambridge, MA: Harvard University Press.

Bijker, W.E. (1995) *Of Bicycles, Bakelites, and Bulbs: Toward a Theory of Sociotechnical Change*, Cambridge, MA: MIT Press.

Bijker, W.E. and Law, J. (eds) (1992) *Shaping Technology/Building Society: Studies in Sociotechnical Change*, Cambridge, MA: MIT Press.

Bijker, W.E., Hughes, T.P. and Pinch, T.J. (eds) (1987) *The Social Construction of Technological Systems: New Directions in the Sociology and History of Technology*, Cambridge, MA: MIT Press.

Bingham, N. (1996) Object-ions: From Technological Determinism Towards Geographies of Relations, *Society and Space* 14(6): 635–657.

Biocca, F. and Lanier, J. (1992) An insider's view of the future of virtual reality, *Journal of Communications* 42(4): 150–172.

Birkerts, S. (1994) *The Gutenberg Elegies: The Fate of Reading in an Electronic Age*, Boston, MA: Faber and Faber.

Bleeker, J. (1992) Vision Culture: Information Management and the Cultural Assimilation of VR, *Afterimage* October.

Bloque 7 (1997) *Temáticas transversales y relacionadas con la red*. Discussion papers, 2nd Encounter for Humanity and against Neoliberalism, Spain.

Blumenberg, H. (1993) Light as a Metaphor for Truth: At the Preliminary Stage of Philosophical Concept Formation (trans. J. Anderson), in Levin, D.M. (ed.) *Modernity and the Hegemony of Vision*, Berkeley, CA: University of California Press.

Boddy, T. (1992) Underground and Overhead: Building the Analogous City, in Sorkin, M. (ed.) *Variations on a Theme Park: The New American City and the End of Public Space*, New York: Hill and Wang.

Bogard, W. (1996) *The Simulation of Surveillance: Hypercontrol in Telematic Societies*, Cambridge: Cambridge University Press.

Bogue, D. (1989) *Deleuze and Guattari*, London: Routledge.

Borch-Jacobsen, M. (1994) The alibis of the subject, in Shamdasani, S. and Münchow, M. (eds) *Speculations after Freud: psychoanalysis, philosophy and culture*, London: Routledge: 77–96.

Borges, J.L. (1971) *The Aleph and Other Stories: 1933–1969*, New York: Bantam Books.

Borges, J. L. (1974) *The book of imaginary beings*, Harmondsworth: Penguin.

Borgmann, A. (1995) The Nature of Reality and the Reality of Nature, in Soulé, M. and Lease, G. (eds) *Reinventing Nature: Responses to Postmodern Deconstruction*, Washington, DC: Island Press.

Boundas, C.V. and Olkowski, D. (1994) *Gilles Deleuze and the Theatre of Philosophy*, London: Routledge.

Boyer, C. (1992) Cities for Sale: Merchandising History at South Street Seaport, in Sorkin, M. (ed.) *Variations on a Theme Park: The New American City and the End of Public Space*, New York: Hill and Wang.

Boyer, C. (1996) *Cybercities: Visual Perception in the Age of Electronic Communication*, New York: Princeton University Press.

Bradley, D. (1995) Situating cyberspace, *Public* 11: 9–19.

Braidotti, R. (1994) *Nomadic Subjects: embodiment and sexual difference in contemporary feminist theory*, New York: Columbia University Press.

Brail, S. (1996) The Price of Admission: Harassment and Free Speech in the Wild, Wild West, in Cherny, L. and Weise, E.R. (eds) *Wired Women: Gender and New Realities in Cyberspace*, Seattle, WA: Seal Press.

Brandt, J. (1997) *Geopoetics. The Politics of Mimesis in Poststructuralist French Poetry and Theory*, Stanford, CA: Stanford University Press.

Brewster, D. (1832) *Letters on Natural Magic, addressed to Sir Walter Scott, Bart*, London: John Murray.

Brewster, D. (1856) *The Stereoscope: Its History, Theory, and Construction*, London: John Murray.

Briggs, A. (1968) *Victorian Cities*, Harmondsworth: Penguin Books.

Brill, T. B. (1980) *Light: Its Interaction with Art and Antiquities*, New York: Plenum Press.

Brook, J. and Boal, I. A. (eds) (1995) *Resisting the Virtual Life: the culture and politics of information*, San Francisco: City Lights.

Brosseau, M. (1995) The City in Textual Form: Manhattan Transfer's New York, *Ecumene* 2: 89–114.

Bruckman, A. (1996) Gender Swapping on the Internet, in Ludlow, P. (ed.) *High Noon on the Electronic Frontier: Conceptual Issues in Cyberspace*, Cambridge, MA: MIT Press.

Bryson, N. (1983) *Vision and Painting: The Logic of the Gaze*, New Haven, CT: Yale University Press.

Buck-Morss S. (1989) *The Dialectics of Seeing: Walter Benjamin and the Arcades Project*, Cambridge, MA: MIT Press.

Bukatman, S. (1993a) Gibson's Typewriter, *South Atlantic Quarterly* 92: 627–645.

Bukatman, S. (1993b) *Terminal Identity: The virtual subject in postmodern science fiction*, Durham and London: Duke University Press.

Bukatman S. (1995) The Artificial Infinite, in Cooke, L. and Wollen P. (eds) *Visual Display: Culture Beyond Appearances*, Seattle, WA: Bay Press: 254–289.

Burgess, J., Harrison, C. and Goldsmith, B. (1990) Pale Shadows for Policy: The role of qualitative research in environmental planning, in Burgess, R. (ed.) *Studies in Qualitative Methods*, Vol. 2, London: JAI Press.

Burgess, J., Limb, M. and Harrison, C. (1988a) Exploring Environmental Values through the Medium of Small Groups: 1. Theory and practice, *Environment and Planning A* 20: 309–326.

Burgess, J., Limb, M. and Harrison, C. (1988b) Exploring Environmental Values through the Medium of Small Groups: 2. Illustrations of a group at work, *Environment and Planning A* 20: 456–476.

Burgin, V. (1988) Geometry and Abjection, in Tagg, J. (ed.) *The Cultural Politics of Postmodernism*, Binghamton, NY: Department of Art and Art History, SUNY Binghamton.

Burnstein, D. and Kline, D. (1995) *Road Warriors: Dreams and Nightmares Along the Information Highway*, New York: Dutton.

Burrows, R. (1997) Virtual culture, urban social polarisation and social science fiction, in Loader, B. (ed.) *The Governance of Cyberspace: Politics, Technology and Global Restructuring*, London: Routledge: 38–45.

Butler, J. (1990) *Gender Trouble*, London: Routledge.

Butler, J. (1993) *Bodies that Matter*, London: Routledge.

Byrne, D. (1996) Chaotic cities or complex cities, in Westwood, S. and Williams, J. (eds) *Imagining Cities*, London: Routledge: 50–70.

Callon, M. (1991) Techno-economic networks and irreversibility, in Law, J. (ed.) *A Sociology of Monsters: Essays on power, technology and domination*, London: Routledge.

Calvino, I. (1992) *Six Memos For The New Millennium*, London: Jonathan Cape.

Campion Smith, B. (1996) Highway 407: Tolls but no jams, *Toronto Star* 29 July: 4–8.

Carey, J. (1989) *Communication as Culture: Essays on Media and Society*, London: Unwin Hyman.

Carlyle, T. (1835) [1829] Signs of the Times, in Cross, M. (ed.) *Selections from the Edinburgh Review Comprising the Best Articles in that Journal from its Commencement to the Present Time*, Paris: Baudry's European Library.

Carr, B. (1996) Crossing Borders: Labor Internationalism in the Era of NAFTA, in Otero, G. (ed.) *Neoliberalism Revisited: Economic Restructuring and Mexico's Political Future*, Boulder, CO: Westview Press: 209–232.

Carruthers, M. (1990) *The Book of Memory: A Study of Memory in Medieval Culture*, Cambridge, MA: Cambridge University Press.

Carter, P. (1987) *The Road To Botany Bay: an essay in spatial history*, London and Boston: Faber and Faber.

Casey, C. (1995) *Work, Self and Society after Industrialism*, London: Routledge.

Castells, M. (1985) High Technology, Economic Restructuring, and the Urban-Regional Process in the United States, in Castells, M. (ed.) *High Technology, Space and Society*, London: Sage.

Castells, M. (1989) *The Informational City: Information Technology, Economic Restructuring and the Urban-Regional Process*, Oxford: Blackwell.

Castells, M. (1996) *The Information Age: Economy, Society and Culture Vol. 1: the Rise of the Network Society*, Oxford: Blackwell.

Castells, M. and Hall, P. (1994) *Technopoles of the World*, London: Routledge.
Castle, T. (1995) *The Female Thermometer: Eighteenth-Century Culture and the Invention of the Uncanny*, New York: Oxford University Press.
Cavell, S. (1971) *The World Viewed*, New York: Viking.
Certeau, M. de (1984) *The Practice of Everyday Life* (trans. Steven Randall), Berkeley, CA: University of California Press.
Ceruzzi, P. (1991) When Computers were Human, *Annals of the History of Computing* 13(3): 237–244.
Chaitkin, S. (1996) Untitled, unpublished paper, Vassar College.
Chant, C. (ed.) (1989) *Science, Technology and Everyday Life 1870–1950*, London: The Open University.
Chartier, R. (1997) *On the Edge of the Cliff: History, Language, Practices*, Baltimore, MD: Johns Hopkins University Press.
Chermayeff, S. and Alexander, C. (1963) *Community and Privacy: Toward a New Architecture of Humanism*, Harmondsworth: Penguin.
Chernaik, L. (1996) Spatial Displacements: transnationalism and the new social movements, *Gender, Place and Culture* 3: 3: 251–275.
Cherryh, C. J. (1983) *Downbelow Station*, London: Methuen.
Chitty, G. (1995) 'A great entail': The historic environment, in Wheeler, M. (ed.) *Ruskin and Environment: The Storm Cloud of the Nineteenth Century*, Manchester: Manchester University Press.
Clark, J. (1995) *Managing Consultants: Consultancy as the Management of Impressions*, Milton Keynes: Open University Press.
Clark, T.J. (1985) *The Painting of Modern Life: Paris in the Art of Manet and his Followers*, Princeton, NJ: Princeton University Press.
Clarke, D. B. and Doel, M. A. (1994) Transpolitical geography, *Geoforum* 25: 505–524.
Cleaver, H. (1994) Introduction. In *¡Zapatistas! Documents from the New Mexican Revolution*, 11–24, Brooklyn, NY: Autonomedia.
Cleaver, H. (1996) *Zapatistas and the Electronic Fabric of Struggle*, Internet: http://www.eco.utexas.edu/faculty/Cleaver/zaps.html
Clegg, S. and Palmer, G. (eds) (1996) *The Politics of Management Knowledge*, London: Sage.
Clerc, S. (1996) Estrogen Brigades and 'Big Tits' Threads: Media Fandom Online and Off, in Cherny, L. and Weise, E.R. (eds) *Wired Women: Gender and New Realities in Cyberspace*, Seattle, WA: MIT Press.
Coddington, D. (1993) *Turning Pain into Gain: The Plain Person's Guide to the Transformation of New Zealand 1984–93*, Auckland: Alister Taylor Publishers.
Cohen, B.M. (1996) *The Edge of Chaos: Financial Booms, Bubbles and Crashes*, New York: John Wiley.
Collier, G. with Lowery Quaratiello, E. (1994) *Basta! Land and the Zapatista Rebellion in Chiapas*, Oakland, CA: Food First Books.
Collins, H.M. (1990) *Artificial Experts: Social Knowledge and Intelligent Machines*, Cambridge, MA: MIT Press.
Collins, J. (1995) *Architectures of Excess: Cultural Life in the Information Age*, New York: Routledge.

Conley, T. (1993) Translator's Foreword: A Plea for Leibniz, in Deleuze, G. *The Fold: Leibniz and the Baroque*, Minneapolis: University of Minnesota Press: ix–xx.

Cooper, R. (1992) Formal Organisation as Representation: Remote Control, Displacement and Abbreviation, in Reed, M. and Hughes M. (eds) *Rethinking Organisation: New Directions in Organisational Theory and Analysis*, London: Sage: 254–272.

Cosgrove, D.E. and Daniels, S.J. (eds) (1988) *The Iconography of Landscape: essays on the symbolic representation, design, and use of past environments*, Cambridge: Cambridge University Press.

Couch, H.N. and Geer, R.M. (1961) *Classical Civilization, Greece* (2nd edn), Englewood Cliffs, NJ: Prentice-Hall.

Coveney, P. and Highfield, R. (1995) *Frontiers of Complexity*, New York: Fawcett Columbine.

Coyne, R. (1994) Heidegger and Virtual Reality, *Leonardo* 27(1): 65–73.

Crang, P. (1994) It's showtime: on the workplace geographies of display in a restaurant in southeast England, *Society and Space* 12: 675–704.

Crary, J. (1994) *Techniques of the Observer: On Vision and Modernity in the Nineteenth Century*, Cambridge, MA: MIT Press.

Crary, J. and Kwinter, S. (eds) (1992) *Incorporations. Zone* 6, New York: Zone Books.

Crawford, M. (1992) The World in a Shopping Mall, in Sorkin, M. (ed.) *Variations on a Theme Park: The New American City and the End of Public Space*, New York: Hill and Wang.

Crawford, M. (1995) Contesting the Public Realm: Struggles over Public Space in Los Angeles, *Journal of Architectural Education* 49: 1: 4–9.

Crawford, R. (1996) Computer-assisted crises, in Gerbner, G. Mowlana, H. and Schiller, H. (eds) *Invisible Crises: What Conglomerate Control of Media Means for America and the World*, Boulder, CO: Westview: 47–81.

Cronon, W. (ed.) (1995) *Uncommon Ground: Toward Reinventing Nature*, New York: W.W. Norton.

Crowther, P. (1989) *The Kantian Sublime: From Morality to Art*, Oxford: Clarendon Press.

Crowther, P. (1993) *Critical Aesthetics and Postmodernism*, Oxford: Clarendon Press.

Csicsery-Ronay Jr, I. (1991a) Cyberpunk and Neuromanticism, in McCaffery, L. (ed.) *Storming the Reality Studio*, Durham, NC: Duke University Press.

Csicsery-Ronay Jr, I. (1991b) The SF of Theory: Baudrillard and Haraway, *Science Fiction Studies* 18: 387–404.

Cubitt, S. (1996) Reviews of Featherstone, M. and Burrows, R. (eds) *Cyberspace, Cyberbodies, Cyberpunk* and Shields, R. (ed.) *Cultures of Internet*, *Sociology* 30(4): 832–835.

Curtis, P. (1996) MUDding: Social Phenomena in Text-based Virtual Realities, in Ludlow, P. (ed.) *High Noon on the Electronic Frontier: Conceptual Issues in Cyberspace*, Cambridge, MA: MIT Press.

Czitrom, D. (1982) *Media and the American Mind: From Morse to McLuhan*, Chapel Hill: University of North Carolina Press.

Daly, G. (1991) The Discursive Construction of Economic Space: Logics of Organization and Disorganization, *Economy and Society* 20(1): 79–102.

Daniels, S. and Rycroft, S. (1993) Mapping the Modern City: Allan Sillitoe's Nottingham Novels, *Transactions of the Institute of British Geographers* New Series 18: 460–480.

Datel, R. E. (1990) Southern Regionalism and Historic Preservation in Charleston, South Carolina, 1920–1940, *Journal of Historical Geography* 16(2): 197–215.

Davies, S. (1994) They've got an eye on you, *The Independent* 2 November.

Davies, S. (1995) *Big Brother: Britain's Web of Surveillance and the New Technological Order*, London: Pan.

Davis, E. (1993) Techgnosis, Magic, Memory, in Dery, M. (ed.) *Flame Wars: The Discourse of Cyberculture*, (South Atlantic Quarterly, Fall), Durham, NC: Duke University Press.

Davis, L. (1987) *Resisting Novels: Ideology and fiction*, New York and London: Methuen.

Davis, M. (1992) Fortress Los Angeles: The militarization of urban space, in Sorkin, M. (ed.) *Variations on a Theme Park: The New American City and the End of Public Space*, New York: Hill and Wang.

Davis, M. (1992) Beyond Blade Runner: Urban Control, the Ecology of Fear, *Open Magazine*, Westfield, NJ.

Davison, G. (1978) *The Rise and Fall of Marvellous Melbourne*, Carlton, Victoria: Melbourne University Press.

De Boer, M. (1993) Public Interiors, in Kloos, M. (ed.) *Public Interiors*, Amsterdam: Architectura & Natura Press.

de Lauretis, T. (1987) *Technologies of Gender: Essays on Theory, Film and Fiction*, Bloomington, IN: Indiana University Press.

Dean, M. (1994) *Critical and Effective Histories: Foucault's Methods and Historical Sociology*, London: Routledge.

Debord, G. (1977) *Society of the Spectacle*, Detroit: Black and Red.

Debord, G. (1991) *Comments on the Society of the Spectacle*, London: Verso [Sheffield: Pirate Press].

DeLanda, M. (1991) *War in the Age of Intelligent Machines*, New York: Zone Books.

DeLanda, M. (1994) Virtual Environments and the Emergence of Synthetic Reason, in Dery, M. (ed.) *Flame Wars: The Discourse of Cyberculture*, Durham, NC: Duke University Press: 263–285.

DeLanda, M. (1996) Markets and Anti-Markets, in Aronowitz, S. *et al.* (eds) *Technoscience and Cyberculture*, New York: Routledge: 181–194.

Deleuze, G. (1983) Plato and the simulacrum, *October* 27: 45–56.

Deleuze, G. (1988a) Postscript on the societies of control, *October* 59, 3–7.

Deleuze, G. (1988b) *Spinoza: Practical Philosophy*, San Francisco: City Lights Books.

Deleuze, G. (1990a) *Expressionism in Philosophy*, New York: Zone Books.

Deleuze, G. (1990b) *The Logic of Sense*, New York: Columbia University Press.

Deleuze, G. (1991) *Bergsonism*, New York: Zone Books.

Deleuze, G. (1992) *The Fold: Leibniz and the baroque*, Minneapolis: University of Minnesota Press.

Deleuze, G. (1993) *The Deleuze Reader* (Boundas, C.V. (ed.)) New York: University of Columbia Press.

Deleuze, G. (1994) *Difference and Repetition*, London: Athlone.

Deleuze, G. and Guattari, F. (1984) *Anti-Oedipus: Capitalism and Schizophrenia* (trans. Robert Hurley, Mark Seem and Helen R. Lane), London: Athlone Press.

Deleuze, G. and Guattari, F. (1987) *A Thousand Plateaus: Capitalism and Schizophrenia* (trans. Brian Massumi), Minneapolis: University of Minnesota Press.

Department of the Environment/MAFF (1995) *Rural England: a nation committed to a living countryside*, London: HMSO.

Derrida, J. (1981) *Dissemination*, Chicago: University of Chicago Press.

Derrida, J. (1982) Choreographies, *Diacritics* 12 (2): 66–76.

Derrida, J. (1994) *Specters of Marx: the work of mourning, the state of the debt, and the New International*, London: Routledge.

Derrida, J. (1995) *On the Name*, Stanford: Stanford University Press.

Dery, M. (1992) Cyberculture, *South Atlantic Quarterly* 91: 501–523.

Dery, M. (1996) *Escape Velocity: cyberculture at the end of the century*, New York: Grove.

Deutsche, R. (1991) Boys Town, *Society and Space* 9(1) 5–30.

Deutsche, R. (1996) *Evictions: Art and Spatial Politics*, Cambridge, MA: MIT Press.

Dibbell, J. (1996 [1993]) A Rape in Cyberspace; or How an Evil Clown, a Haitian Trickster Spirit, Two Wizards, and a Cast of Dozens Turned a Database into a Society, in Ludlow, P. (ed.) *High Noon on the Electronic Frontier: Conceptual Issues in Cyberspace*, Cambridge, MA: MIT Press.

Dieberger, A. (1996) Browsing the WWW by interacting with a technical virtual environment – A framework for experimenting with navigational metaphors, in *Hypertext '96: The Seventh ACM Conference on Hypertext*, New York: ACM Press. Also at http://www.lcc.gatech.edu/faculty/dieberger/HT96.paper.html

Dienst, R. (1994) *Still Life in Real Time*, Durham, NC: Duke University Press.

Doel, M. A. (1995) Bodies without organs: schizoanalysis and deconstruction, in Pile, S. and Thrift, N. (eds) *Mapping the subject: geographies of cultural transformation*, London: Routledge: 226–240.

Doel, M. A. (1996) A hundred thousand lines of flight: a machinic introduction to the nomad thought and scrumpled geography of Gilles Deleuze and Félix Guattari, *Environment and Planning D: Society and Space* 14: 421–439.

Doel, M. A. (forthcoming) *Poststructuralist Geographies: The diabolical art of spacial science*, Edinburgh: Edinburgh University Press.

Donaldson, H. (1994) Recent Developments in the Regulatory Environment. Speech to the New Zealand Telecommunications Summit, Wellington, 2 March.

Douglas, S. (1986) Amateur Operators and American Broadcasting: Shaping the Future of Radio, in Corn, J. (ed.) *Imagining Tomorrow: History, Technology and the American Future*, Cambridge, MA: MIT Press.

Driver, F. (1984) Power, space and the body: A critical reassessment of Foucault's *Discipline and Punish, Society and Space* 3: 425–446.

Droege, P. (ed.) (1997) *Intelligent Environments*, North Holland: Amsterdam.

Druckrey, T. (ed.) (1994a) *Electronic Culture: Technology and Visual Representation*, New York: Aperture.

Druckrey, T. (1994b) Introduction, in Druckrey, T. (1994a) (ed.) *Electronic Culture: Technology and Visual Representation*, New York: Aperture: 13–26.

Dyrkton, J. (1996) Cool runnings: the contradictions of cyberreality in Jamaica, in Shields, R. (ed.) *Cultures of Internet: Virtual Spaces, Real Histories, Living Bodies*, London: Sage: 49–57.

Eco, U. (1986a) Function and Sign: Semiotics of Architecture, in Gottdiener, M. and Lagopoulos, A. Ph. (eds) *The City and the Sign: An Introduction to Urban Semiotics*, New York: Columbia University Press. Originally appeared in Italian in 1973.

Eco, U. (1986b) *Travels in Hyperreality* (trans. W. Weaver), San Diego: Harcourt Brace Jovanovich. Originally appeared in Italian in 1975.

Eco, U. (1989) *The Open Work* (trans: Anna Cancogni), Cambridge, MA: Harvard University Press.

Economist (1991) Telecommunications: The New Boys, 5 October.

Economist (1996) The Hitchhiker's Guide to Cybernomics: A Survey of the World Economy, 28 September.

Elliott, E. and Kiel, L.D. (eds) (1997) *Chaos Theory in the Social Sciences. Foundation and Application*, Ann Arbor: University of Michigan Press.

Elloriaga, J. (1997) *Del sumense al construyamos: el zapatismo a partir de sus cuatro Declaraciones de la Selva Lacandona*, Internet: http://spin.com.mx/~floresu/fzln/

Emery, F. (ed.) (1969) *Systems Thinking*, London: Routledge.

Engwall, L. (1992) *Mercury Meets Minerva*, Oxford: Pergamon.

Escobar, A. (1994) Welcome to Cyberia – Notes on the Anthropology of Cyberculture, *Current Anthropology* 35(3): 211–232.

Esteva, G. (1994) *Crónica del Fin de una Era*, Mexico City: Editorial Posada.

European Commission (1988) *The Future of Rural Society*, Brussels.

Fabrice, F. (1990) Information Landscapes, in Ambron, S. and Hooper, K. (eds) *Learning with interactive Multimedia*, Microsoft Press.

Falk, T. and Abler, R. (1980) Intercommunications, distance and geographical theory, *Geografiska Annaler* Series B: 59–67.

Featherstone, M. and Burrows, R. (eds) (1995) *Cyberspace, Cyberbodies, Cyberpunk: Cultures of Technological Embodiment*, London: Sage.

Ferguson, N. (ed.) (1997) *Virtual History: Alternatives and Counterfactuals*, London: Picador.

Fernback, J. and Thompson, B. (1995) *Computer-mediated Communications and the American Collectivity: the dimensions of community within cyberspace*. Paper presented at the annual convention of the International Communications Association, Alberquerque, NM, May.

Fischer, C. (1988) 'Touch Someone': The Telephone Industry Discovers Sociability, *Technology and Culture* 29(1): 32–61.

Fischer, C. (1992) *America Calling: A social history of the telephone to 1940*, Berkeley, CA: University of California Press.

Fischer, C. (1997) Technology and Community: Historical Complexities, *Sociological Inquiry* 67(1): 113–118.

Fishman, T C. (1995) The Bull Market in Fear, *Harper's Magazine* October: 55–62.

Fiske, J. (1989) *Reading the Popular*, Boston: Unwin Hyman.

Flanagan, B. (1996) Cause to Celebrate?, *Metropolitan Home* 28(4): 54–60.

Flynn, E. A. and Schweickart, P. P. (eds) (1986) *Gender and Reading: Essays on readers, texts, and contexts*, Baltimore and London: Johns Hopkins University Press.

Foucault, M. (1977) *Discipline and Punish*, New York: Pantheon.

Foucault, M. (1980) *Power/Knowledge: selected interviews and other writings, 1972–1977* (ed. and trans. C. Gordon (*et al.*)), New York: Pantheon.

Foucault, M. (1982) *The Archaeology of Knowledge*, New York: Pantheon.

Foucault, M. (1986) Of Other Spaces, *Diacritics* 16(1): 22–27.

Francaviglia, R. (1996) *Main Street Revisited: Time, Space, and Image-Building in Small-Town America*, Iowa City: University of Iowa Press.

Friedberg, A. (1993) *Window Shopping: Cinema and the Postmodern*, Berkeley, CA: University of California Press.

Frieden, B. and Sagalyn, L. (1989) *Downtown Inc.: How America Rebuilds Cities*, Cambridge, MA: MIT Press.

Friedland, R. and Boden, D. (eds) (1994) *NowHere: space, time and modernity*, London: University of California Press.

Froehling, O. (1997) A War of Ink and Internet, *Geographical Review* 87(2): 291–307. Special Issue on Geographical Views of Computer Networking.

Frow, J. (1997) *Time and Commodity Culture. Essays in Cultural Theory and Post-modernity*, Oxford: Oxford University Press.

Garbade, K.D. and Silber, W.L. (1978) Technology, communication and the performance of financial markets: 1840–1975, *Journal of Finance* 33(3): 819–832.

Gelerntner, D. (1991) *Mirror Worlds: The Day Software Puts the Universe in a Shoebox . . . How It Will Happen and What It Will Mean*, New York: Oxford University Press.

Gell, A. (1992) *The Anthropology of Time*, Oxford: Berg.

Genosko, G. (1994) *Baudrillard and Signs: signification ablaze*, London: Routledge.

Gibson, A. (1996) *Towards a Postmodern Theory of Narrative*, Edinburgh: Edinburgh University Press.

Gibson, W. (1984) *Neuromancer*, London: Grafton.

Gibson, W. (1986/7) *Count Zero*, London: Grafton.

Gibson, W. (1986/8) *Burning Chrome*, London: Grafton.

Gibson, W. (1988/9) *Mona Lisa Overdrive*, London: Grafton.

Gibson, W. (1991) Academy Leader, in Benedikt, M. (ed.) *Cyberspace, First Steps*, Cambridge, MA: MIT Press: 27–29.

Gibson, W. (1996) *Idoru*, New York: Putnam.

Giedion, S. (1948) *Mechanization Takes Command: a contribution to anonymous history*, New York: Oxford University Press.

Godwin, J. (1979) *Athanasius Kircher: A Renaissance Man and the Quest for Lost Knowledge*, London: Thames and Hudson.

Goheen, P.G. (1994) Negotiating access to public space in mid-nineteenth century Toronto, *Journal of Historical Geography* 20: 430–449.

Goldhill, S. (1996) Refracting Classical Vision: Changing Cultures of Viewing, in

Brennan, T. and Martin, J. (eds) *Vision in Context: Historical and Contemporary Perspectives on Sight*, New York: Routledge.

Gómez Peña, G. (1995) The Subcommandante of Performance, in Katzenberger, E. (ed.) *First World, Ha Ha Ha! The Zapatista Challenge*, San Francisco: City Lights: 89–98.

Goodchild, P. (1996) *Deleuze and Guattari: An Introduction to the Politics of Desire*, London: Sage.

Goodwin, B.C. (1963) *The Temporal Organisation of Cells*, London: Academic Press.

Goodwin, B.C. (1979) Generative and cognitive models of biological pattern formation, in Cullen, I.G. (ed.) *Analysis and Decision in Regional Policy. London Papers in Regional Science 9*, London: Pion: 20–25.

Goss, J. (1996) Disquiet on the Waterfront: Reflections on Nostalgia and Utopia in the Urban Archetypes of Festival Marketplaces, *Urban Geography* 17(3): 221–247.

Gowdy, V. (1994) Alternatives to prison, *The Futurist* January–February, 53.

Graham, S. (1996) Imagining the Real-Time City: Telecommunications, urban paradigms and the future of cities, in Westwood, S. and Williams, J. (eds) *Imagining Cities: Scripts, signs and memories*, London: Routledge.

Graham, S. (1998a) The end of geography or the explosion of place? Conceptualising space, place and information technology, *Progress in Human Geography* 22(2).

Graham, S. (1998b) The Spaces of surveillant-simulation: New technologies, digital representations, and material geographies, *Environment and Planning D: Society and Space* 16: 483–503.

Graham, S. and Marvin, S. (1996) *Telecommunications and the City: Electronic Spaces, Urban Places*, London: Routledge.

Graham, S., Brooks, J. and Heery, D. (1996) Towns on the Television: Closed Circuit TV in British Towns and Cities, *Local Government Studies* 22 (3): 3–27.

Gregory, D. (1993) *Geographical Imaginations*, Oxford: Blackwell.

Gregson, N. (1995) And now it's all consumption?, *Progress in Human Geography* 19(1): 135–141.

Grewal, I. and Kaplan, C. (eds) (1994) *Scattered Hegemonies*, Minneapolis: University of Minnesota Press.

Griffith, V. (1996) I know that face, *Financial Times*, 14 May: 10.

Grimes, J. and Warf, B. (1997) Counter-Hegemonic Discourses and the Internet, *Geographical Review* 87(2): 259–274. (Special Issue on Geographical Views of Computer Networking.)

Grint, K. and Woolgar, S. (1995) On Some Failures of Nerve in Constructivist and Feminist Analyses of Technology, in Grint, K. and Gill, R. (eds) *The Gender-Technology Relation: Contemporary Theory and Research*, London: Taylor & Francis.

Grint, K. and Woolgar, S. (1997) *The Machine at Work. Technology, Work and Organisation*, Cambridge: Polity Press.

Griscom, J. (1845) *The Sanitary Condition of the Laboring Population of New York*, New York: Harper.

Grosz, E. (1990) Inscriptions and body-maps: representations and the corporeal, in Threadgold, T. and Cranny-Francis, A. (eds) *Feminine/Masculine and Representation*, London: Allen & Unwin: 62–74.

Grosz, E. (1994) *Volatile Bodies: toward a corporeal feminism*, Indianapolis: Indiana University Press.

Grosz, E. (1995a) *Space, Time and Perversion*, London: Routledge.

Grosz, E. (1995b) Women, *chora*, dwelling, in Watson, S. and Gibson, K. (eds) *Postmodern Cities and Spaces*, Oxford: Blackwell: 47–58.

Gruen, V. (1957) Main Street 1969, *American Planning and Civic Annual*: 16–22.

Gruen, V. (1964) *The Heart of Our Cities. The Urban Crisis: Diagnosis and Cure*, New York: Simon and Schuster.

Guattari, F. (1995) *Chaosmosis: an ethico-aesthetic paradigm*, Sydney: Power Publications.

Guattari, F. (1996) *The Guattari Reader* (ed. G. Genosko), Oxford: Blackwell.

Habermas, J. (1989) *The Structural Transformation of the Public Sphere*, Cambridge: Polity Press.

Habermas, J. (1992) The Normative Content of Modernity, *The Philosophical Discourse of Modernity, Twelve Lectures*, Cambridge, MA: MIT Press.

Hall, K. (1996) Cyberfeminism, in Herring, S. (ed.) *Computer-Mediated Communication: Linguistic, Social and Cross-Cultural Perspectives*, Amsterdam: John Benjamins.

Hall, P. and Preston, P. (1988) *The Carrier Wave: New Technology and the Geography of Innovation*, London: Unwin.

Hall, S. (1993) *Mapping the Next Millennium*, New York: Vintage.

Halperin, D. (1995) *Saint Foucault*, Oxford: Oxford University Press.

Hannah, M. (1997) Imperfect panopticism: Envisioning the construction of normal lives, in Benko, G. and Stohmayer, U. (eds) *Space and Social Theory*, Oxford: Blackwell: 344–359.

Hannerz, U. (1987) The world in creolisation, *Africa* 57(4): 546–559.

Hannerz, U. (1989) Notes on the global ecumene, *Public Culture* 1(2): 66–75.

Hannerz, U. (1992) *Cultural Complexity: Studies in the Social Organization of Meaning*, New York: Columbia University Press.

Hannerz, U. (1996) *Transnational Connections: Culture, People, Places*, London: Routledge.

Hansard (1986) New Zealand Parliamentary Debates, State-Owned Enterprises Bill, Introduction, 30 September. Vol. 474: 4724–4739.

Hansard (1987) New Zealand Parliamentary Debates, State Enterprises Restructuring Bill, Report of Commerce and Marketing Committee, 9 June. Vol. 477: 9326–9336.

Hansard (1990) New Zealand Parliamentary Debates, Finance Bill, 20 March. Vol. 506: 802–861.

Haraway, D. (1989) *Primate Visions: Gender, Race and Nature in the World of Modern Science*, New York: Routledge.

Haraway, D. (1991a) *Simians, Cyborgs, and Women: the Reinvention of Nature*, London: Free Association Books.

Haraway, D. (1991b) Cyborg at Large, Interview; and The Actors are Cyborg, Nature is Coyote and the Geography is Elsewhere, both in Penley, C. and Ross, A. (eds) *Technoculture*, Minneapolis: University of Minnesota Press.

Haraway, D. (1997) *Modest.Witness@Second.Millennium.FemaleMan Meets OncoMouse*, London: Routledge.

Harding, S. (1992) *Whose Science, Whose Knowledge*, Ithaca: Cornell University Press.

Hardt, M. and Negri, A. (1994) *Labor of Dionysus: a critique of the state form*, Minneapolis: University of Minnesota Press.

Harpold, T. (1991) Contingencies of the Hypertext Link, *Writing on the Edge* 2(2) Spring: 126–137. Also on-line as http://www.lcc.gatech.edu/faculty/harpold/papers/contingencies/index.html

Harpold, T. (1996) Author's Note to web version of 'Contingencies of the Hypertext Link', http://www.lcc.gatech.edu/faculty/harpold/papers/contingencies/index.html

Harvey, D. (1982) *The Limits to Capital*, Oxford: Blackwell.

Harvey, D. (1989) *The Condition of Postmodernity: An Enquiry into the Origins of Cultural Change*, Oxford: Basil Blackwell.

Harvey, D. (1990) Between space and time: reflections on the geographical imagination, *Annals of the Association of American Geographers* 80(3): 418–834.

Harvey, D. (1993) From space to place and back again: reflections on the condition of postmodernity, in *Mapping the Futures: Local Cultures, global change*, London: Routledge.

Harvey, D. (1996) *Justice, Nature and the Geography of Difference*, Oxford: Blackwell.

Harvey, N. (1996) Rural Reforms and the Zapatista Rebellion: Chiapas 1988–1995, in Otero, G. (ed.) *Neoliberalism Revisited: Economic Restructuring and Mexico's Political Future*, Boulder, CO: Westview Press: 187–208.

Hayden, D. (1984) *Redesigning the American Dream: The Future of Housing, Work and Family Life*, New York: W.W. Norton.

Hayles, N. K. (1993) Virtual bodies and flickering signifiers, *October* 66: 69–91.

Hayles, N.K. (1995) Simulated Nature and Natural Simulations: Rethinking the Relation Between the Beholder and the World, in Cronon, W. (ed.) *Uncommon Ground: Toward Reinventing Nature*, New York: W.W. Norton.

Hayles, N. K. (1996) Narratives of artificial life, in Robertson, G. *et al.* (eds) *Future Natural: Nature, Science, Culture*, London: Routledge: 146–164.

Headrick, D.R. (1981) *The Tools of Empire: Technology and European Imperialism in the Nineteenth Century*, New York: Oxford University Press.

Heaney, S. (1990) *Selected Poems 1966–1987*, New York: The Noonday Press, Farrar Straus and Giroux.

Heelas, P. *et al.* (eds) (1996) *De-traditionalization*, Oxford: Blackwell.

Heim, M. (1993) *The Metaphysics of Virtual Reality*, New York: Oxford University Press.

Held, R. and Durlach, N. (1991) Telepresence, Time Delay and Adaptation, in Ellis, S. (ed.) *Pictorial Communication in Virtual and Real Environments*, New York: Taylor and Francis.

Hepworth, M. (1989) *The Geography of the Information Economy*, London: Belhaven.

Hepworth, M. and Ducatel, K. (1992) *Transport in The Information Age: Wheels and Wires*, London: Belhaven Press.

Herbert, F. (1965/84) *Dune*, London: New English Library.

Herring, S. (1996) Posting in a Different Voice: Gender and Ethics in Computer-Mediated Communication, in Ess, C. (ed.) *Philosophic Perspectives in Computer-Mediated Communication*, Albany, NY: State University of New York Press.

Hetherington, K. (1996) Identity formation, space and social centrality, *Theory Culture and Society* 13: 33–52.

Hillis, K. (1994) The Power of Disembodied Imagination: Perspective's Role in Cartography, *Cartographica* 31(3): 1–20.

Hillis, K. (1998) Human.language.machine, in Pile, S. and Nast, H. (eds) *Mapping The Body*, London: Routledge.

Hillis, K. (forthcoming) *Digital Sensations: Identity, Embodiment and Space in Virtual Reality*, Minneapolis: University of Minnesota Press.

Hiss, T. (1990) *The Experience of Place*, New York: Knopf.

Hobsbawm, E.J. (1968) *Industry and Empire*, Harmondsworth: Penguin Books.

Hofstadter, A. and Kuhns, R. (eds) (1976) *Philosophies of Art and Beauty*, Chicago: University of Chicago Press.

Holquist, M. (1990) *Dialogism: Bakhtin and his world*, London: Routledge.

Hoy, D. (1986) *Foucault: A Critical Reader*, Oxford: Blackwell University Press.

Huczynski, A. (1993) *Management Gurus: What Makes Them and How to Become One*, London: Routledge.

Hughes, T. (1983) *Networks of Power: Electrification in Western Society, 1880–1930*, Baltimore: Johns Hopkins University Press.

Hugo, V. (1963) *Les Misérables*, Paris: Garnier Frères. Originally published in 1862.

Huxtable, A. (1997) *The Unreal America: Architecture and Illusion*, New York: New Press.

Information Society Forum (1995) *Information Society Forum Papers*, http://www.ispo.cec.be/infoforum/pub.html

Ingold, T. (1995) Building, dwelling, living: How animals and people make themselves at home in the world, in Strathern, M. (ed.) *Shifting Contexts: Transformations in Anthropological Knowledge*, London: Routledge: 57–80.

Jackson, P. (1989) *Maps of Meaning*, London and New York: Routledge.

Jackson, P. and Holbrook, B. (1995) Multiple Meanings: shopping and the cultural politics of identity, *Environment and Planning A* 27(12): 1913–1923.

Jackson, P. and Holbrook, B. (1996) Shopping around: focus group research in North London, *Area* 28(2): 133–139.

Jackson, R. (1981) *Fantasy: The literature of subversion*, London and New York: Routledge.

Jacobs J. (1961) *The Death and Life of Great American Cities*, New York: Vintage Books.

Jacobsen, R. (1994) Virtual worlds capture spatial reality, *GIS World* December: 36–39.

James, W. (1907) *On Pragmatism*, London: Longmans.

Jameson, F. (1991) *Postmodernism, or the Cultural Logic of Late Capitalism*, London: Verso.

Jameson, F. (1995) Marx's purloined letter, *New Left Review* 209: 75–109.

Jammer, M. (1969) *Concepts of Space: The History of Theories of Space in Physics* (2nd edn), Cambridge, MA: Harvard University Press.

Janelle, D.G. (1968) Central place development in a time-space framework, *The Professional Geographer* 20: 5–10.

Jenkins, H. (1992) *Textual Poachers: Television fans and participatory culture*, London and New York: Routledge.

Johannessen. N. (ed.) (1991) *'Ring Up Britain': The Early Years of the Telephone in the United Kingdom*, London: British Telecommunications PLC.

Johnson, C. (1993) *System and Writing in the Philosophy of Jacques Derrida*, Cambridge: Cambridge University Press.

Jonas, H. (1982) *The Phenomenon of Life*, Chicago: University of Chicago Press.

Joseph, R. (1993) *The Politics of Telecommunications Reform: A Comparative Study of Australia and New Zealand.* University of Wollongong Science and Technology Analysis Research Programme.

Joyce, M. (1996) (Re)Placing the Author: A Book in the Ruins, in Nunberg, G. (ed.) *The Future of the Book*, Brepols: University of California Press.

Kellner, D. (ed.) (1994) *Baudrillard: A Critical Reader*, Oxford: Blackwell.

Kelly, K. (1994a) Hive Mind, *Whole Earth Review* 82.

Kelly, K. (1994b) *Out of Control: The New Biology of Machines*, London: Fourth Estate.

Kendall, L. (1996) MUDder? I Hardly Know 'Er! Adventures of a Feminist MUDder, in Cherny, L. and Weise, E.R. (eds) *Wired Women: Gender and New Realities in Cyberspace*, Seattle, WA: Seal Press.

Kendall, L. (forthcoming) *Hanging Out in the Virtual Pub: Identity and Relationships Online*. Unpublished Ph.D thesis, Department of Sociology: University of California, Davis.

Kern, S. (1983) *The Culture of Time and Space 1880–1918*, Cambridge, MA: Harvard University Press.

Keyes, R. (1973) *We, the Lonely People; searching for community*. New York, Harper and Row.

King, G. (1996) *Mapping Reality: An Exploration of Cultural Cartographies*, Basingstoke: Macmillan.

Kircher, A. (1646) *Ars Magna Lucis et Umbrae*, Rome: Ludovici Grignani.

Kirsch, S. (1995) The Incredible Shrinking World – Technology and the Production of Space, *Society and Space* 13(5): 529–555.

Kitto, H.D.F. (1964) *The Greeks*, Harmondsworth: Penguin.

Knorr-Cetina, K. (1997) Sociality with objects: Social relations in postsocial knowledge societies, *Theory, Culture and Society* 14: 1–30.

Koch, R. (1995) The Case of Latour, *Configurations* 3(3): 319–347.

Kogut, B. and Bowman, E.H. (1996) Redesigning for the 21st century, *Financial Times*, 22 March: 13–14.

Koyré, A. (1957) *From The Closed World to The Infinite Universe*, Baltimore, MD: Johns Hopkins University Press.

Kristeva, J. (1991) *Strangers to Ourselves* (trans. L.S. Roudiez), New York: Columbia University Press.

Kroker, A. and Weinstein, M.A. (1994) *Data Trash: the theory of the virtual class*, New York: St Martin's Press.

Kunstler, J. (1993) *The Geography of Nowhere: The Rise and Decline of America's Man-Made Landscape*, New York: Simon and Schuster.

Kunze, D. (1995) The Thickness of the Past: The Metonymy of Possession,

Intersight3 The Online Journal of the School of Architecture and Planning, State University of New York University at Buffalo http://www.arch.buffalo.edu/~intrsght/archives/intersight3/kunze/kunze.html

Lacan, J. (1977) *Écrits: a selection*, London: Tavistock/Routledge.

Lacan, J. (1979) *The Four Fundamental Concepts of Psycho-analysis*, Harmondsworth: Penguin.

Laidlaw, M. (1993) Virtual Reality: our new romance with plot devices, *South Atlantic Quarterly* 92: 647–668.

Laidlaw, M. (1996) The Egos at i-D, in *Wired* 4.08, August 1996: 122–127, 186–189.

Landes, D.S. (1969) *The Unbound Prometheus: Technological Change and Industrial Development in Western Europe from 1750 to the Present*, Cambridge: Cambridge University Press.

Lanier, J. and Biocca, F. (1992) An insider's view of the future of virtual reality, *Journal of Communications* 42 (4): 150–172.

Larkham, P. (1996) *Conservation and the City*, London: Routledge.

Larner, W. (1996) The 'New Boys': Restructuring in New Zealand 1984–95, *Social Politics: Special Edition on Gender Inequalities in Global Restructuring* 3(1): 32–56.

Larner, W. (1997a) The Legacy of the Social: Market Governance and the Consumer, *Economy and Society* 26(3): 373–399.

Larner, W. (1997b) 'A Means to an End': Neo-liberalism and State Processes in New Zealand, *Studies in Political Economy* 52: 7–38.

Larner, W. (1997c) Hitching a Ride on a Tiger's Back: Globalization and Spatial Imaginaries. Paper presented to the American Sociological Association, 9–13 August, Toronto.

Latour, B. (1987) *Science in Action*, Cambridge, MA: Harvard University Press.

Latour, B. (1988a) 'Opening one eye while closing the other' . . . a note on some religious paintings, in Fyfe, G. and Law, T. (eds) *Picturing Power: Visual Depiction and Social Relations*, London: Routledge: 15–38.

Latour, B. (1988b) The politics of explanation: an alternative, in Woolgar, S. (ed.) *Knowledge and Reflexivity*, London: Sage: 155–177.

Latour, B, (1990) The Enlightenment without the Critique: A Word on Michel Serre's Philosophy, in Griffiths, A. (ed.) *Contemporary French Philosophy*, Cambridge: Cambridge University Press.

Latour, B. (1991) Technology is society made durable, in Law, J. (ed.) *A Sociology of Monsters*, London: Routledge: 103–132.

Latour, B, (1993) *We Have Never Been Modern*, London: Harvester Wheatsheaf.

Latour, B. (1996a) *Aramis, or the Love of Technology*, Cambridge, MA: Harvard University Press.

Latour, B. (1996b) Social Theory and the Study of Computerised Work Sites, in Orlikowski, W., Walsham, G., Jones M. and DeGross, J. (eds) *Information Technology and Changes in Organisational Work*, London: Chapman & Hall.

Latour, B. (1997) On Actor-Network Theory: A Few Clarifications, http://www.keele.ac.uk/depts/stt/stt/ant/latour.htm

Latour, B. and Woolgar, S. (1979, 1986) *Laboratory Life*, Princeton, NJ: Princeton University Press.

Law, J. (1994) *Organising Modernity*, Oxford: Blackwell.

Law, J., and Mol, A. (1995) Notes on materiality and sociality, *Sociological Review* 24: 641–671.

Law, J. and Mol, A. (1996) Decision's. Paper presented to the Centre for Social Theory and Technology Seminar, November, 1996.

Le Bot, Y. (1997) *Subcomandante Marcos: El sueño Zapatista*, Mexico: Plaza & Janés.

Lee, R. and Wills, J. (eds) (1997) *Geographies of Economies*, London: Arnold.

Lenoir, T. and Ross, C. L. (1996) The Naturalized History Museum, in Galison, P. and Stump, D. (eds) *The Disunity of Science: Boundaries, Contexts, and Power*, Stanford: Stanford University Press.

Lévy, P. (1996) Collective Intelligence and its Objects: Many-to-Many Communication in a Meaning World, http://www.design-inst.nl/doors/doors3/transcripts/levy.html

Levy, S. (1992) *Artificial Life*, New York: Pantheon.

Liebes, T. and Katz, E. (1990) *The Export of Meaning: Cross-Cultural Readings of Dallas*, Oxford: Oxford University Press.

Light, J. (1996) Editorial: Developing the Virtual Landscape, *Environment and Planning D: Society and Space* 14(2): 127–131.

Light, J. (1997) The Changing Nature of Nature, *Ecumene: A Journal of Environment, Culture, Meaning* 4(2): 181–195.

Lilley, R. and Knapper, P. (1993) The corrections-commercial complex, *Crime and Delinquency* 39(2) April: 150–166.

Lippard, L. (1983) *Overlay*, New York: Pantheon Books.

Lister, M. (1995) (ed.) *The Photographic Image in Digital Culture*, Routledge: London.

Littlewoods (1997) *Spring/Summer Catalogue*, Liverpool: Littlewoods Home Shopping Group Company.

Livingstone, D. (1992) *The Geographical Tradition*, Oxford: Blackwell.

Lofland, L. (1973) *A World of Strangers: order and action in urban public space*, New York: Basic Books.

Lowe, P., Murdoch, J. and Ward, N. (1995) Networks in Rural Development: beyond exogenous and endogenous models, in van der Ploeg, J. D. and Van Dijk, G. (eds) *Beyond Modernization: the impact of endogenous rural development*, Assen: Van Gorcum.

Ludwig, M. (1996) Virtual Catastrophe: Will Self-Reproducing Software Rule the World?, in Leeson, H. (ed.) *Clicking In: Hot Links to a Digital Culture*, Seattle, WA: Bay Press: 238–246.

Luhmann, N. (1997) *Die Gesellschaft der Gesellschaft*, Frankfurt am Main: Suhrkamp.

Lukás, G. (1971) *History and Class Consciousness: studies in Marxist dialectics*, London: Merlin.

Lupton, D. (1994) Panic computing: the viral metaphor and computer technology, *Cultural Studies* 8: 556–568.

Lupton, D. (1995) The embodied computer/user, in Featherstone, M. and Burrows, R. (eds) *Cyberspace, Cyberbodies, Cyberpunk: Cultures of Technological Embodiment*, London: Sage: 97–112.

Lynd, R. and Lynd, H. (1956) *Middletown: A Study in American Culture*, New York: Harcourt.

Lyon, D. (1994) *The Electronic Eye: The Rise of Surveillance Society*, Cambridge: Polity.

Lyotard, J.-F. (1984) *The postmodern condition: a report on knowledge*, Manchester: Manchester University Press.

Lyotard, J.-F. (1990) *Duchamp's TRANS/formers*, Venice: Lapis.

Lyotard, J.-F. (1994) *Libidinal Economy*, Bloomington, IN: Indiana University Press.

McCaffery, L. (ed.) (1991) *Storming the Reality Studio*, Durham, NC: Duke University Press.

McDonough, J. P. (forthcoming). Designer selves: construction of technologically mediated identity within graphical, multiuser virtual environments, *Journal of the American Society of Information Science*.

McDowell, L. (1991) Life Without Father and Ford: The New Gender Order of Post-Fordism, *Transactions of the Institute of British Geographers* 16: 400–419.

MacKenzie, D. and Wajcman, J. (1985) *The Social Shaping of Technology: How the Refrigerator Got its Hum*, Milton Keynes: Open University Press.

McKenzie, E. (1994) *Privatopia: homeowner associations and the rise of residential private government*, New Haven, CT: Yale University Press.

McKie, R. (1994) Never mind the quality, just feel the collar, *The Observer*, 13 November: 1.

McLuhan, M. (1964) *Understanding Media: the extensions of man*, New York: McGraw-Hill.

McRae, S. (1996) Coming Apart at the Seams: Sex, Text and the Virtual Body, in Cherny, L. and Weise, E.R. (eds) *Wired Women: Gender and New Realities in Cyberspace*, Seattle, WA: Seal Press.

Maffesoli, M. (1996) *The Time of the Tribes*, London: Sage.

Makin, A. (1994) With Clear You're More than Just a Number, *Proceedings of TUANZ '94 Conference*, Auckland.

Malin, H. (1998) Contour and Consciousness, *Eastgate Review of Hypertext*, Watertown, MA: Eastgate Systems.

Malmgren, C. D. (1991) *Worlds Apart: Narratology of science fiction*, Bloomington and Indianapolis, IN: Indiana University Press.

Malmgren, C. D. (1993) Self and Other in SF: Alien encounters, *Science Fiction Studies* 20: 15–33.

Manovich, L. (1992) Virtual Cave Dwellers: Siggraph '92, *Afterimage* October.

Marcos (1994) *¡Zapatistas! 1994*, Brooklyn, NY: Autonomedia.

Marcos (1997a) 7 Preguntas a quien corresponda. Imágenes del Neoliberalismo en el México de 1997, *La Jornada* 24 de Enero de 1997.

Marcos (1997b) *Statement of Subcomandante Marcos* to the Freeing the Media Teach-In organised by the Learning Alliance, Paper Tiger TV and FAIR in cooperation with the Media and Democracy Congress, 31 January and 1 February 1997.

Marcus, G. (1989) *Lipstick Traces: A Secret History of the Twentieth Century*, Cambridge, MA: Harvard University Press.

Martin, M. (1991) *Hello Central?: Gender, Technology, and Culture in the Formation of Telephone Systems*, Montreal and Kingston: McGill-Queen's University Press.

Marvin, C. (1988) *When Old Technologies were New: thinking about electric communications in the late-nineteenth century*, New York: Oxford University Press.

Marx, K. (1993) *Grundrisse: Foundations of the Critique of Political Economy* (trans. Martin Nicolaus), New York: Penguin.
Marx, L. (1964) *The Machine in the Garden: Technology and the Pastoral Ideal in America*, New York: Oxford University Press.
Marx, L. (1994) The Idea of 'Technology' and Postmodern Pessimism, in Marx, L. and Smith, M.R. (eds) *Does Technology Drive History? The dilemma of technological determinism*, Cambridge, MA: MIT Press
Marx, L. and Smith, M. R. (1994) *Does Technology Drive History? The dilemma of technological determinism*, Cambridge, MA: MIT Press.
Mason, R. and Morris, M. (1986) *Post Office Review*, Wellington: Government Printer.
Massey, D. (1993) Power-geometry and a progressive sense of place, in Bird, J., Curtis, B., Putnam, T., Robertson, G. and Tickner, L. (1993) *Mapping The Futures: Local Cultures, Global Change*, London: Routledge: 59–69.
Massey, D. (1997) Spatial disruptions, in Golding, S. (ed.) *Eight Technologies of Otherness*, London: Routledge.
Massey, J. (1996) Keeping the customer satisfied, *Information Age*, July/August: 22–24.
Massumi, B. (1992) *A user's guide to* Capitalism and Schizophrenia*: deviations from Deleuze and Guattari*, London: MIT Press.
Massumi, B. (1996) The evolutionary alchemy of reason. http://www.telefonica.es/fat/emassumi.html
Matthew Gallery, University of Edinburgh (1996) *Strangely Familiar: Narratives of Architecture in the City*. Exhibition.
Maturana, H. (1991) Response to Jim Birch, *Journal of Family Therapy* 13: 375–393.
Maturana, H. and Varela, F. (1980) *Autopoiesis and Cognition: the Recognition of the Living*, Dordrecht: Reidel.
Meier, R.C. (1962) *A Communications Theory of Urban Growth*, Cambridge, MA: MIT Press.
Merleau-Ponty, M. (1962) *Phenomenology of Perception* (trans., Smith C.) London: Routledge and Kegan Paul.
Meyrowitz, J. (1984) *No Sense of Place: The Impact of Electronic Media on Social Behavior*, New York: Oxford University Press.
Michie, R.C. (1997) Friend or foe? Information technology and the London Stock Exchange since 1700, *Journal of Historical Geography* 23(3): 304–326.
Miller, D. (1995) *Acknowledging Consumption: A Review of New Studies*, London: Routledge.
Miller, D. (1997) *Capitalism: An Ethnographic Approach*, Oxford: Berg.
Miller, D. (1998) A theory of virtualism, in Carrier, J. and Miller, D. (eds) *Virtualism and its Discontents*, Oxford: Berg.
Miller, L. (1995) Women and Children First: Gender and the Settling of the Electronic Frontier, in Brook, J. and Boal, I.A. (eds) *Resisting the Virtual Life: The Culture and Politics of Information*, San Francisco: City Lights.
Miller, P. and Rose, N. (1995) Production, Identity, and Democracy, *Theory and Society* 24: 427–467.
Ministry of Commerce (1995) Telecommunications Reform in New Zealand: 1987–

1995, *Telecommunications Information Leaflet No. 5*, Communications Division: Wellington.

Mitchell, D. (1995) The End of Public Space? People's Park, Definitions of the Public, and Democracy, *Annals of the Association of American Geographers* 85(1): 108–133.

Mitchell, W. J. (1992) *The Reconfigured Eye: visual truth in the post-photographic era*, London: MIT Press.

Mitchell, W. J. (1996) *City of Bits: Space, Place, and the Infobahn*, Cambridge, MA: MIT Press.

Moeschin, F. (1931) *Amerika vom Auto aus*, Zurich: Erlenbach.

Mol, A. and Law, J. (1994) Regions, Networks, and Fluids – Anaemia and Social Topology, *Social Studies of Science* 24(4): 641–671

Morford, H. (1867) *Paris in '67; or The Great Exposition, its Side-shows and Excursions*, New York: G. W. Carleton and Co.

Moulthrop, S. (1991) Reading from the Map: Metonymy and Metaphor in the Fiction of Forking Paths, in Delany, P. and Landow, G. (eds) *Hypermedia and Literary Studies*, Cambridge, MA: MIT Press.

Mowshowvitz, A. (1996) Social control and the network marketplace, in Lyon, D. and Zuriek, E. (eds) *Computers, Surveillance, and Privacy*, Minneapolis: University of Minnesota Press: 79–103.

Mumford, L. (1934) *Technics and Civilisation*, New York: Harcourt, Brace and Company.

Murdoch, J. and Pratt, A. C. (1993) Rural Studies: Modernism, Postmodernism and the 'Post-rural', *Journal of Rural Studies* 9 (4): 411–427.

National Business Review (1993) Telecom Puts Away Rainy Day Profits as Staff Gets Drenched, 19 February.

Nesbitt, M. (1992) In the absence of the parisienne . . . , in Colomina, B. (ed.) *Sexuality and Space*, Princeton, NJ: Princeton Architectural Press.

New Zealand Herald (1993) Telecom Shares Hit Peak with Record Earnings, 17 February.

Nietzsche, F. (1968) *Twilight of the Idols* and *The Anti-Christ*, Harmondsworth: Penguin.

Nixon, N. (1992) Cyberpunk: Preparing the ground for revolution or keeping the boys satisfied?, *Science Fiction Studies* 19: 219–235.

Norris, C. and Armstrong, G. (1997) *Categories of Control: The Social Construction of Suspicion and Intervention in CCTV Systems*. Report to the ESRC.

Norris, C., Moran, J. and Armstrong, G. (1996) *Algorithmic Surveillance – the future of automated visual surveillance*. Mimeo.

Novak, M. (1991) Liquid Architectures in Cyberspace, in Benedikt, M. (ed.) *Cyberspace: First Steps*, Cambridge, MA: MIT Press: 225–254.

Novak, M. (1996) *Trans Terra Form: Liquid Architectures And The Loss of Inscription*, http://www.t0.or.at/~krcf/nlonline/nonMarcos.html or http://flux.carleton.ca/SITES/PROJECTS/LIQUID/Novak1.html

Nowotny, H. (1994) *Time: The Modern and Postmodern Experience*, Cambridge: Polity Press.

Nye, D.E. (1990) *Electrifying America: Social Meanings of a New Technology, 1880–1940*, Cambridge, MA: MIT Press.

Nye, D.E. (1994) *American Technological Sublime*, Cambridge, MA: MIT Press.

Oldenburg, R. (1989) *The Great Good Place: cafés, coffee shops, community centers, beauty parlors, general stores, bars, hangouts, and how they get you through the day*, New York: Paragon House.

Olson, C. (1947) *Call Me Ishmael*, San Francisco: City Lights.

Openshaw, S. (1994) Computational human geography: towards a research agenda, *Environment and Planning A* 26(4): 499–505.

Ormrod, S. (1995) Feminist Sociology and Methodology: Leaky Black Boxes in Gender/Technology Relations, in Grint, K. and Gill, R. (eds) *The Gender-Technology Relation: Contemporary Theory and Research*, London: Taylor & Francis.

Osborne, P. (1995) *The Politics of Time*, London: Verso.

O'Tuathail, G. (1994) Shadow Warriors and the Electronic Jury: Mexico and Chiapas Revolt in the Geo–Financial Panopticon, *Ecumene* 4(3): 300–317.

Owen, J.J. (1996) Chaos theory, marxism and literature, *New Formations* 29: 84–112.

Pall Mall Gazette, 20 January 1893.

Parkes, D. and Thrift, N.J. (1980) *Times, Spaces, Places. A Chronogeographic Perspective*, Chichester: John Wiley.

Pascal, A. (1987) The Vanishing City, *Urban Studies* 24: 597–603.

Pascale, T. (1991) *Managing on the Edge*, Harmondsworth: Penguin.

Patton, P. (1994) Anti-Platonism and Art, in Boundas, C. V. and Olkowski, D. (eds) *Gilles Deleuze and the Theatre of Philosophy*, London: Routledge: 141–156.

Pepperell, R. (1995) *The Post-human Condition*, Oxford: Intellect.

Perkin, H. (1969) *Origins of Modern English Society*, London: Routledge.

Perlez, J. (1996) Few Forints to Spend, but Hungarians Like the Mall, *New York Times*, 146, 12/24: D1.

Perniola, M. (1995) *Enigmas: the Egyptian Moment in Society and Art*, London: Verso.

Perrella, S. (1996) *Being@Home . . . as Becoming Information and Hypersurface*, http://www.mediamatic.nl/doors/Doors2/Perrella/Perrella-Doors2-E.html

Perry, C.R. (1977) The British experience 1876–1912: the impact of the telephone during the years of delay, in Pool, I. de Sola (ed.) *The Social Impact of the Telephone*, Cambridge, MA: MIT Press: 69–96.

Perry, C.R. (1992) *The Victorian Post Office: The Growth of a Bureaucracy*, Woodbridge, Suffolk: The Boydell Press.

Pesce, M. (1995) *VRML: Browsing and Building Cyberspace*, Indianapolis: New Riders Publishing.

Philo, C. (1992) Foucault's geography, *Society and Space* 10: 137–161.

Philo, C. (1993) Postmodern Rural Geography? A reply to Murdoch and Pratt, *Journal of Rural Studies* 9 (4): 429–436.

Pickles, J. (1991) Geography, GIS and the surveillant society, *Papers and Proceedings of Applied Geography Conference* 14: 80–91.

Pickles, J. (1995) Representations in an Electronic Age, in Pickles, J. (ed.) *Ground Truth: The Social Implications of Geographic Information Systems*, New York: Guilford Press.

Pike, R.M. (1989) Kingston adopts the telephone: the social diffusion and use of the telephone in urban central Canada, 1876 to 1914, *Urban History Review* 18(1): 32–47.

Pile, S. (1994) Cybergeography: 50 years of *Environment and Planning A, Environment and Planning A,* 26: 1815–1823.

Pile, S. and Thrift, N. (1996) Mapping the subject, in Pile, S. and Thrift, N. (eds) *Mapping the Subject: Geographies of Cultural Transformation*, London: Routledge: 13–51.

Pimentel, K. and Texeira, K. (1993) *Virtual Reality: through the new looking-glass*, New York: Intel/Windcrest, McGraw Hill.

Pineda, F. (1996) La guerra de baja intensidad, in Barreda, A. *et al*. (eds) *Chiapas 2*, Mexico D.F.: Instituto de Investigaciones Económicas: 173–196.

Plant, S. (1995) The future looms: weaving women and cybernetics, in Featherstone, M. and Burrows, R. (eds) *Cyberspace, Cyberbodies, Cyberpunk: Cultures of Technological Embodiment*, London: Sage: 45–64.

Plant, S. (1996) On the Matrix: Cyberfeminist Simulations, in Shields, R. (ed.) *Cultures of Internet: Virtual Spaces, Real Histories, Living Bodies*, London: Sage.

Plant, S. (1997) *Zeros and Ones: Digital Women and The New Technoculture*, London: Fourth Estate.

Platt, H.L. (1991) *The Electric City: Energy and the Growth of the Chicago Area, 1880–1930*, Chicago: University of Chicago Press.

Pool, I. de Sola (ed.) (1977) *The Social Impact of the Telephone*, Cambridge, MA: MIT Press.

Porta, G. (1658) *Natural Magick*, London: Thomas Young and Samuel Speed.

Post 84, BT Archives (BTA) (1984) Telecommunications: telephones, private companies 1879–1915.

Post 86, BT Archives (BTA) (1986) Telecommunications: telephones, inland 1880–1938.

Poster, M. (1990) *The Mode of Information: poststructuralism and social context*, Cambridge: Polity.

Poster, M. (1995) *The Second Media Age*, Cambridge: Polity.

Postman, N. (1992) *Technopoly: the surrender of culture to technology*, New York: Knopf.

Potter, E. (forthcoming) Making gender/making science, in Spanier, B. (ed.) *Making a Difference*, Bloomington, IN: Indiana University Press.

Putnam, R. (1995) Bowling Alone: Declining Social Capital, *Journal of Democracy* 6(1): 65–78.

Rabinow, P. (1996) *Essays on the Anthropology of Reason*, Princeton, NJ: Princeton University Press.

Radway, J. (1984) *Reading the Romance: Women, patriarchy, and popular literature*, Chapel Hill, NC: University of North Carolina Press.

Ramasubramanian, L. (1996) Building communities: GIS and participatory decision making, *Journal of Urban Technology* 3(10): 67–79.

Ray, C. (1996a) *The Dialectic of Local Development: the case of the E.U. LEADER I rural development programme*, Working Paper 23, Centre for Rural Economy, University of Newcastle upon Tyne.

Ray, C. (1996b) *Local Rural Development in the Western Isles, Skye and Lochalsh, and*

Brittany, unpublished PhD thesis, Welsh Institute of Rural Studies, University of Wales.

Ray, C. and Woodward, R. (1998) 'Voluntary Organisations: status and role', *Environment & Société* 20: 27–33.

Reid, E.M. (1996) Text-based Virtual Realities: Identity and the Cyborg Body, in Ludlow, P. (ed.) *High Noon on the Electronic Frontier: Conceptual Issues in Cyberspace*, Cambridge, MA: MIT Press.

Relph, E. (1987) *The Modern Urban Landscape*, Baltimore, MD: Johns Hopkins University Press.

Rheingold, H. (1991) *Virtual Reality*, New York: Simon and Schuster.

Rheingold, H. (1993) *The Virtual Community: homesteading on the electronic frontier*, Menlo Park, CA: Addison-Wesley.

Richards, C. (1995) Virtual bodies, *Public* 11: 35–39.

Richardson, R. (1994) Back officing front office functions – organisational and locational implications of new telemediated services, in Mansell, R. (ed.) *Management Of Information And Communication Technologies*, London: ASLIB: 309–335.

Robberson, T. (1995) Mexican Rebels Using A High-Tech Weapon; Internet Helps Rally Support, *The Washington Post*, 1995 February 20.

Robins, K. (1991) Into the image: visual technologies and vision cultures, in Wombell, P. (ed.) *PhotoVideo: photography in the age of the computer*, London: Rivers Oram: 52–77.

Robins, K. (1995) Collective Emotion and Urban Culture, in Healey, P. *et al.* (eds) *Managing Cities: The New Urban Context*, Chichester: John Wiley.

Robins, K. (1996) *Into the Image: Culture and Politics in the Field of Vision*, London: Routledge.

Robins, K. and Hepworth, M. (1988) Electronic Spaces: new technologies and the future of cities, *Futures*, 20 (2): 155–176.

Robson, B. (1973) *Urban Growth: An Approach*, London: Methuen.

Ronell, A. (1994) Video/Television/Rodney King: Twelve Steps Beyond the Pleasure Principle, in Bender, G. and Druckney, T. (eds) *Culture on the Brink: Ideologies of Technology*, Dia Center for the Arts 9, Seattle, WA: Bay Press.

Rose, G. (1993) *Feminism and Geography: the limits of geographical knowledge*, Cambridge: Polity Press.

Rose, N. (1996) *Inventing Our Selves: Psychology, Power and Personhood*, Cambridge: Cambridge University Press.

Rosello, M. (1994) The Screener's Maps: Michel de Certeau's 'Wandersmänner' and Paul Auster's Hypertextual Detective in Landow, G. (ed.) *Hyper/Text/Theory*, Baltimore, MD: Johns Hopkins University Press.

Rosen, F. (1996) Zapatistas' New Political Organization prompts Realignments on the Left, *NACLA Report on the Americas* XXIX(5) March/April: 2–3.

Ross, A. (1991) *Strange Weather: Culture, science, and technology in the age of limits*, London and New York: Verso.

Ross, J. (1995a) *Rebellion from the Roots: Indian uprising in Chiapas*, Monroe, ME: Common Courage Press.

Ross, J. (1995b) The EZLN, a History: Miracles, Coyunturas, Communiqués, in *Shadows of Tender Fury- The Letters and Communiqués of Subcommandante Marcos and the Zapatista Army of National Liberation*, New York: Monthly Review Press.

Roszak, T. (1994) *The Cult of Information*, Berkeley, CA: University of California Press.

Rubin, B. (1979) Aesthetic Ideology and Urban Design, *Annals of the Association of American Geographers* 69(3): 339–361.

Ryan, M-L. (1994) Immersion vs. interactivity: virtual reality and literary theory, *Postmodern Culture* 5 (1). http://jefferson.village.virginia.edu/pmc/

Sack, R. (1976) Magic and Space, *Annals of the Association of American Geographers* 66(2): 309–322.

Said, E. (1978) *Orientalism*, Harmondsworth: Penguin.

Samuelson, D. N. (1993) Modes of Extrapolation: The formulas of hard SF, *Science Fiction Studies* 20: 191–232.

Sandoval, C. (1991) US Third World Feminism: The Theory and Method of Oppositional Consciousness in the Postmodern World, *Genders* 10: 1–24.

Sauer, C.O. (1963) The Education of a Geographer, in Leighly, J. (ed.) *Land & Life, A Selection from the Writings of Carl Ortwin Sauer*, Berkeley, CA: University of California Press.

Sauer, C.O. (1968) *Northern Mists*, San Francisco: Turtle Island Foundation.

Schiller H. (1989) *Culture Inc: The Takeover of Corporate Expression*, New York: Oxford University Press.

Schivelbusch, W. (1986) *The Railway Journey: the industrialization of space and time in the nineteenth century*, Berkeley, CA: University of California Press.

Schmidt, G. (1996) *Der Indio-Aufstand in Chiapas*, München: Droemersche Verlagsanstalt Th. Knaur Nachf.

Schroeder, R. (1997) Networked Worlds: Social Aspects of Multi-User Virtual Reality Technology, *Sociological Research Online* 2: http://www.socresonline.org.uk/socresonline/2/4/5.html

Schwartz, R. (1994) *Vision: Variations on some Berkeleian Themes*, London: Blackwell.

Schwartz, V. (1994) The Morgue and the Musée Grévin: Understanding the Public Taste for Reality in Fin-de-Siècle Paris, *Yale Journal of Criticism* 7(2): 151–173.

Scottish Office (1995) *Rural Scotland: people, prosperity and partnership*, Edinburgh: HMSO.

Select Committee on Telephone Charges (1920) BPP, vol. 8.

Select Committee on Telephones (1898) BPP, vol. 12.

Select Committee on the Telegraphs Bill (1892) British Parliamentary Papers (BPP), vol. 17.

Select Committee on the Telephone Service (1895) BPP, vol. 13.

Select Committee on the Telephone Service (1921) BPP, vol. 7.

Select Committee on the Telephone Service (1922) BPP, vol. 6.

Serres, M. (1982) *Hermes: Literature, Science, Philosophy*, Baltimore, MD: Johns Hopkins University Press.

Serres, M. (1995) *Genesis*, Michigan: University of Michigan Press.

Serres, M. with Latour, B. (1995) *Conversations on Science, Culture, and Time*, Michigan: University of Michigan Press.

Sexton, R. (1995) *Parallel Utopias: Sea Ranch and Seaside: The quest for community*, San Francisco: Chronicle Books.

Shade, L.R. (1996) Is there free speech on the net? Censorship in the global information infrastructure, in Shields, R. (ed.) *Cultures of Internet: Virtual Spaces, Real Histories, Living Bodies*, London: Sage: 11–32.

Shapin, S. (1994) *A Social History of Truth: Civility and Science in Seventeenth-Century England*, Chicago: University of Chicago Press.

Shapin, S. and Schaffer, S. (1985) *Leviathan and the Airpump*, Princeton, NJ: Princeton University Press.

Shapiro, A. (1985) *Housing the Poor of Paris, 1850–1902*, Madison: University of Wisconsin Press.

Shields, R. (ed.) (1996) *Cultures of internet: virtual spaces, real histories, living bodies*, London: Sage.

Shirtcliffe, P. (1993) Letter to Philip Bowyer, General Secretary, PTTI, 9 March.

Silverstone, R. (1994) *Television and Everyday Life*, London: Routledge.

Silverstone, R. *et al.* (eds) (1992) *Consuming Technologies: Media and Information in Domestic Spaces*, London: Routledge.

Slouka, M. (1996) *War of the Worlds: Cyberpsace and the High-Tech Assault on Reality*, London: Abacus.

Smith, J. (1992) *The Frugal Gourmet Whole Family Cookbook*, New York: William Morrow.

Soja, E. (1989) *Postmodern Geographies*, London: Verso.

Soja, E. (1993) History, geography, modernity, in During, S. (ed.) *The Cultural Studies Reader*, New York: Routledge.

Soja, E. (1996) *Thirdspace: Journeys to Los Angeles and Other Real-and-Imagined Places*, Cambridge, MA: Blackwell.

Sontag, S. (1977) *On Photography*, New York: Farrar, Straus and Giroux.

Sorkin, M. (1992a) Scenes from the Electronic City, *I.D. Magazine* May: 70–77

Sorkin, M. (1992b) See You in Disneyland, in Sorkin, M. (ed.) *Variations on a Theme Park: The New American City and the End of Public Space*, New York: Hill and Wang.

Sorkin, M. (1993) Meeting Spaces, *Progressive Architecture* 74(4): 106–107.

Spinoza, Benedict/Baruch (1955) *The Ethics* (trans. Elwes), New York: Dover Press.

Springer, C. (1991) The Pleasure of the Interface, *Screen* 32: 303–323.

Spufford, F. and Uglow, J. (eds) (1996) *Cultural Babbage. Technology, Time and Invention*, London: Faber and Faber.

Squires, J. (1994) Private lines, secluded places: privacy as political possibility, *Society and Space* 12: 387–401.

Squires, J. (1996) Fabulous Feminist Futures and the Lure of Cyberculture, in Dovey, J. (ed.) *Fractal Dreams: New Media in Social Context*, London: Lawrence & Wishart: 194–216.

Stabile, C.A. (1994) *Feminism and the Technological Fix*, Manchester: Manchester University Press.

Stallabrass, J. (1995) Empowering Technology: The Exploration of Cyberspace, *New Left Review* 211: 3–32.

Starr, P. (1994) Seductions of Sim, *American Prospect* 17: 19–29.

Star, S.L. (1989) The structure of ill-structured solutions: boundary objects and heterogeneous distributed problem solving, in Gasser, L. and Huhn, N. (eds) *Distributed Artifical Intelligence*, New York: Morgan Kauffman: 37–54.

Star, S. and Ruhleder, K. (1996) Steps towards an Ecology of Infrastructure: Design and Access for Large Information Spaces, *Information Systems Research* 7(1): 111–134.

Stephenson, N. (1992) *Snowcrash*, New York: Bantam.

Stephenson, N. (1993) *Snow Crash*, New York: Spectra.

Stephenson, N. (1994) Spew, *Wired* 2(10): 91–95, 142–147.

Steuer, J. (1992) Defining virtual reality: dimensions determining telepresence, *Journal of Communications* 42 (4): 73–93.

Stone, A. R. (1991) Will the Real Body Please Stand Up? Boundary stories about virtual cultures, in Benedikt, M. (ed.) *Cyberspace: First steps*, Cambridge, MA: MIT Press: 81–118.

Stone, A.R. (1994) Preface, in Druckrey, T. (ed.) *Electronic Culture: Technology and Visual Representation*, New York: Aperture: 6–10.

Stone, A. R. (1995) *The War of Desire and Technology at the Close of the Mechanical Age*, Cambridge, MA: MIT Press.

Stone, R. J. (1990) Virtual reality in telerobotics, in *Computer graphics*, proceedings of the Conference held in London, November 1990, Pinner: Blenheim Online: 32.

Strathern, M. (1992) Reproducing anthropology, in Wallman, S. (ed.) *Contemporary Futures. Perspectives on Social Anthropology*, London: Routledge: 172–189.

Strathern, M. (1996) Cutting the Network, *Journal of the Royal Anthropological Institute* NS2: 517–535.

Strum, S. and Latour, B. (1987) The Meanings of the Social: From Baboons to Humans, *Social Science Information* 26: 783–802.

Sui, D. (1997) Reconstructing urban reality: From GIS to electropolis, *Urban Geography* 18(1): 74–89.

Swett, C. (1995) *Strategic Assessment: The Internet*. Report. Office of the Assistant Secretary of Defense for Special Operations and Low-Intensity Conflict (Policy Planning). Internet: http://www.fas.org/cp/swett.html

Swyngedouw, E. (1993) Communication, mobility and the struggle for power over space, in Giannopoulos, G. and Gillespie, A. (eds) (1993) *Transport And Communications in the New Europe*, London: Belhaven: 305–325.

Tabbi, J. (1995) *Postmodern Sublime: Technology and American Writing from Mailer to Cyberpunk*, Ithaca, NY: Cornell University Press.

Talbot, H. (1997a) *Telematics for the Rural North*, Sunderland: Northern Informatics.

Talbot, H. (1997b) *Rural Telematics in England: strategic issues*. Research Report, Centre for Rural Economy, University of Newcastle upon Tyne.

Tarr, J.A. and Dupuy, G. (eds) (1988) *Technology and the Rise of the Networked City in Europe and North America*, Philadelphia: Temple University Press.

Tarr, J.A., Finholt, T. and Goodman, D. (1987) The city and the telegraph: urban telecommunications in the pre-telephone era, *Journal of Urban History* 14(1): 38–80.

Taylor, C. (1994) *Multiculturalism*, Princeton, NJ: Princeton University Press.

Telecom NZ (1991) Annual Report.

Telecom NZ (1993) Annual Report.

Telecom NZ (1994) Annual Report.

Telephone Development Association (1930) *The Stranglehold on Our Telephones: A Practical Remedy*, London: The Telephone Development Association.

Terry, J. and Calvert, M. (1997) Introduction: Machines/Lives, in Terry, J. and Calvert, M. (eds) *Processed Lives: gender and technology in everyday life*, London: Routledge.

Thompson, J.B. (1990) *Ideology and Modern Culture: Critical Social Theory in the Era of Mass Communication*, Cambridge: Polity Press.

Thompson, J.B. (1995) *The Media and Modernity: A Social Theory of the Media*, Cambridge: Polity Press.

Thrift, N. (1995) The Hyperactive World, in Johnston, R. Taylor, P. and Watts, M. (eds) *Geographies of Global Change*, Oxford: Blackwell: 18–35.

Thrift, N. (1996a) New urban eras and old technological fears: reconfiguring the goodwill of electronic things, *Urban Studies* 33(8): 1463–1493.

Thrift, N. (1996b) *Spatial Formations*, London: Sage.

The Times (1883–1909) various issues.

Todorov, T. (1973) *The Fantastic: A structural approach to a literary genre*, Ithaca, NY: Cornell University Press.

Toffler, A. (1980) *The Third Wave*, New York: William Morrow.

Tomas, D. (1991) Old Rituals for a New Space: Rites de Passage and William Gibson's Cultural Model of Cyberspace, in Benedikt, M. (ed.) *Cyberspace, First Steps*, Cambridge, MA: MIT Press: 31–47

Tuan, Y. (1978) Literature and Geography: Implications for geographical research, in Ley, D. and Samuels, M. (eds) *Humanistic Geography: Prospects and Problems*, London: Croom Helm.

Turkle, S. (1984) *The Second Self: Computers and the Human Spirit*, New York: Simon & Schuster.

Turkle, S. (1995) *Life on the Screen: Identity in the Age of the Internet*, New York: Simon and Schuster.

Ullman, E. (1996) Come In CO: The Body on the Wire, in Chermy, L. and Weise, E.R. (eds) *Wired Women: Gender and Nw Realities in Cyberspace*, Seattle, WA: Seal Press: 3–23.

Ulmer, G. L. (1989) *Teletheory: Grammatology in the Age of Video*, New York: Routledge.

Urry, J. (1994) Time, leisure and social identity, *Time and Society* 3: 131–150.

Van der Ploeg, J.D. and Van Dijk, G. (1995) *Beyond Modernization: the impact of endogenous rural development*, Assen: Van Gorcum.

Vidler, A. (1978) The Scenes of the Street: Transformations in Ideal and Reality 1750–1871, in Anderson, S. (ed.) *On Streets*, Cambridge, MA: MIT Press.

Villiers de l'Isle Adam, A. (1879) *L'Eve Future*, Paris: M. de Brunhoff.

Virilio, P. (1986) *Speed and Politics* (trans. Mark Polizzotti), New York: Semiotext(e).

Virilio, P. (1987) The Overexposed City, *Zone* 1(2): 14–31.

Virilio, P. (1992) The third interval: a critical transition, *Art & Design* 7 (1/2): 78–85.

Virilio, P. (1993) The Third Interval: A critical transition, in Andermatt-Conley, V. (ed.) *Rethinking Technologies*, London: University of Minnesota Press: 3–10.

Virilio, P. (1994a) The Third Interval: A critical transition, in Andermatt-Conley, V. (ed.) *Rethinking Technologies*, London: University of Minnesota Press: 3–10.

Virilio, P. (1994b) *The vision machine*, Bloomington, IN: Indiana University Press.

Virilio, P. (1995) Red alert in cyberspace, *Radical Philosophy* 74: 2–4.

VNX Matrix (1998) All New Gen, in Broadhurst Dixon, J. and Cassidy, E.J. (eds) *Virtual Futures: Cybererotics, Technology, and Post-Human Pragmatism*, London: Routledge.

Voloshinov, V. (1973) *Marxism and the Philosophy of Language* (trans. L. Matejka and I. R. Titunik), New York: Seminar Press.

Wakeford, N. (1995) Sexualised Bodies in Cyberspace, in Chernaik, W. and Deegan, M. (eds) *Beyond the Book: Theory, Text and the Politics of Cyberspace*, London: London University Press.

Wakeford, N. (1997a) Cyberqueer, in Medhurst, A. and Munt, S.R. (eds) *Lesbian and Gay Studies: A Critical Introduction*, London: Cassell.

Wakeford, N. (1997b) Networking Women and Grrls with Information/ Communication Technology: Surfing Tales of the World Wide Web, in Terry, J. and Calvert, M. (eds) *Processed Lives: gender and technology in everyday life*, London: Routledge.

Wakeford, N. (1998) Urban culture for virtual bodies: comments on lesbian 'identity' and 'community' in San Francisco Bay Area cyberspace, in Ainley, R. (ed.) *New Frontiers of Space, Bodies and Gender*, London: Routledge.

Wakeford, N. (forthcoming) *Networks of Desire: Gender, Sexuality and Computing Culture*, London: Routledge.

Waldrop, M.M. (1993) *Complexity*, New York: Viking.

Waller, P.J. (1983) *Town, City and Nation: England 1850–1914*, Oxford: Oxford University Press.

Walter, E. V. (1988) *Placeways*, Chapel Hill, NC: The University of North Carolina Press.

Wark, M. (1994a) Third Nature, *Cultural Studies* 8(1): 115–132.

Wark, M. (1994b) *Virtual geography: living with global media events*. Indianapolis, IN: Indiana University Press.

Warren, S. (1996) Popular Cultural Practices in the 'Postmodern City', *Urban Geography* 17(6): 545–567.

Watson, R. (1995) When Words are the Best Weapon, *Newsweek* 27 February: 36–40.

Wehling, J. (1995) *Netwars and Activist Power on the Internet*, Internet: http://www.teleport.com/~jwehling/Netwars.html

Welsh Office (1995) *A Working Countryside for Wales*, London: HMSO.

Wertheim, M. (1997) *Pythagoras' Trousers. God, Physics and the Gender Wars*, London: Fourth Estate.

Westwood, S. and Williams, J. (eds) (1997) *Imagining Cities: Scripts, Signs, Memories*, London: Routledge.

Whyte, W. (1988) *City: Rediscovering the Center*, New York: Doubleday.

Wigley, M. (1993) *The architecture of deconstruction: Derrida's haunt*, London: MIT Press.

Williams, R.H. (1973) *The Country and the City*, New York: Oxford University Press.

Williams R.H. (1990) *Notes on the Underground: An Essay on Technology, Society, and Imagination*, Cambridge, MA: MIT Press.

Williams, R.H. (1991) *Notes on the Underground*, Cambridge, MA: MIT Press.

Wilson, E. (1995) The Invisible Flâneur, in Watson, S. and Gibson, K. (eds) *Postmodern Cities and Spaces*, Oxford: Blackwell.

Wilson, E. (1997) Nostalgia and the City, in Westwood, S. and Williams, J. (eds) *Imagining Cities: Scripts, Signs, Memory*, New York: Routledge.

Wilson, K. (1986) The Videotext Revolution: Social Control and the Cybernetic Commodity of Home Networking, *Media, Culture And Society* 8: 7–39.

Wincapaw, C. (forthcoming) Lesbian and Bisexual Women's Electronic Mailing Lists as Sexualised Spaces, *Journal of Lesbian Studies*.

Winckler, M. (1991) Walking Prisons: The Developing Technology of Electronic Controls, *The Futurist* July–August: 34–36.

Winner, L. (1996) Who Will Be in Cyberspace, *The Information Society* 12: 63–71.

Winston, B. (1995) Tyrrell's Owl: The Limits of the Technological Imagination in an Epoch of Hyperbolic Discourse, in Adam, B. and Allan, S. (eds) *Theorising Culture*, London: UCL Press: 225–235.

Winter, J. (1993) *London's Teeming Streets 1830–1914*, London: Routledge.

Winterson, J. (1992) *Sexing the Cherry*, London: Vintage.

Wired (1995) Godzone, November: 164–167, 230–236.

Wolmark, J. (1993) *Aliens and Others: Science fiction, feminism and postmodernism*, Hemel Hempstead: Harvester Wheatsheaf.

World Economic Forum (1996) World Competitiveness Report, Lausanne.

Wright, G. (1980) *Moralism and the Model Home: Domestic Architecture and Cultural Conflict in Chicago: 1873–1913*, Chicago: University of Chicago Press.

Wright, R. (1996) Art and science in Chaos: contested readings of scientific visualisation, in Robertson, G. *et al.* (eds) *Future natural. Nature, Science, Culture*, London: Routledge: 218–236.

Yeatman, A. (1990) *Bureaucrats, Technocrats, and Femocrats: Essays on the contemporary Australian state*, Sydney: Allen and Unwin.

Youngblood, G. (1989) The new renaissance: art, science and the universal machine, in Loveless, R. L. (ed.) *The Computer Revolution and the Arts*, Tampa, FL: University of South Florida Press.

¡Zapatistas! Documents from the New Mexican Revolution (1994) Brooklyn, NY: Autonomedia.

Zimmerman, J. (1986) *Once upon the Future: A Woman's Guide to Tomorrow's Technology*, New York: Pandora.

Zisek, S. (1989) *The Sublime Object of Ideology*, London: Verso.

Zisek, S. (1993) *Tarrying with the Negative*, Durham, NC: Duke University Press.

Index

Note: Page numbers in **bold** type refer to **figures**
Page numbers followed by n refer to notes